GW01179624

Schering Foundation Workshop
BIOSCIENCE ⇌ SOCIETY

Goal of this Workshop:
Do current and anticipated developments in bioscience require a new covenant between science and society?

Schering Foundation Workshop

Series Editors: G. Stock
U.-F. Habenicht

Held and published on behalf of
Schering AG, Berlin

Sponsored by:
Schering AG, Berlin
The Senate of the City of Berlin

Schering Foundation Workshop

BIOSCIENCE ⇌ SOCIETY

D.J. Roy, B.E. Wynne, R.W. Old, Editors

Report of the Schering Workshop on
BIOSCIENCE ⇌ SOCIETY
Berlin 1990, November 25–30

Rapporteurs: R.A. Gordon, S.S. Jasanoff, J.D. Moreno, D.W. Sharp

Program Advisory Committee:
D.J. Brahams, H.G. Gassen, R.A. Gordon, D.J. Roy,
R.C. Solomon, W. Van den Daele

Consulting Organizer:
S. Bernhard

John Wiley & Sons
Chichester New York Brisbane Toronto Singapore

Copy Editors: Danny Lee Lewis, John Berwick

Photographs: E.P. Thonke

With 4 photographs, 15 figures, and 13 tables

Copyright © 1991 by John Wiley & Sons Ltd
Baffins Lane
Chichester
West Sussex PO19 1UD, England

ISBN 0 471 93152 7

Typesetting with $\TeX$: Danny Lee Lewis Buchproduktion, Berlin
Printed and bound in Great Britain by
Biddles Ltd, Guildford

Table of Contents

Preface

To mark the occasion of the 100th anniversary of Schering research two events took place:

a) a Workshop on "Bioscience ⇌ Society", which was held
   November 24–30, 1990,
   and
b) a Round Table Discussion, held on
   December 1, 1990,
   where the results of the workshop were discussed by a special invited panel.

The results of the Workshop are presented here as the first volume in a newly created publication series called "Schering Foundation Workshop".

The Round Table Discussion will be published* separately as the first volume in a newly created publication series called "Schering Lecture Series".

We hope that these publications will contribute to a better understanding of science and to a better recognition of its merits.

G. Stock

---

* This series will be available on request by the Schering Research Foundation, Müller-strasse 178, D–1000 Berlin 65.

Bioscience ⇌ Society
edited by D.J. Roy, B.E. Wynne, and R.W. Old, pp. 1–8
© 1991, John Wiley & Sons

Introduction

> It is the great glory as it is the great threat of science that everything which is in principle possible can be done if the intention to do it is sufficiently resolute. Scientists may exalt in the glory, but in the middle of the twentieth century the reaction of ordinary people is more often to cower at the threat.
> – *P.B. Medawar* (Medawar 1990, p. 15)

BIOSCIENCE ⇌ SOCIETY

How the scientific community should relate to the broader human community in free societies is an enduring question, rejuvenated with each new generation and each major scientific innovation. This relationship was a central issue in debates over recombinant DNA technology in the 1970s, and it is in the field of the biosciences that this relationship has been most frequently scrutinized over the last twenty years.

During the 1970 debates on the then newly discovered methods for splicing genes from one organism to another, Maxine Singer accented the failure of scientists to educate the public adequately about recombinant DNA, and she observed: "It should not be surprising if deep fears and ambiguities arise in the minds and hearts of those who suddenly learn the depths of modern insights into the nature of living things" (Singer 1977, p. 30).

THE BERLIN MEETING

The ambiguities of contemporary advances in the biosciences, summarized in the frequent coupling of "peril and promise", "glory and threat", "dangers and benefits", perdure long after the suddenness of insight has passed, and they shaped the context of the meeting documented in this book. Fifty people from ten countries and various scientific, humanist, and professional disciplines met in Berlin from November 25 to December 1, 1990 to pursue the overarching workshop question: "Do current and anticipated developments in bioscience require a new covenant between science and society?" The workshop was held on the occasion of the 100th anniversary of the foundation of the first pharmaceutical laboratory at Schering AG, Berlin, and was sponsored by the Senate of the City of Berlin and by Schering. The conference participants worked according to the "Dahlem Workshop Model*", originated by Dr. Silke Bernhard and tested in over sixty conferences during the past sixteen years.

This book uses twenty-three background papers and four working group reports to explore four questions:

- Does bioscience threaten human integrity?
- Does bioscience threaten ecological integrity?
- What's wrong with the interaction between bioscience and society?
- What actions are required to improve the present uneasy relationship between bioscience and society?

If the Berlin meeting was an inquiry into the question about the need for a new covenant, the four working group questions were planks thrown down in front of us over uncertain and unstable ground, planks the inquiry could use to stop and take its bearings as it advanced upon the lead question.

As in all conferences patterned on the Dahlem Workshop Model, the organizers asked about half of the participants to write background papers on urgent issues in a given field. These papers, representing different disciplinary points of view, were sent to all participants long before the conference, with the request that they formulate written questions and comments for the authors. On the first day of the conference, each working group, aided by the background papers and the participants' written questions, prepared a list of the most important issues the group would discuss throughout the remainder of the conference. The objective of the Dahlem method is to foster dialogue between, not only within, the working groups. The group sessions were accordingly scheduled to allow the participants in each working group to attend at least one meeting of each of the three other groups. On the last day of the conference, all participants discussed the four group reports in a plenary session. As a result, the group reports, though textually the responsibility of each report's author, contain and reflect ideas and views from the workshop's entire roster of participants.

---

* For an explanation of the Dahlem Workshop Model see page 392.

The intense intergroup discussion encouraged by the Dahlem method deliberately (and successfully) searches out cross-disciplinary misunderstandings, divergent discourses, and incompatible premises. It cannot pretend to resolve such deeply embedded conflicts, but it did make them apparent, often uncomfortably so. The disciplines represented spanned an extremely wide range, including those which question conventional notions about the nature and the basis of the authority of science. The papers therefore properly reflect sometimes divergent perspectives, as well as conflicting value positions. As editors, we have attempted not to stifle such rigorous intellectual struggle, but only to challenge authors to identify and question their own taken-for-granted concepts and premises.

A NEW COLLEGIALITY?

Over a period of twenty years, from the early 1940s to the mid-1960s, critical scientific experiments, discoveries, and advances in molecular biology and other disciplines in the biosciences largely occurred in the silence of laboratories undisturbed by public debate and controversy. The "moves of the game", if not the "rules of the game", were set while the people were preoccupied elsewhere. There were few, if any, passionate expressions of concern or warnings of danger during the two decades following World War II. An extensive literature of awakening began to appear in the 1960s and early 1970s, even before the 1973 Gordon Conference announcement of the recombinant DNA innovation. Scientists, such as R. Sinsheimer and S. Luria, and people working in disciplines such as law, philosophy, ethics, sociology, and theology arose to draw the attention of the general public to momentous happenings in the biological sciences.

As people began to learn about the depths of scientific insights and technological interventions into the nature of living things, seminars, symposia, and a wide range of books and essays took up the themes of genetic engineering, genetic diagnosis, genetic therapy, the reproductive technologies, and the endangering of the environment. The language of many essays in this prophetic literature, often inflated and apocalyptic, was designed to stir and grasp the attention of an apparently uninterested people. Salvador Luria, a scientist who contributed to this literary genre, called for a leadership that would find the way to awaken the public and their elected representatives from complacency (Luria 1969).

That awakening has begun. We are no longer in the middle of the twentieth century, a moment in the history of the biosciences-society relationhip when

Medawar's opening quotation may have held true. So-called ordinary people are now, as we enter the last decade of this century, much less likely to cower at the threat of science, and scientists are much more at the forefront of critical reflection on the glory of science than Medawar's simple opposition of extreme attitudes would lead anyone to suspect. Since Asilomar, innumerable institutes of ethics, working groups at regional, provincial, or state levels of government, and national commissions of inquiry throughout the world have convened scientists from many disciplines and people from many walks of life, to devise, in consort, realistic alternatives to the notion that power is its own ethic.

Power is ambivalent, particularly the power to do things that never could be done before. The category of the Novum may harbor life-giving goods, imaginable but highly uncertain dangers, unintended and unforeseeable perils, and possible distant mischief as well. Integrating science and society in the process of establishing the goals, the norms, and the conditions for the exercise of new power is itself an image, as well as the most promising safeguard, of human and ecological integrity. We have come to recognize more clearly over the last twenty years that this process, though not infallible, is self-corrective if it is sustained.

Of course, science and society can never be fully integrated, if this means an absence of tension. Science is in its essence instrumental and manipulative. There is nothing wrong with this when devoted to the proper ends, under proper conditions. However, instrumentalism and reductionism too often slide from means into normative ends, as science tends towards scientism – the uncritical use of scientific methods for control of complex issues where values are at stake.

To the extent that scientific claims on truth are ultimately confirmed or denied by some form of implementation in society, and often require the reorganization of society and its values to effect those validations, the "integration" of science and society cannot be self-corrective in the sense of approximating ever more closely to a pre-established ideal. Sustaining struggle and dialogue under conditions of openness can therefore be seen as a necessary, but not sufficient, condition of progress. The negotiation of acceptable scientific knowledges and techniques, and of acceptable objectives for these, is a more open-ended struggle than the classical Popperian notion of self-correction recognizes.

We need better critical understanding of the institutional factors in the control and practice of science which inherently encourage scientists themselves to misunderstand and oversell, not only what science can achieve technically, but also more subtly, what authority science can and should exercise over human values.

People from widely different educational, professional, and cultural backgrounds are slowly moving out of complacency and beyond confrontation towards an emerging new collegiality of responsibility for the implementation of scientific and technological innovations, and for the commonweal of this planet. That passage, though, from complacency to confrontation, and then from confrontation to collegiality, is tentative, difficult, and far from complete and secure.

A NEW COVENANT?

Our discussions in Berlin, summarized in this book, were far from easy. They were not always congenial either, and some participants even wondered from time to time about the discussions's coherence. It is unsettling to come to a conference, confident of one's competence, articulate in one's expertise, and accustomed to a shared disciplinary universe of discourse – and then to find oneself passing through the experience of Babel.

The moderator of one working group expressed extreme puzzlement over the apparent inability of a group of highly intelligent and educated people to move beyond divisive "gut reactions" and simple everyday language statements – puzzlement over the group's difficulty in producing articulate and mutually acceptable reasoned statements – on the controversial questions under deliberation. An example was the issue of whether research on the human embryo should be permitted at all, and if so, under what conditions and within what limits. One participant reacted vigorously with the response: "But what we have been doing all along during the past few days, all this intellectual struggling we've been going through, is precisely what this new covenant is all about."

Indeed, do we not all too frequently find ourselves in situations like that described in the story of the tower of Babel? We no longer have, readily available, mutually comprehensible languages or mutually convincing arguments and agreements about what we should or should not do on the frontiers of scientific and technological innovation. The process of breaking through encrusted biases and moving beyond narrow personal and professional perspectives, and doing this by attentively listening to others while we vigorously proclaim our own views, and as we grope towards each other through a common mist of misconstrued meaning (Steiner 1975, p. 65) – this is the process of a new convenant between the biosciences and society.

However, this interpersonal metaphor has limitations beyond which we must also extend. The personalized model of the endless struggle to achieve the conditions of free communication has implicit social and institutional corollaries which require further thought, critical debate, and action. Are the dominant interests and institutions of science, for example, really struggling to communicate freely and openly, or are they balancing competing interests which include power, profit, credibility, and survival as well as some (perhaps parochial!) notion of "the public good"?

Going from the interpersonal metaphor to the institutional dimensions also accentuates the begged question: if we talk about a *new* covenant, what is implied about an existing or previous one? Because there was relative public quiescence until the 1970s, does that imply a *consensus,* and a "covenant" which all parties to it understood and espoused? We will only create the institutional conditions of a more progressive coevolution of bioscience and society if we more fully examine the past rhetorical constructions of science, and the impetus towards scientism.

The growing post-Asilomar involvement of people untrained in science with scientists in the construction of a consensus about what we should do, about what we must prohibit, about what we as a society can tolerate, is a search for norms and limits. These can no longer be simply deduced from traditionally established principles. A society, as Nicola Chiaromonte has put it in his *The Paradox of History,* is not merely a collection of individuals and cannot be reduced either to the sum total of the political and juridical institutions on which it rests or to its economic and cultural structures. Society also consists of the beliefs on which members of a community agree or clash (Chiaromonte, undated, p. 134).

In the classicist notion, culture was unitary, universal, and permanent: there was one set of beliefs, ideals, and norms, and these were the standard for all human beings in all places and all times. Those who did not espouse or aspire to these standards were barbarians. We, however, live in a post-classicist world. Our science is not a ready-made achievement stored for all time in a great book, but an ongoing process that no library, let alone any single mind, is expected to encompass (Lonergan 1988, p. 241). Our ethics, likewise, is not a simple inheritance of principle, completed and ready for universal application. We are in the cultural situation of having to struggle towards a commonality of belief and meaning. We should not, in that struggle, rely on the illusion that we can attain a new permanent universal normative consensus.

The principle of the new covenant between the biosciences and society is not a text of codes, norms, roles, or guidelines. These, as useful as they may well be, are only products, often of only provisional validity, of the real principle of the bioscience-society relationship. That principle is a process that could, in the light of Karl-Otto Apel's work, be called "Dialog als Methode" (Apel 1973). This is the process that generates both science and ethics in a post-classicist world. However, this critical dialogue needs to be sustained and *given effect* through the institutional forms of scientific research and development. Otherwise the dialogue is reduced to unauthentic scholasticism. Whether the needed institutional forms already exist, or what they might be, must also be subject to critical dialogue.

After his visit to East Berlin, Marcus Pembrey, one of the conference participants, told some of us about two painted graffiti he had seen on a remnant of the Berlin Wall. They were the imperatives "Save the Earth!" and "Get human!" The covenants of antiquity, as described in the stories of Noah, Abraham, Moses, and David, encompassed the profoundly human aspirations for a stable natural order, for abundant posterity, for liberation from enslavement, and for survival. These ancient experiences and aspirations of human beings are not foreign to us today, as too abundant a posterity threatens the survival of many, and as human intervention into the natural order of things threatens the ecology of our planet. The graffiti on the Berlin Wall captured the underlying imperatives that occasioned the Berlin meeting on the relationship between the biosciences and society. "Save the Earth!" and "Get human!" To do that, scientists, professionals, and women and men from all the diverse walks of life – all of us – need a relationship as open and as demanding as that of the ancient covenants.

A new covenant between society and its sciences will be uniquely exacting. New scientific capabilities emancipate us from biological constraints, but a simple turning to the past no longer allows us rapidly and surely to discern what new forms of bondage these freedoms may entail. It is equally inescapable to ask in what ways these same capabilities emancipate us from earlier moral and social constraints, and what new disciplines and uncertainties these freedoms harbor in the moral and social domain. We stand, quite surely, at the borders of a new era of responsibility for future generations. So we turn to an ancient word, covenant, and open it to house the largely still unexplored critical process through which we must all pass if we are to protect and enhance the inheritance of people decades distant from us on this planet.

D.J. Roy   B.E. Wynne   R.W. Old

REFERENCES

Apel, K.-O. (1973). *Transformation der Philosophie. Vol 2: Das Apriori der Kommunikationsgemeinschaft.* Frankfurt am Main: Suhrkamp Verlag.

Chiaromonte, N (undated). *The Paradox of History. Stendhal, Tolstoy, Pasternak and others.* London: Weidenfeld and Nicolson.

Lonergan, B. (1988). Dimensions of meaning. In: *Collected Works of Bernard Lonergan,* eds. F.E. Crowe and R.M. Doran, vol. 4. Toronto: University of Toronto Press.

Luria, S.E. (1969). Modern biology: A terrifying power. *The Nation,* October 20, 1969: 409.

Medawar, P. (1990). *The Threat and the Glory. Reflections on Science and Scientists.* New York: Harper Collins.

Singer, M. (1977). The involvement of scientists. In: *Research and Recombinant DNA,* an Academy Forum, March 7–9, 1977. Washington, D.C.: National Academy of Sciences.

Steiner, G. (1975). *After Babel. Aspecs of Language and Translation.* London / Oxford / New York: Oxford University Press.

Bioscience ⇌ Society
edited by D.J. Roy, B.E. Wynne, and R.W. Old, pp. 9–27
© 1991, John Wiley & Sons

Key Techniques in Contemporary Life Sciences

Robert W. Old

Department of Biological Sciences
University of Warwick
Coventry CV4 7AL, U.K.

Abstract. The recent spectacular confidence of biologists is firmly based on the successes of molecular biologists in developing a full range of techniques which endow researchers with the "read, write, cut, copy, and paste" functions for the chromosomes of living organisms. Application of even just the technology of today affects every biological discipline: it is limited by the need for further knowledge of complex biological processes, and of ethics. Gene technology and immunochemistry will drive progress in the two great biological questions for the next century: How do development and ageing of complex organisms occur? and, How does the brain work?

INTRODUCTION

Contemporary Life Sciences is such a broad subject area that it is with humility that one approaches the task of briefly reviewing its key techniques. Many disciplines are subsumed, ranging from organic chemistry to paleontology. One must be selective, and no doubt important techniques will be overlooked. There is insufficient space here to give details of many techniques. These can be found elsewhere (Harlow and Lane 1988; Old and Primrose 1989; Sambrook et al. 1989).

Improvements in microscopy, particularly electron microscopy, will always be crucial in biological investigations. Indeed, it is of interest to note in the context of the current debate about the Human Genome Project that there is the possibility that, during the lifetime of the enormous sequencing effort envisaged, electron microscopy may make it possible to "read" a DNA molecule directly, and hence possibly supercede the current relatively laborious and slow sequencing technology. Also, techniques such as X-ray crystallography and NMR, which can be applied to structural studies of biological macromolecules, must rank as key techniques in biology for the foreseeable future.

It is obvious that there has recently been a revolution in biology, and that this revolution stems from the ability to analyze and manipulate the genetic material of living organisms. My review is focused on this revolutionary technology, and the distinct field of immunology. The gene technology has already affected almost every field of biology, including crop plants, farm animals, diagnosis of inherited and infectious disease, cancer studies, prospects for totally new approaches to medical treatments (for example, exploiting antisense nucleic acids), new drugs, new insights into evolution and anthropology, forensic science, new routes to vaccines, and totally new biological catalysts. Even this impressive list does not convey the full scope of current technology, let alone that envisioned in the near future.

It is possible that we are on the threshold of an industrial revolution that, when fully realized, will be comparable with previous industrial revolutions. The first industrial revolution arose from the application of new sources of energy for the mass production of goods. The second industrial revolution developed from the electronics of telecommunications and the extension of electronic information processing to commercial and industrial purposes. Both revolutions were the result of man's ability to manipulate the world of physics.

The new revolution would be different from the previous ones in an important respect, because this revolution concerns the world of biology. Practitioners of the new biology not only have the ability to analyze living organisms in immense detail, but can also intervene in the genetic make-up of organisms. These abilities to understand and manipulate the biological world are all the more powerful because, of course, we ourselves and the ecosystem on which we depend are possible subjects of this technology. Whatever the wider implications of the new biology eventually turn out to be, it is safe to speculate that gene technology and modern immunochemical techniques will play an important part in advancing other biological disciplines, such as neuroscience. If it is true to say that the 20th century has been the century of physics in science, then the 21st century is going to be the century of biology, with gene technology and immunochemistry leading the way.

HYBRIDIZATION OF NUCLEIC ACIDS

The double helical structure of DNA that was discovered by Watson and Crick in 1953 comprises two antiparallel strands of nucleic acid, in which the bases of one strand of the helix are hydrogen-bonded to bases in the other strand. The base adenine always pairs with thymine; guanine always pairs with cytosine. The two strands of DNA are said to be *complementary* because their sequences of bases are such that they can associate. The two strands of a double helix readily come apart under conditions which disrupt hydrogen bonds, such as heating. This is a reversible process. Under suitable conditions of temperature, pH, and salt concentration, two sequences can come together stably through base pairing *if they are complementary in sequence*. This base pairing can also occur if one or both strands of nucleic acid consist of RNA. (Like DNA, RNA contains bases, but with uracil replacing thymine. Uracil, like thymine, base pairs with adenine.) The formation of any duplex nucleic acid from complementary single strands is called, in rather loose jargon, *hybridization* of the nucleic acids. The specificity of hybridization is very great, because the two strands must be perfectly (or almost perfectly) complementary in sequence. Therefore, by tagging one strand of nucleic acid in some way, for example by making it highly radioactively labelled, a *probe* is created which can be used to associate stably with, and thus detect, the complementary sequence.

There are many variations on this basic principle of the hybridization probe. The specificity and sensitivity of hybridization probes are so remarkable in practice, for example, in detecting a single gene among the whole human genome, that the technology daily performs minor technological miracles in the whole range of molecular biological investigations.

I wish to make a final comment on this subject. The reader should be aware that the molecular biological use of the term hybridization has no connection whatsoever with the original use of the word to describe the product of cross-breeding two strains of plants or animals.

SIMPLE DNA MANIPULATION AND CLONING

In the 1970s a range of techniques was developed which allowed precise cutting and joining of DNA molecules in the test tube. The ability to cut and join DNA led to the ability to construct new combinations of DNA segments from diverse sources. Virtually any DNA could be made to replicate and pass from generation to generation in a bacterium such as *Escherichia coli* by inserting the DNA into a vector. A vector is simply a small, amenable piece of DNA which has the ability to replicate in its own right in the host cell. The first vectors were developed for *E. coli* and were plasmids (extra minichromosomes found in the cell), or virus DNAs. Thus any piece of DNA can be inserted into a vector to create a *recombinant* vector, and *transformation* procedures are available for getting the DNA into the living host cell where it will replicate and be propagated. The technology was soon extended from bacteria to yeasts, other eukaryotic microbes, and to cultured plant and animal cells. As an alternative to using a vector DNA it is possible to select cells in which incoming DNA has integrated into a host cell chromosome. The incoming DNA is then transferred to succeeding cell generations as part of the normal genetic make-up of the chromosome. This latter approach is employed in creating transgenic plants and animals (see below).

An important consequence follows from the use of a vector to carry the foreign DNA: this is because simple methods are available for purifying the recombinant vector molecule from cells. Thus not only does the vector provide the ability for the recombinant DNA to be propagated in the host, and hence the capability for it to be grown in bulk cultures, but the vector also permits the easy preparation of the recombinant DNA, from which the foreign component can be readily released if desired.

EXPRESSION OF MANIPULATED DNA IN HOST CELLS

It is frequently the aim of a gene manipulator to ensure that manipulated DNA sequences are expressed in the host cell, i.e., that a manipulated gene actually directs the synthesis of the protein which it encodes. In general this is achieved by fusing the manipulated coding sequence to DNA sequences that occur naturally in the host cell and whose function is to direct gene expression in that cell by insuring that the processes of transcription and translation occur efficiently. This is what is called the "Homologous Regulon" approach in the words of a well-known early patent in the field.

Without wanting to belittle the technical problems involved in achieving expression and recovery of functional proteins from transformed cells, we can claim that relatively mature technology currently exists for the expression of any desired protein, with a good expectation of success, in any one of a number of host organisms.

TRANSGENIC PLANTS

Transgenic plants are plants into which manipulated DNA has been introduced, so that the introduced DNA is passed to subsequent generations as part of the genetic constitution of the plant. The technology is dependent upon a variety of techniques for introducing the DNA into plant cells, so that it becomes integrated into a plant chromosome. Techniques based upon the plant pathogen *Agrobacterium tumefaciens* are applicable to a range of, mostly dicot (i.e., broad-leaved, not cereals), plant species. Another approach is based upon microscopic metal spheres which are coated with DNA and literally fired from a gun into host cells.

Progress in this field depends crucially upon the ability to recover whole plants from transformed cells, and this is a limitation which has restricted the reliable production of transgenic plants to a few species such as tomato, tobacco, potato, and sugar beet. The important crops maize and rice are yielding to technology, but wheat and barley have proved more difficult.

The prospects of introducing genes conferring insect resistance, fungal reistance, or herbicide resistance into crop plants have already begun to be realized. Agrochemical companies see an important marketing opportunity in the case of herbicide resistant crop varieties; that is because the cultivation of such a variety is combined with the use of herbicide to control weeds. An ecologically more favorable outcome may be foreseen in the use of genetically engineered insect and fungal resistance, which has the prospect of diminishing the use of applied agrochemicals. Transgenic plants with other desirable properties such as altered oil composition are envisioned. However, most of the desirable attributes of crop plants are not well understood at the molecular genetic level and may also require the concerted action of many genes, so that future progress will depend on advances in plant biology on a broad front. The transgenic technology can be adapted to tagging and isolating plant genes and generally advancing plant molecular biology.

Plants are the cheapest form of biomass. This factor will make genetically engineered plants favored hosts for expressing genes encoding intermediate or low value proteins. Mammalian antibody genes have already been successfully expressed in plants: this is a pointer to the future.

TRANSGENIC MAMMALS

Several methods have been developed for introducing manipulated DNA into the germ line of mammals. A prerequisite was the availability of techniques for removing fertilized eggs or early embryos from the mother, culturing them briefly in vitro, and then returning them to the uterus of foster mothers, where embryogenesis could proceed. This opened the way for 1) the mixing of cells from different embryos, i.e., chimera production; 2) introducing pluripotent stem cells such as ES cells into developing embryos, where they could populate a range of cell lineages including the germ line; 3) microinjecting DNA into eggs, so that it integrates into a chromosome; and 4) infection by recombinant virus vectors. Among these methods, the one which has had the greatest general impact is the direct microinjection of DNA into the newly fertilized egg. The resulting adult animals can be bred to establish transgenic progeny lines. This methodology was developed in mice, and has been extended to farm animals, including sheep and cattle.

Transgenic mammals provide the test-tube for studying gene regulation during embryogenesis, for studying the action of oncogenes in the progression of tumors and for studying the intricate interactions of cells in the immune system. In other words, the whole animal is the ultimate assay system for studying genes which control complex biological processes (Hanahan 1989).

Importantly, transgenic mammals can be used as hosts for the expression of valuable recombinant proteins, especially in instances where the protein carries modifications, such as glycosylation, which are important for its function (pharmokinetic properties may be affected), and which are characteristically performed by animal cells but not other host systems. The genetic farming concept is comparable with the practice of raising valuable antisera in animals, except that a single injection into a feritlized egg substitutes for multiple somatic injections. An example of this approach which is already in progress is the creation of transgenic sheep, in which a human blood-clotting factor IX gene is expressed under the control of milk protein regulatory DNA sequences. The blood-clotting factor is secreted into the milk of lactating transgenic animals. Further work is required to improve the production performance of such animals.

Transgenic mammal technology has the prospect of improving animal resistance to diseases, but conversely, questions arise about the welfare of animals genetically engineered for the high-level production of foreign proteins.

An important application of transgenic technology is the creation of animal models for human genetic disorders: these animals are a test system for possible treatments. Most simple genetic disorders are the result of the inactivation of a particular gene. This means that techniques for "knocking out" particular genes in transgenic animals are required. This can be achieved by targeted gene inactivation procedures wh;ich are available for cultured pluripotential

ES cells. Once these cultured cells have been manipulated and screened for the inactivation of the gene in question, such cells can be introduced into a mouse embryo, where they will enter inter alia the germ line. Heterozygous and homozygous mutant progeny can be obtained by breeding from such animals.

A question raised by transgenic animal technology is whether it should be possible to obtain patent protection for strains of animals created by this technology. This question is still not resolved in most jurisdictions.

ANALYZING DNA SEQUENCES

Knowledge of the sequence of a DNA region may be an end in its own right, for example in understanding mutations responsible for a human genetic disorder. In any event, sequence information is a prerequisite for any substantial DNA manipulation.

The technology currently favored is based upon the "dideoxy" or chain terminator approach of Sanger. The original technology employs radioactive labelling during the sequencing reactions, followed by detection of the products, resolved by gel electrophoresis, using autoradiography on X-ray film. The biochemical reactions are amenable to automation, but the gel electrophoresis, interpretation of the autoradiograph, and inputting of the data into a computer file all present problems. This has been partly overcome by the development of automated scanners to read and interpret the autoradiograph, but a more radical solution is to replace the radioactive label with fluorescent tags. By the expedient of using different tags for the four sequencing reactions, the sequencing reaction products can be mixed and resolved in a single electrophoretic separation. Products are detected as they pass a laser scanner: this is readily amenable to automated reading (Brumbaugh et al. 1988).

Once substantial amounts of data have been accumulated, very considerable computer capacity is required for a full analysis of the data base. The scale of the task for the human genome can be appreciated by realizing that the 3×10^9 nucleotides of DNA requires 5 CD discs to file it. The human genome project will be discussed by other participants; here I wish to raise just the simple point of accuracy. It is clear to any molecular biologist that the current sequence data base is very imperfect. Extrapolating from one's own limited knowledge of particular entries, it is clear that errors are common. Is this important? What level of accuracy is acceptable in the human genome project? Most proponents of large-scale sequencing projects are technological optimists who expect substantial future improvements in speed and accuracy.

ANALYZING VERY LARGE DNA REGIONS AND CHROMOSOME JUMPING

"Conventional" cloning vectors typically accept DNA fragments up to about 45 Kb. As a rule of thumb, about 1000 Kb corresponds to a recombination frequency of 1% in humans, which represents a map distance which is difficult to better in genetic studies in man. The problems of mapping, and otherwise analyzing, such large distances have been reduced by progress in four areas:

1) Restriction endonucleases have been discovered which cut DNA at very infrequent sites. For example, the enzyme *Not I* cleaves DNA at the octanucleotide target sequence GCGGCCGC. Such long target sequences are rare, and would be expected to occur by chance (assuming a random DNA sequence, with all four bases equally common) about once every 64 Kb. In fact mammalian DNA is deficient in the dinucleotide sequence CG, making target sites (including this dinucleotide) rarer than expected. Additional specificities in restriction endonuclease target sites may be created by combining the specificity of certain methylases with that of restriction endonucleases.
2) Gel electrophoretic systems have been developed in recent years which can resolve DNA fragments up to 2000 Kb in size. This is in the size-range of whole chromosomes in yeast *Saccharomyces cerevisiae.* These electrophoretic systems depend upon the behavior of DNA in electric fields which invert, or otherwise vary, periodically.
3) DNA can be cloned in specialized yeast vectors as artificial chromosomes (YACS, yeast artifical chromosomes). Fragments of several hundred Kb can be cloned in this way.
4) Chromosome jumping is a long-distance version of chromosome walking. These techniques (Fig. 1) allow the experimenter to start with a known DNA sequence at one location on a chromosome, and then move in steps along the chromosome by cloning adjacent or nearby DNA fragments. The jumping technique depends upon the circularization of very large DNA fragments, followed by the cloning of the region of the closure site, thus bringing together DNA sequences that were originally hundreds of Kb apart in the chromosome.

DNA SYNTHESIS

A notable breakthrough in 1977 was the expression of a gene that had been synthesized chemically (Itakura et al. 1977). The technology for chemical

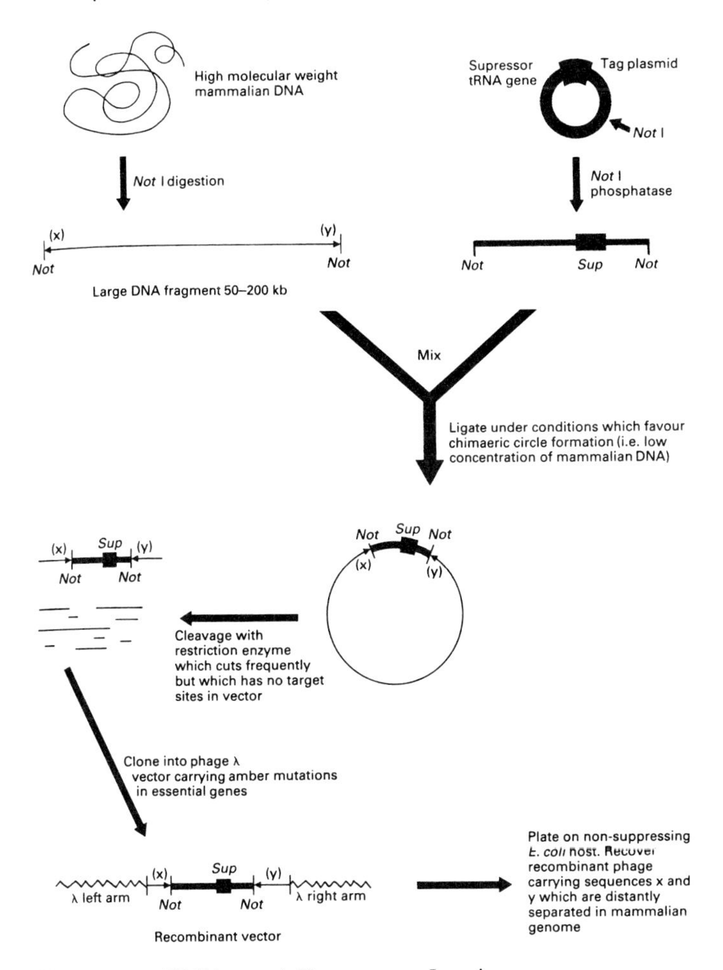

Fig. 1. Chromosome Walking and Chromosome Jumping.
A. Chromosome walking. It is desired to clone DNA sequences of gene B, which has been identified genetically, but for which no probe is available. Sequences of a nearby gene, A, are available in cloned fragement 1. In a large random genomic DNA library, many overlapping cloned fragments are present. Clone 1 sequences can be used as a probe to identify overlapping clones 2, 3, and 4. Clone 4 can, in turn, be used as a probe to identify clone 5, and so on. It is therefore possible to walk along the chromosome until gene B is reached.
B. Jumping library construction. The example illustrates the use of *Not I*-digested mammalian DNA. Large DNA fragments are circularized in vitro, and distant sequences that are thus brought together can be cloned by use of a suppressor gene "tag".

synthesis of DNA has developed dramatically since then, so that today automated machines are widely available which are capable of synthesizing any desired DNA sequence, in a matter of hours, in the range up to about 100 nucleotides. Sequences longer than this can be assembled by joining shorter units. The consequences of this capability are important.

Chemically synthesized short oligonucleotides are required for site-directed mutagenesis methods, which allow the alteration of existing DNA sequences at the will of the experimenter. If desired, complete, entirely novel, genes can be synthesized.

The first generation of biotechnology-based therapeutic agents were essentially natural proteins (e.g., human insulin, growth hormone, interferons, tissue plasminogen activator), whose supply was limited. Second generation products currently under development include modified agents which are based upon natural proteins, but which are altered for a pharmacologically desirable effect. For the future we have the prospect of entirely novel proteins. The freedom of the molecular biologist in this field is almost unlimited. Consider that a typical protein consists of about 300 amino acid residues. Since there are 20 different amino acids in proteins, the number of possible proteins of this length is 20^{300}, or about 10^{390}. This is a superastronomical number. For comparison, the number of fundamental particles in the observable universe is estimated at about 10^{80}. So we certainly could not even begin to synthesize and test every possible protein: we would very soon run out of matter! It is also evident that the range of proteins tested throughout the course of evolution is an infinitesimal fraction of the total possible. Although the experimenter can synthesize and express a gene for any protein, he does not have the theory to design a functional protein ab initio. Until protein structure and activity can be predicted from a knowledge of an amino acid sequence, biochemistry must be regarded as a very incomplete science.

THE POLYMERASE CHAIN REACTION

The PCR is a technique which has shaken molecular biology to its technical foundations. Like many great ideas, the PCR concept is simple, and it is remarkable that PCR was discovered only relatively recently, in 1985 (Erlich 1989). The ease with which PCR can be performed, and its tremendous range of applications, make this a truly outstanding technical development.

The technique depends upon the ability of a small piece of DNA, a chemically synthesized "primer" oligonucleotide of 15–50 nucleotides, to find and bind to a complementary target sequence in a long stretch of DNA. Two primers are used, and these define the region of duplex DNA that will be amplified by the reaction (Fig. 2). By repeating the amplification cycle 25–30 times, amplifications of over a millionfold are achieved in practice within a few hours. There are two important aspects to the PCR reaction, one is the enormous amplification achieved, the other is that only a particular specified region of DNA is amplified. The procedure is simple and very readily automated. The technique is extremely sensitive; even a single molecule can be amplified. With such sensitivity, contamination of the sample or reagents can be a serious problem. Once the amplified DNA is in hand there is sufficient of it for analysis by almost any of the methods of modern molecular biology.

Many variations of the basic procedure exist. These often exploit the fact that molecules covalently attached to the 5′ ends of the primers will be incorporated at the ends of the amplified product.

In many situations where cloning of a DNA was once necessary, the PCR obviates this need. PCR speeds up the prenatal diagnosis of genetic diseases (in some circumstances its sensitivity allows the analysis of fetal DNA present in trace amounts in blood samples withdrawn from the mother, hence avoiding more invasive sampling procedures), it can be used to detect viruses and other pathogens, and it can be applied to tissue typing. The extreme sensitivity of the technique can be put to use in other ways. Since a single intact DNA molecule of a few hundred nucleotides can be amplified, a DNA sample that is degraded may still yield a result. A region of mitochondrial DNA from a woolly mammoth, dead for 40,000 years, has been analyzed in this way (Paabo et al. 1989). The reaction can be applied to biological evidence from the scenes of crimes, but scrupulous avoidance of contamination is necessary.

DETECTION OF GENETIC VARIATION IN MAN

There are an estimated 4,000 single gene defects known in man. Defective genes responsible for inherited disorders range from the rather common (cystic fibrosis has a carrier frequency of about 5% of the Caucasian population, and a disease frequency of about 1 in 2000 live births) to the extremely rare. Many of these disorders are serious for the affected individual and his family, and although treatments may be available, there is usually no cure. Prenatal diagnosis in affected kinships is an important option in such circumstances.

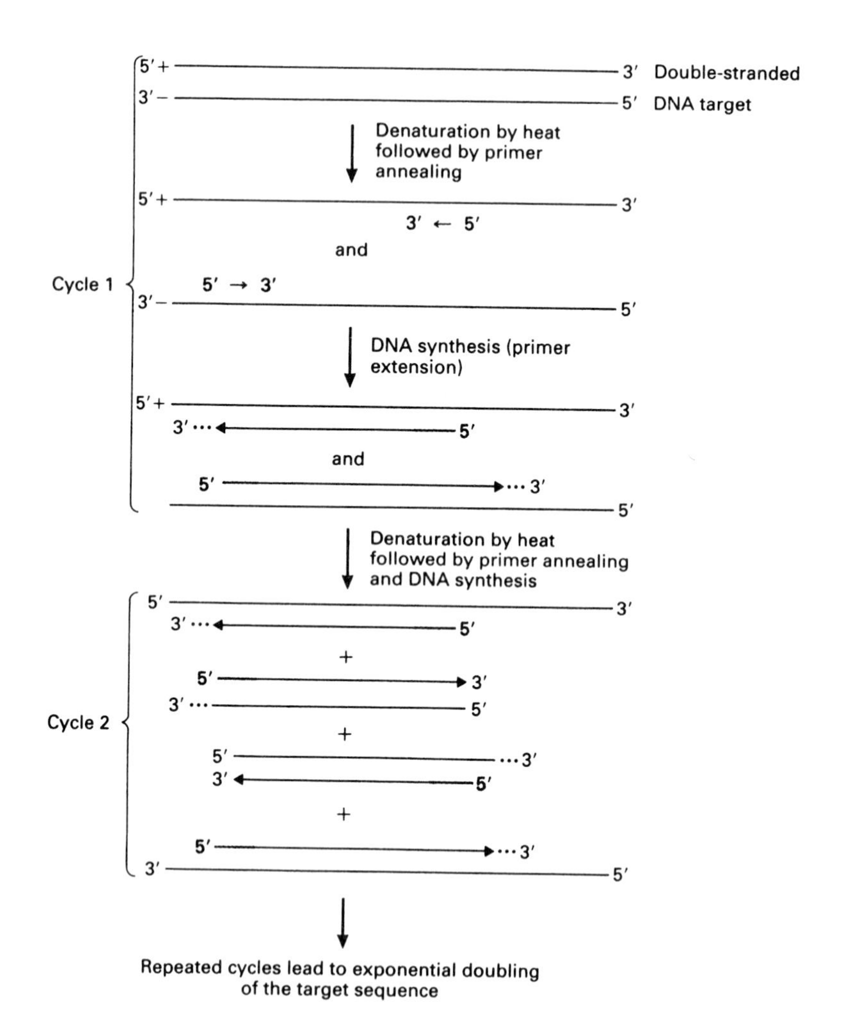

Fig. 2. The Polymerase Chain Reaction. Primers bind to heat-denatured DNA, and are extended in a reaction catalyzed by the thermostable DNA polymerase, *Taq* DNA polymerase. Note that although the extended DNA is not a full copy of the long starting DNA, it will be amplified if it extends beyond the primer sites which define the region to be amplified.

Much research is devoted to identifying genes resonsible for inherited disorders in man. The recent success in identifying the cystic fibrosis gene is a landmark in this field (Rommens et al. 1989), especially in view of the fact that the approach was rather general in its applicability. Chromosome jumping played an important part in this work. The whole gamut of DNA techiques can be applied to human DNA analysis once a DNA region has been identified, PCR being almost inevitably involved.

The discovery of the gene defect in an inherited disorder should lead to an understanding of its molecular basis and provides at least the prospect of new pharmacological or gene therapies. A distinction must be drawn between somatic gene therapy and germ line gene therapy. In the former it is envisioned to insert a functional gene into somatic cells of the body – blood-forming tissue is the target of much research. If stem cells of the blood-forming lineages can be transformed with a functional gene, and if these cells can repopulate the bone marrow of the affected individual (in a bone marrow transplant with his own transformed cells), then there is the prospect of somatic gene therapy for certain disorders. Introduction of genes into the germ line of man, using techniques such as those used in creating transgenic animals, suffer from the ethical problem that the alternation, with any unforeseen detrimental effects, can be transmitted to progeny. Germ line therapy conjures up images of controlling human evolution.

Testing for genetic defects in man is accompanied by many ethical and counselling problems. For example, consider the case of identifying a healthy person as being positive for the dominant inherited disorder Huntington's disease. This is a serious disease which usually becomes apparent in later life, after the time at which an individual may have already produced a family. Little can be done to influence the progress of the disease, and no cure is available. In such circumstances individuals at risk may not wish to have a test, and if they do so wish, very careful counselling is required.

Testing can reveal the genotype of individuals at gene loci known to influence predisposition to heart disease or cancer. These are facts which may be used for discrimination by insurance companies. The recent experience with HIV does give grounds for concern, with insurance companies asking such questions as "Have you ever been tested for HIV?" An affirmative answer prompts a request for the result of the test, and an HIV-positive individual is definitely uninsurable for a whole range of health and life-related insurance risks. One result of this scenario may be the avoidance of tests. However, in the future it seems very probable that genotypic information will be demanded by insurance companies as part of an individual's medical profile, just as phenotypic information is routinely the stuff of current medical assessment (e.g., cholesterol level, hypertension, etc.).

"DNA fingerprinting" is a term brought to the fore by the discoveries of Jeffreys and co-workers (Jeffreys et al. 1985). They exploited the discovery of so-called minisatellite sequences which occur as tandemly repeated blocks of a repeating unit. These blocks of repeats are scattered throughout the human genome and are very variable from individual to individual; in other words, humans are extremely polymorphic at such genetic loci. This variation is revealed by digesting genomic DNA with restriction enzymes, and probing Southern blots with a probe containing the repeat unit. Because a large number of variable bands are produced in such an analysis, the pattern is unique to the individual. Fingerprints obtained by such methods can be prepared from biological evidence from the scenes of crimes, and thus, used to incriminate or exonerate suspects. In one early case the admission of avoiding the test led to suspicion falling on a man who was subsequently convicted.

The bands revealed in DNA fingerprints correspond to genetic loci, and so any band in an individual must be evident in the fingerprint of either his mother or father. In fact, conclusions of unprecented accuracy can be drawn about kinship, leading to the adoption of this procedure for evidence in immigration disputes in the U.K.

IN VITRO FERTILIZATION AND EMBRYO TRANSFER

In the 1960s techniques were developed in experimental animals for recovering eggs, fertilizing them in the test tube, and implanting them in the uterus of unrelated foster mothers. This was the technology that made transgenic animal technology possible (see above). The technology was extended to a range of farm animals, including sheep, pigs, and cattle. In later years, IVF was developed for humans, where it can be applied to overcome certain causes of infertility. With domesticated animals the related technology of transporting frozen embryos interantionally is valued for its economy and avoidance of disease transmission. Embryo multiplication is being actively researched. One technique is to separate individual cells from the 4-cell embryo from which 4 complete, identical embryos can be produced.

Following the birth of the first baby resulting from human IVF in 1978, Louise Brown, there has been growing public debate of issues concerning IVF and human embryo research. This culminated in recent legislation in the

U.K. to regulate human embryo research. Research is permitted, under certain conditions, for a period of up to 14 days.

MONOCLONAL ANTIBODIES

Fused cell techniques allow the fusion of two dissimilar cells to yield a novel cell containing the whole, or parts, of the genetic material of both parents. These techniques are the basis of protoplast fusion procedures for creating novel plant hybrids. However, cell fusion is most notable as the basis of monoclonal antibody production.

In order to obtain an unlimited supply of a particular antibody molecule, an immortal clone of cells (B lymphocytes, the antibody-secreting cells of the immune system) is created. The B cells are not naturally immortal, that is, they do not grow indefinitely in culture, but they can be made immortal by fusing them with an immortal, malignant myeloma cell. The fused ("hybridoma") cells produce antibodies, and furthermore they are clonogenic, i.e., single fused cells can grow into a clone which can be selected and screened to isolate a hybridoma producing the desired, monoclonal, antibody molecule. The technology is mostly applicable only to mouse antibodies, but manipulation of the mouse antibody genes can be performed so as to make them resemble their human counterparts.

Monoclonal antibodies have many applications throughout biology (Sikora and Smedley 1984). The property that makes monoclonal antibodies so useful is the fact that all the molecules in a preparation are identical (and the preparation can be obtained in unlimited amounts if desired). Their reaction with an antigen is exactly the same each time. They are reagents which can be used to detect and quantify complex biological molecules with great precision. By linking other molecules to antibodies, the specificity of the antibody can be harnessed for imaging tumors, or targeting drugs to tumors.

SINGLE DOMAIN ANTIBODIES

This important recent advance employs a simple expression system in bacteria for the production of single domain antibodies. This exploits the finding that a single chain of an antibody molecule can have sufficient specificity and binding activity to be useful and that even an unimmunized animal displays sufficient antibody diversity for the selection of reactive antibodies against many antigens. Libraries of immunoglobulin sequences can be expressed in bacteria, and screened for expression of light chain domains with any desired binding specificity. The single domain approach has been extended by Lerner's group (Huse et al. 1989). They have combined two libraries, so that bacteria express both types of antibody chain, allowing a greater range of specificities than the single domain approach. Again, libraries can be screened for bacteria, producing antibodies with a desired binding specificity. These techniques hold the promise of selecting highly specific monoclonal antibodies within just a few days, in contrast to the conventional approach, which takes several months.

CATALYTIC ANTIBODIES

One of the ways in which enzymes catalyze reactions is to bind to, and hence stabilize, an unstable and energetically unfavored chemical intermediate that must be formed along a reaction pathway. In this way the enzyme lowers the energy barrier between substrate and product.

Antibodies can be raised, which bind to *stable* chemical *analogues* of unstable reaction intermediates. By binding to a true reaction intermediate, the antibody stabilizes it and can therefore catalyze the reaction in which the intermediate is involved. Recent research has confirmed that such antibodies can be catalytic. The reaction rate enhancements achieved so far do not reach the high values typical of enzymes, but can match the rates achieved when enzymes act on unnatural substrates.

This technology allows the creation of regioselective, stereospecific catalysts for reactions which do not occur in living organisms and for which no enzyme catalyst is known (Shokat et al. 1989).

REFERENCES

Brumbaugh, J.A., L.R. Middendorf, D.L. Grone, and J.L. Ruth (1988). Continuous, on-line DNA sequencing using oligodeoxynucleotide primers with multiple fluorophores. *Proceedings of the National Academy of Science U.S.A* **85:** 5610–5614.

Erlich, H.A. (1989). *PCR Technology: Principles and Application for DNA Amplification.* New York: Stockton Press.

Hanahan, D. (1989). Transgenic mice as probes into complex systems. *Sceince* **246:** 1265–1275.

Harlow, E. and D. Lane (1988). *Antibodies: A Laboratory Manual.* New York: Cold Spring Harbot Laboratory Press.

Huse, W.D., L. Sastry, S.A. Iverson, A.S. Kang, M. Alting-Mees, D. Burton, S.J. Benkovic, and R.A. Lerner (1989). Generation of a large combinatorial library of the immunoglobulin repertoire in phage lambda. *Science* **246:** 1275–1281.

Itakura, K., T. Hirose, R. Crea, A.D. Riggs, H.L. Heyneker, F. Bolivar, and H.W. Boyer (1977). Expression in *Escherichia coli* of a chemically synthesized gene for the hormone somatostatin. *Science* **198:** 1056–1063.

Jeffreys, A.J., J.F. Brookfield, and R. Semenoff (1985). Positive identification of an immigration test case using human DNA fingerprinting. *Nature* **317:** 577–579.

Old, R.W., and S.B. Primrose (1989). *Principles of Gene Manipulation,* 4th edition. Oxford: Blackwell Scientific Publications.

Paabo, S., R.J. Higuchi, and A.C. Wilson (1989). Ancient DNA and the polymerase chain reaction. *Journal of Biological Chemistry* **264:** 9709–9712.

Sambrook, J., E.F. Fritsch, and T. Maniatis (1989). *Molecular Cloning: A Laboratory Manual,* 2nd edition. New York: Cold Spring Harbor Laboratory Press.

Sikora, K., and H.M. Smedley (1984). *Monoclonal Antibodies.* Oxford: Blackwell Scientific Publications.

Shokat, K.M., C.J. Leumann, R. Sugasawara, P.G. Schultz (1989). A new strategy for the generation of catalytic antibodies. *Nature* **338:** 269–271.

Rommens, J.M., M.C. Iannuzzi, B.-S. Kerem, M.L. Drumm, G. Melmer, M. Dean, R. Rozmahel, J.L. Cole, D. Kennedy, N. Hidaka, M. Zsiga, M. Buchwald, J.R. Riordan, L.-C. Tsui, and F. Collins (1989). Identification of the cystic fibrosis gene: Chromosome walking and jumping. *Science* **245:** 1059–1065.

GLOSSARY

antibody: A protein made by, and often secreted from, B lymphocytes of a vertebrate animal in response to a foreign antigen with which the antibody can bind. Antibodies are immunoglobulins.

antisense: A strand of DNA or RNA that is complementary to a mRNA or other single stranded (e.g., viral) RNA. It has the potential to base-pair with its complementary sequence and possibly interfere with the normal functioning of the mRNA in vivo.

complementary: A nucleic acid sequence is said to be complementary to another if it is able to form a hydrogen-bonded duplex with it, according to the rules of base-pairing, A with T (or U), and C with G.

ES (embryo-derived stem) cells: A cell line, derived from early mouse embryos, which can be cultured in an undifferentiated state in vitro. These cells are pluripotential when introduced by injection into an early embryo. Their descendants form many cell lineages including germ cells.

Kb (Kilobase): A one thousand base sequence of a nucleic acid molecule.

library (gene library): A large random collection of cloned DNA fragments that collectively includes all the genetic information of the source from which the DNA was originally derived.

minisatellite: Short DNA sequences (<100 bp) which are repeated many times and dispersed at multiple sites throughout the chromosomes of an organism.

monoclonal antibody: An antibody preparation which contains only a single type of antibody molecule. A clone of cells producing a single antibody type may be prepared by fusing normal lymphocytes with myeloma cells to create a hybridoma.

oligonucleotide: A short nucleic acid molecule (can be synthesized chemically).

plasmid: An extra-chromosomal circular DNA molecule capable of autonomous replication in a cell.

pluripotent (stem cell): A cell that is capable of multiplying and differentiating into a wide variety of differentiated cell types.

primer: An oligonucleotide designed to be complementary to a region of single-stranded DNA. The enzyme DNA polymerase absolutely requires the end of a primer at which to begin polymerisation.

polymorphism: The existence, commonly, of more than one DNA sequence at a particular genetic locus, in a population.

recombinant: A DNA molecule containing a new combination of sequences.

Regulon (Homologous): DNA sequences adjacent to the coding region of a gene which are recognised by cellular machinery responsible for initiating transcription and translation. Such sequences are homologous when they are derived from the same species of host cell as that in which they are intended to function.

restriction endonucleases (= restriction enzymes): Enzymes which recognise and catalyse cleavage of double-stranded DNA sequences at specific sites on each strand, generating either blunt ends or ends with single stranded protrusions. Usually isolated from bacteria.

site-directed mutagenesis: A technique in which a cloned gene is *specifically* mutated in vitro so that a particular, desired change in its sequence is created.

Southern blot (transfer): A technique which combines the resolving power of gel electrophoresis with the detection sensitivity of nucleic acid hybridisation. DNA fragments separated in an agarose gel are transferred, usually by blotting via capillary action, to a sheet of nitrocellulose paper (or some other sheet) to which the DNA can be permanently fixed in a denatured (i.e., single stranded) state. The blotted DNA is then detected by hybridisation with a labelled (often radioactively labelled) "probe" consisting of a nucleic acid which is complementary to sequences to be detected on the sheet. The hybridised probe is revealed by autoradiography.

transcription: The process in which a single-stranded RNA molecule, complementary to a DNA template, is synthesised by RNA polymerase.

transformation: In genetic engineering this is the term used to describe the introduction of exogenous DNA into a host cell where it is maintained.

translation: The process of protein synthesis carried out by ribosomes under the direction of messenger RNA (mRNA). The sequence of bases in the mRNA is translated into an amino acid sequence according to the triplet genetic code.

Bioscience ⇌ Society
edited by D.J. Roy, B.E. Wynne, and R.W. Old, pp. 27–40
© 1991, John Wiley & Sons

Does Society Have a Claim on My Body? Legally? Morally?*

Diana Brahams

Barrister-at-Law
15, Old Square, Lincoln's Inn
London WC2A 3UH, U.K.

Abstract. The integrity of a society, and indeed its survival, will in part depend on the regulation and preservation of the rights of individuals, both as between individuals and between the individual and society at large. Where the personal wishes or interests of an individual and/or his family unit and society as a whole come into conflict, there will be a balancing of interests which will be upheld by law.

Society, usually acting in its capacity as a nation state, may override individual rights and wishes in order to preserve the accepted fabric and greater good, e.g., payment of taxes. Where its survival is at risk – for example, in times of war or famine – society may exact greater sacrifices from individuals. To preserve its moral integrity, and at times its very existence, a society may impose negative or positive restraints or requirements on individuals which may affect their liberty, for example, to read obscene literature, to travel freely, to have an abortion, to be entitled to all forms of medical treatment, or to assist death. Society may even require individuals to place their bodies at its disposal, as in conscription for military service, or that an individual's body be examined after death by autopsy, for example, to determine the cause of death. Autopsy may be required even when it is against the stated wishes of the individual when alive or

* The text of this paper contains material which has been the subject of earlier publication in *The Lancet,* with whose kind permission it is reproduced here.

against the wishes of relatives after the person's death. How far should society demand that the tissue or organs of a dead fetus or deceased individual be available for transplantation into other members of society in order to alleviate their medical condition? Should society restrain living individuals from donating or selling their organs for transplantation or for research?

INTRODUCTION

In all civilized societies, there are laws which regulate the interaction of citizens both with the state and with each other. In Britain, there traditionally were considerable feudal claims over an individual's body, from the king down to his nobles, and then on downward to the meanest serf. These feudal dues could range from money or taxes in kind, to working on a lord's land, to the first night with a bride (the *droit de seigneur*), to produce from a tenant's farm or small-holding, to work on the lord's land, and to fight in the king's army or that of a nobleman, who might in turn have to render up a fighting force to put at the king's disposal. Certainly, there can be no greater claim on the body of an individual than the placing of his life at the disposal of the state or its delegate or emissary.

The enforcement of such provisions and societal claims has been backed by law, and continues to be backed by law, which at the same time limits the legal ability of a lucid adult individual to consent to an injury inflicted on himself which was not therapeutically intended. These two apparently conflicting legal aims are usually reconcilable; the common law principle, which rendered deliberate maiming of an individual unlawful, had its origins in the king's interest (the head and policy maker of society) in preserving as many healthy and able-bodied fighting men as possible in order to defend the realm against attack or subversion. Maiming for the purpose of begging may also have been contrary to public policy and morals generally, however.

Consent to a surgical operation for a purpose recognized as valid by the law is effective, and this includes a sex change operation and, presumably, cosmetic surgery and organ transplants. It is otherwise when the purpose is one condemned by the law. In 1604 Chief Justice Coke tells us that "a young strong and lustie rogue to make himself impotent, thereby to have the more colour to begge or to be relieved without putting himself to any labour, caused his companion to strike off his left hand." Both of them were convicted of mayhem. Maiming, even with consent, was unlawful because it deprived the king of a fighting man. In early Victorian times, soldiers had to bite cartridges as part of their military drill (biting the bullet). A soldier got a dentist to extract his front teeth so that he could avoid the drill and render himself unfit. They were both found guilty of committing a crime. Recently, a court approved the decision of a known child molester and abuser to consent to chemical castration for a period of time in order to reduce his urge to reoffend.

Fighting in the street or at home, unless regulated by the appropriate rules or for purposes considered proper, is unlawful even if both sides consent. Where the safety of the realm is at stake or potentially so, however, society may actually require an individual to cause harm to others or to justify his refusal. Though formal conscription to the army was only introduced in Britain during the First World War (1914–1918), in earlier times the infamous "press gangs" roamed the country and seaports in search of young men whom they could force into military or naval service. Recruitment to the services was aided by emotional appeal or blackmail, inducing the individual man to volunteer. Posters with slogans such as "Your Country Needs You" and "What Did You Do in the War, Daddy?" were particularly successful in this regard.

Desertion from the forces in wartime was a capital offence, though conscientious objection was grudgingly accepted as an option for refusal to serve. Other alternatives for fit men of fighting age were war work requiring either special skills or regarded as especially vital to the war effort. Women were drafted into jobs of vital importance to the nation, many of which were formerly performed by the men who had gone to war. Many countries continue to have conscription in peacetime, but Britain abolished this practice in the late 1950s. However, even in Britain, fighting for Queen and Country continues to be a claim that society may choose to make, superceding the autonomy of its citizens for the collective good.

Where medical and food resources are low, there may be rationing. In war, triage is accepted in situations where there are shortages of personnel, equipment, and drugs. The most badly wounded casualties, or those least likely to benefit from scarce medical resources, may be sacrificed so that those with a better chance of survival and response to treatment can be given preference. Supplies of limited scarce drugs and skilled medical personnel in times of war are likely to be diverted to the military hospitals first, presumably with the aim of preserving the fighting force to protect the nation and of encouraging those who have volunteered to have confidence that society will look after their welfare.

ABORTION AND HUMAN EMBRYO RESEARCH IN BRITAIN

These few examples set out above clearly demonstrate that society has made demands on the bodies of its citizens for centuries. It has also had the power to regulate to a considerable extent what may be done with or to them. The underlying reason for such societal restraints are likely to reflect Judeo-Christian thinking and moral norms, as well as society's undoubted pragmatism. Suicide was regarded as self-murder, and only legalized in 1961. Abortion of a nonquick fetus was not a crime according to common law in Britain, but all abortion was criminalized by statute in the early nineteenth century, subject to the right to kill the fetus in order to save the mother or protect her health.

It took the Abortion Act 1967 (which does not apply to Northern Ireland) to legalize abortion under certain conditions. Overnight, on the passing of the act, the position changed, and women were allowed much greater freedom to choose whether or not to continue with a pregnancy. In 1990 the new, much debated Human Embryology and Fertilization Act has, inter alia, reduced the time span for social abortions to 24 weeks, but extended until birth the time for aborting fetuses thought to be seriously handicapped or whose continued presence endangers the mother's life (the latter was always the position). The act also regulates experimentation on human embryos, which is allowed until 14 days (i.e., the primitive streak phase). Spare embryos for the purpose of research may be created, but all clinics and laboratories engaging in such human embryo research must be licensed and inspected.

Early in 1988, reports emerged of experiments which involved the transplantation of tiny portions of aborted fetal brain tissue into the brain of patients suffering from Parkinson's disease. How far should such experiments be restricted, subjected to controls or guidance issued by society or the medical profession, hereby regarded as representing society?

The possibility that a woman might deliberately conceive to create a fetus which she would abort to supply appropriate tissue or spare parts for transplantation into a sick relative was a risk which was felt to be morally unacceptable by most people. Hurriedly, guidelines were issued. Thus, doctors and scientists who in the past had gone quietly about the business of collecting aborted fetal remains for research purposes were now required to seek the aborting mother's permission to use her fetal material in this way, even though the alternative was consignment to the incinerator or dustbin.

BODY PARTS AS PROPERTY

Patients going into hospital to have a diseased organ removed rarely give any thought as to what will become of it. Most of us assume that such material is valueless in monetary terms, and that after pathological examination it will be speedily disposed of (probably by incineration). But suppose our doctor or hospital employees decide instead to sell it for animal food? Does our tacit consent to destruction allow for this? Or suppose our organ contains potentially valuable biological properties, which are later patented by the doctors for their own financial benefit? Could we call them to account? Who (if anybody) owns the organ, once it is removed from the patient's body? Has the patient any rights over it, and does he retain any, notwithstanding his tacit agreement to its destruction? Is the patient entitled to sue for conversion? A long-running dispute in California between a patient and patentees over the ownership of an immortal and valuable cell line cultured from a diseased spleen gives cause for thought.

JOHN MOORE'S CLAIM

In 1976 John Moore, a surveyor on the Alaskan pipeline, was suffering from hairy cell leukemia. Dr. David W. Golde, (a hematologist-oncologist working at UCLA), removed Moore's spleen, which was grossly enlarged from about half a pound in weight to over fourteen pounds. In consequence the patient's health rapidly improved. After surgery was completed, Dr. Golde took a sample from the spleen, and isolated and cultured an immortal cell line capable of producing a variety of products, including the lymphokine GM-CSF (granulocyte-macrophage colony stimulating factor) which is being tested as a treatment for AIDS. In 1979 Golde took steps to patent his discovery with the University of California, and in 1983 the university applied for a patent on the cell line, naming Golde and his research assistent Shirley Quan as inventors. The patent was granted in 1984 (Annas 1988). In the meantime, the patient had moved to Seattle, but returned to consult Dr. Golde at six-month intervals. Reportedly, he stated that he would not have known about the cell line had Dr. Golde not contacted him in September 1983 to tell him that he had "missigned the consent form, circling '[I] do not', rather than '[I] do', grant the University all rights in any cell line." The patient consulted a lawyer, and in September 1984 proceedings were commenced against the University of California, Golde, Quan, the Genetics

Institute Inc, and Sandoz. Allegations included inter alia conversion, lack of informed consent, and breach of fiduciary duty, an account, and other relief.

At first instance, Moore's claim was struck out as disclosing no cause of action (by way of preliminary issue), but this decision was reversed in July 1988, when by a majority of 2 to 1 the California Court of Appeals held that Moore was entitled to bring an action in conversion. (An appeal has been lodged).

The appeal court concentrated on the issue of conversion, a tort of "strict" liability whose principles are based in English common law. Conversion requires neither knowledge nor intent on the part of the defendant. In his discussion of the issues (Annas 1988), Professor George Annas considers that Moore's case raises three questions regarding conversion: a) Were the spleen cells Moore's property? b) Did Golde wrongfully take them? c) Did Moore suffer damage in consequence? Annas concludes that technically the court came to the right decision, but that he would like to see the U.S. laws amended to prohibit the patenting of human cells and the sale of human tissue and cells for any purpose. This looks to be unlikely in the light of further developments. Moore's case moved on a stage further when, in the summer of 1990, on appeal, the High Court of California ruled that a patient did not hold the rights to scientific and medical patents derived from tissues removed (presumably lawfully) from his body. The legal inference is that passivity or silence on the part of the source body of the material will not, where this is commercially exploited, lead to the passing of ownership to the researcher or hospital (*John Moore vs. The Regents of the University of California,* 202 Cal. App. 3d. 1230; [1988] Cal. App. Lexis).

Interestingly, both parties seem to regard the court's decision as a victory for their own corner, though Moore's is pyrrhic and moral; thus far his claim to the financial profit on his cell line has failed. However, most importantly, the judge held that a patient's consent must be obtained before research is carried out. Although the ruling is binding only within the state of California, it has set a precedent: the California court has ruled that doctors must tell their patients in advance of an operation if they intend to use their genetic material for research purposes, and must list the potential uses to which it is to be put, and in this way the patient would have the opportunity to negotiate an appropriate fee with a drug or development company. It remains to be seen how this commercialization of body parts works out in practice.

THE POSITION UNDER ENGLISH LAW

It is perfectly possible to patent a human cell line and other biological riches under English law. The standard criteria are novelty, utility, and the taking of an "inventive step". Further, DNA techniques would enable positive identification of the donor of the cells or tissue to be made. The issue of ownership of the cells or tissue then falls to be considered.

Traditionally, English common law has held that there is no "ownership" in a corpse (Handyside's Case 1750). However, the extent of this rule (which originates from the fact that burial law was once the exclusive preserve of the ecclesiastical courts) is doubtful (see the Human Tissue Act 1961), and current English law "acknowledge[s] (at least for certain purposes), it is not dictated by logic or practical considerations and there is no reason why the executors should not at least have possession of a corpse" (Smith and Hogan Criminal Law, 6th Edition, p. 524), and bailment of a dead body would seem possible (Palmer on Bailment, 1979).

In *Doodeward vs. Spence* (1908, 6 CLR 406), the High Court of Australia held that a two-headed stillborn fetus (born ten years earlier) could be the subject of ownership, and that the English authorities related primarily to the condition of the human body at death, at which time it could accurately be said to be "nullius in rebus", i.e., incapable of becoming the subject of ownership. Accordingly, medical exhibits and scientific samples can be the subject of ownership. It would seem to follow, therefore, that there may be ownership of an organ or cell line taken from a living person (and a dead person) which can qualify as "goods" for the purposes of theft or conversion. The preferred view is that English courts would be likely to follow the approach adopted by the later American authorities so that "the custodian of it (the body or its part) has a legal right to its possession for the purposes of preservation and burial [and for any other purpose not contrary to law and permitted by the appropriate individual] and that any interference with that right, by mutilating or otherwise disturbing the body, is an actionable wrong" (*Larson vs. Chase,* 1891, 50 NW 238).

CONVERSION AND THEFT DEFINED

Conversion has been defined as "an act intentionally done inconsistent with the owner's right, though the doer may not know of or intend to challenge the property or possession of the true owner" (Lord Porter in *Caxton Publishing Co. Ltd. vs. Sutherland,* 1939, AC 178 at 202). A person is guilty of theft if he dishonestly appropriates property belonging to another with the intention of permanently depriving the other of it (The Theft Act 1968). In *R. vs. Welsh* (1974, RTR 478, CA), a driver who had provided a specimen of urine for analysis and subsequently poured the specimen down the sink was convicted of stealing the urine. (See also *R vs. Rothery,* 1976, RTR 550).

The question of ownership then falls to be considered. Abandonment is not to be lightly inferred – in *Hibbert vs. McKiernan,* (1948, 2 KB 142) there was a conviction of theft of golf balls lost on club premises and in *William vs. Philipps* (1957, 41 Cr App Rep 5, DC), theft of refuse from a dustbin.

Accordingly, at the point of removal from the body, it would seem that the organs/tissues are not abandoned by the patient and are probably capable of forming the subject of limited ownership, or there are restrictions imposed by society on what may be done with them.

In Moore's case, the patient, when invited to do so, did not agree to consign his organ to research. In Britain, the issue would not normally be raised (save with regard to fetal spare parts or organ donation for transplantation), so that any terms or restrictions would have to be implied from normal practice and what the parties understood this to be. Though some tissue is used for therapeutic or research purposes without express consent being obtained from patients, arguably there is an analogy to be found concerning the use of fetal tissue for research in transplantation therapy for Parkinson patients.

The British Medical Association's 1988 guidelines accept that possession of fetal remains destined for destruction may be subject to limitations: para. 2 states that "the woman from whom the fetal material is obtained must consent to the use of fetal material for research and/or therapeutic purposes." It seems therefore that the patient's consent should not be taken for granteed. It could equally well be argued, however, that abortion raises special issues, and that this guidance does not recognize a legal but an ethical obligation.

Perhaps it is unrealistically idealistic to suggest that it should not be possible to patent cultures emanating from the human body, and that they should all be available for the good of society at large. High costs of research may alas need to be fuelled by the prespect of substancial monetary returns in a free market economy. Accordingly, therefore, the whole issue of disposal of body parts, and what may be done with them, should be discussed with patients and noted on the consent form. It should be made clear that compliance with the doctors' request should be unrelated to the offer of treatment.

It seems likely that most patients will not be looking for profit from their discarded, failing body tissues or organs, but will expect decent, confidential, and prompt disposal.

AVAILABILITY OF ORGANS FOR TRANSPLANT

In Britain, the law does not require patients, or if deceased or incapable, their relatives, to contract out of a duty to provide them for transplantation or research. However, the continuing shortage of organs available for transplantation has caused this issue to be raised on many occasions, and led to an unsavoury commercial trade in 1989 (and before) in London. Kidneys were removed from impoverished Turkish peasants in return for payment of medical expenses and about £2,000–£3,000, and transplanted into unrelated recipients with kidney failure.

This practice of removing organs such as kidneys from unrelated donors for cash has now been outlawed by a new criminal statute, it may well have been unlawful in any event, and there are civil proceedings on foot apparently. The three doctors involved with the kidney transplant scandal were found guilty of serious professional misconduct, with the "ringleader" (the procuring doctor) struck off the medical register.

BRAINS AS ORGANS FOR DONATION

"Cogito ergo sum" – "Je pense donc je suis" – "I think, therefore I am", said Descartes. But is the converse equally true, namely: I lack the capacity to think, therefore I am not (in being/I a person)? At what point do I cease being I? Where is the "I" located? Is the individual persona or human personhood extinguished when the brain dies or ceases to function cognitively? Most ordinary people accept that it is our brain which contains, supplies, and stores our persona, our individuality and being and, some would argue cogently, the whole "I". With this in mind, one can understand the sensation of unease experienced, even by scientists and doctors operating in research projects which involve research on the human brain, either by transplanting substances or tissues into a living brain and/or carrying out research on a brain from a dead person or fetus.

For the more pragmatic among us, once dead or irretrievably defunct, the brain merits no greater status than any other parts of a dead body, and certainly this is the legal position. Accordingly, the anxieties and ethical disquiet which may be generated by proposals for brain research and transplantation after death are logically unsustainable. Our instinctive repugnance, if it is not religiously or philosophically based, seems therefore to be an emotional response rather than a logical one, probably prompted by our concern for showing respect for what once represented the "I" in the person who is now dead.

In fact, several "brain banks" are already functioning in the U.K. (for example, there is one in Cambridge, and a small specialized center is already operating at Charing Cross Hospital). They keep a discreetly low profile, and their existence and the nature of their work are not widely known. More research is urgently needed, and therefore more such banks. A neuropathology center, or "brain bank", was recently set up at the Unit of Mental Handicap at Charing Cross and Westminster Medical School to serve the large North West Thames Region. It will be sourced by some six hospitals. With a view to allaying their own anxieties and reassuring staff and members of the public with regard to the setting up and running of such a center, the consultants in charge were anxious that the legal, moral, and practical issues should be openly discussed.

THE NEED FOR RESEARCH INTO THE CAUSES OF DEMENTIA AND DISEASES OF THE CNS

As the population ages, the problems of dementia, and particularly senile dementia and Alzheimer's disease, are becoming increasingly acute and costly in both human and economic terms. Research into the causes of diseases of the central nervous system, and in particular, into how to delay or prevent the onset or progression of such diseases, is urgently needed. Yet, ironically, one of the most important hindrances to the furtherance of such research is the shortage of human brains.

As one of the founders, Professor Ben I. Sacks explains,

> Without a suitable supply of brains which need to be removed within hours of death it is very difficult to confirm diagnoses made during life and elucidate the causation and mechanisms of disease. Examining gross and microscopic defects and biochemical disorders can only be performed on the brain itself and such investigations will help research into the causes and mechanisms of handicapping conditions and therefore aid in the prevention and alleviation of these disorders, the associated stresses as well as to plan more logically for the needs of such people.

Modern techniques for research into mental handicap require that tissue be very fresh. Although not as time-sensitive as organ transplantation, such research requires considerable organization to ensure that, if death occurs off hospital premises, the body can be brought in quickly for prompt postmortem and the removal of tissue.

It is not only ethical and compassionate but also practical for the issue to have been discussed and agreed with the patient (if he has the capacity) and/or his relatives well before death occurs. It is noteable that instructions contained in a will for burial or disposal of organs are not legally binding on the testator's executors, who may disregard them if they choose; further, if the will, or a copy of it, is not available either before or at the moment of death, any permission contained in it may not be known until well after the 6–8 hour maximum period has elapsed after which the organ will no longer be useful for this type of research.

THE LEGAL STATUS OF DECEASED HUMAN TISSUE IN BRITAIN

Although traditionally English law recognized no ownership in a dead body, this principle is subject to many exceptions. Individual organs or body fluids which are removed and held for research or other purposes may be owned, and ownership rights may be exercised (East Reports). The precise legal status of a newly deceased corpse and its organs is uncertain, but the legal position with regard to taking possession and removal of any organs for research is clarified by the Human Tissue Act of 1961, which is "an Act to make provision with respect to the use of parts of bodies of deceased persons for therapeutic purposes and purposes of medical education and research and with respect to the circumstances in which post mortem examinations may be carried out . . ."

Section 1(1) states that

> if any person, either in writing or orally in the presence of two witnesses during his last illness has expressed a request that his body or any specified part of his body be used after death for therapeutic purposes or for the purposes of medical education or research, the person lawfully in possession of his body after his death, may unless he has reason to believe that the request was subsequently withdrawn, authorise the removal from the body of any part or, as the case may be, the specified part, for use in accordance with the request.

However, s1(2) provides for the situation where the deceased did not express such a request in the presence of two witnesses, and states that without prejudice to s1(1)

> the person lawfully in possession of the body of a deceased person may authorize the removal of any part from the body for use for the said purposes if, having made such reasonable inquiry as may be practicable, he has no reason to believe –
> (a) that the deceased had expressed an objection to his body being so dealt with after his death, and had not withdrawn it; or
> (b) that the surviving spouse or any surviving relative of the deceased object to the body being so dealt with.

In my view, any inquiries need only be such as are reasonable in all the circumstances (not exhaustive and endless). Section 1(2)(b) refers to "any surviving relative" without defining the minimun degree of relationship to the deceased. In my view, the term "relative" for this purpose should be interpreted as one who is reasonably close to the deceased, or who was primarily concerned with the daily life of the deceased. Where the (arguably) close relatives are in disagreement, and the deceased did not made his willingness to dispose of his organs for research plain before witnesses, then s1(2)(b) will not authorize the use and removal of his body or body parts for research or transplantation etc.

The act also provides that "subject to (4) and (5) the removal and use of any part of the body will be lawful." Subsection 4 requires the removal of the organs to be performed by a fully registered medical practitioner who must have satisfied himself by personal examination of the body that life is extinct, and (5) requires the request of the coroner to be obtained if there is a reason to believe an inquest may be required.

The question of who can be said to be in lawful possession of the body has in the past been the subject of debate, but despite "the absence of any specific case authority, it is now accepted that it is the hospital or institutional authority in whose care the body lies which is in lawful possession" (Palmer 1979), and this is made clear by s1(7) of the act. There is also in force a NHS circular (1975 (Gen) 34, May 1975, DHSS London) to this effect. Obviously, if the person dies at home or outside hospital or institutional authority premises, this will not apply, and the next of kin or executors are likely to be in possession unless the body is removed to a hospital. Similarly, by s1(6) no authority for the removal of any part of the body may be given in respect of any body by a person entrusted with it only for the purposes of interment or cremation.

Though s1(8) of the act says that nothing in s1 shall be construed as rendering unlawful any dealing with, or with any part of, the body of a deceased person which is lawful apart from this act, any protocol for procedures and any consent forms drawn up by "bank managers" in so sensitive an area as this should adhere where possible to its provisions and be appproved by an ethical commitee. However, common sense and discretion should be exercized in its interpretation, otherwise valuable and much needed organs will be lost to research. A good example of where the act is unduly restrictive with regard to cadaveric removal of organs for research purposes is s1(4), which provides that the organ only be removed by a registered medical practitioner. In practice, it is often not doctors but qualified technicians and biochemists who are designated to do this, and who usually do this. Provided there has been pronouncement of death by a medical doctor in accordance with s1(4), any breach with regard to the qualification of the person appointed to remove the organ(s) would seem merely technical and will not render the procedure unlawful.

Protocols should take account of religious beliefs and customs which may bear on the question of organ removal and donation, and it should be made clear that no pressure is being exerted either on an individual in anticipation of his decease or his relatives. Where these are used, consent forms should be drafted so as to allow for the body and body parts to be used for research generally (and where appropriate for organ donation and other allied purposes) rather than for a specified and narrow particular purpose, such as mental handicap.

Brain banks should be subject to proper protocols and procedures which insure that any dealings and requests made to prospective donors and their relatives are handled by trained personnel. The staff as well as the public must be confident that proper safeguards are in force.

CONCLUSIONS

In some societies, as in Britain, health care is available to all, regardless of the ability to pay. Yet resources are limited and must be shared out on some sort of equable basis. One would assume on a priority/ability-to-benefit ratio, depending on what resources are available in that field of medicine. The National Health Service has an interest in keeping people out of hospital and in providing them with the best treatment available at the most reasonable cost. Transplantation of organs and tissue can be life-saving, and at least highly beneficial to the recipient, but there is at present no duty imposed by society on individual citizens to donate blood, let alone bone marrow or other organs. The whole system of donor organs is organized on a volunteer, opt-in basis in Britain, and indeed in most countries. The sale of organs or tissue or oocytes on a commercial basis is regarded as immoral and against public policy, though an alternative view may be argued.

The question of whether after-death donation of organs, such as kidneys, liver, pancreas, heart, and cornea, should be mandatory, subject to an opt-out request by the individual or his relatives, has been repeatedly floated in Britain, but so far without success. If this were the case, doctors and health care personnel might find it easier to approach the families of dying patients with a view to organ retrieval than at present.

It would be consistent with the demands often made on the living by society for it to impose an "opt-out" requirement on the dead. At times, depending on the circumstances, it is arguable that society does legally and morally have a claim over my body, alive or dead. Ironically, as the autonomy of the individual within society which in life becomes the watchword for the last decades of the twentieth century, this is likely to be less justifiable after death, particularly as religious taboos give way to medical and scientific demands. At the end of the day, all the rights which society seeks to exercise over the bodies of its individual members must be closely regulated and restrained within "proper" limits. What is "proper" will always be a matter for debate and will shift to accord with society's values.

REFERENCES

Hastings, G.J. (1988). *Whose Waste is it Anyway? The Case of John Moore.* Hastings Center Report, October/November 1988, pp. 37–39.

Mason, J.K. and R.A. McCall Smith (1987). Human Tissue Act 1961. In: *Butterworths Medico-Legal Encyclopaedia,* pp. 274–275.

Bioscience ⇌ Society
edited by D.J. Roy, B.E. Wynne, and R.W. Old, pp. 41–52
© 1991, John Wiley & Sons

Genetic Intervention: Prevention, Restoration, or Optimization?

Barbara Hobom

Arndtstrasse 14
D–6300 Giessen, Germany

Abstract. Through gene technology it has become possible to transfer foreign genes into any other organism. As soon as scientists became able to tinker with genes, they also became aware of the possibility of manipulating the human genome. The first somatic gene therapy in humans has been performed recently by manipulating human white blood cells. Scientists worldwide agree that germ line gene therapies should not be done in the foreseeable future. Indeed, the manipulation of germ line cells to eliminate a heritable disease will not even be necessary if preimplantation diagnosis becomes practicable for the selection of healthy embryos.

GENES ARE INVOLVED IN THE MOST COMMON DISEASES

About 4,500 disorders have been identified which are caused by a single defective gene. Some of these heritable disorders are quite common, like cystic fibrosis, a severe lung obstruction, which occurs at a frequency of about 1 in 2,000 newborns, or Duchenne muscular dystrophy, a degeneration of muscle cells, which afflicts 1 in every 4,000 newborn boys; others are extremely rare with only a handful of cases all over the world, e.g., adenosine deaminase deficiency (ADA-deficiency), a severe immunodeficiency syndrome. It is estimated that about 1% of all children born suffer from a single gene defect or a chromosomal abnormality. As we have learned in recent years, genes are involved in many more diseases than was anticipated. They play a strong role in the most common diseases of Western societies, in cardiovascular disease, diabetes, and cancer, in the susceptibility to infections and possibly even in psychiatric illnesses like manic depression. In these common diseases, multiple genes seem to be involved which, however, are mostly unknown so far. Whereas single gene disorders seem to be curable rather easily by gene technology, it would be much more difficult to alter "bad" predispositions.

Table 1. Some candidate single gene disorders for somatic gene therapy

Adenosin-Deaminase(ADA)-Deficiency
Lesch-Nyhan Syndrome
Haemophilia A and B
Sickle Cell Anaemia
Phenylketonuria
Duchenne Muscular Dystrophie
Cystic Fibrosis
Hypercholesterolaemia

Table 2. Some multiple gene disorders

cardiovascular disease
diabetes
cancer
susceptibility to infections
Alzheimer disease?
manic depressive disorders?

THE PRACTICAL APPLICATION OF GENE TECHNOLOGY IN MAMMALS

Genetically manipulated so-called transgenic animals are being used as invaluable model systems to study human diseases, including cancer, diabetes, and heritable disorders. In addition, farm animals are being developed which display profitable traits, like disease resistance or faster growth (Pursel et al. 1989).

However, human beings are not experimental animals. Whereas one may discard transgenic mice or carp that do not seem perfect after gene transfer, gene transfer into mammals this is not possible with a human being. Therefore, before discussing gene transfer experiments in humans, it is important to go into some detail with the methods of gene transfer used in animals. This may help to point out where the problems are – at least on the physical side.

METHODS FOR THE TRANSFER OF GENES INTO HUMAN CELLS

There are three ways for transferring genes into the mammalian genome that seem applicable to human cells:

1) *Transfer via so-called retroviral vectors.* This method gives the highest yield of stably transferred genes. The method uses truncated retroviruses as vehicles for the transport of a desired gene. With this method the emergence of viable retroviruses originating from the recombination of the truncated virus with endogenous retrovirus genomes has occasionally been observed in mice.
2) *Transfer via microinjection of genes directly into the male or female (pro-) nucleus of a fertilized egg.* This method works well with mice, and has been successful with some other animal species, like pig, sheep, rabbit and carp.
3) *Using embryonal stem (ES) cells which are added to a very young embryo at the stage of a so-called blastomere, which consists of only a few cells.* This method has been applicable only to mice until now, because totipotent embryonal stem cells are, so far, available only from mice.

Genes transferred by one of these methods are delivered to a random site in the recipient chromosome. Consequently, the foreign gene may integrate into an existing gene, which by this means would be severely damaged or lose its function. Thus, although methods for genetically manipulating the human genome do exist, they are far from perfect and even potentially very dangerous (Friedmann 1989). However, this may change in the future. Mario Capecchi of the University of Utah in Salt Lake City has developed a method to direct genes precisely to a specific site (Mansour et al. 1988). The method of gene targeting is now being used extensively for delivering genetic material to a specific gene (Chisaka and Capecchi 1991).

RESTORATION OF GENETIC DEFECTS

In principle, there are two ways of correcting a defective gene: by somatic gene therapy, and by germ line gene therapy. In somatic gene therapy a healthy gene is introduced into selected body cells of the patient, blood cells, for instance, in order to provide the individual with the protein function which is lacking in his organism. The patient may still pass his mutant gene to his offspring, because his germ line cells will stay unchanged in this kind of treatment. In contrast, in germ line therapy the healthy gene is transferred to the fertilized egg so that all the cells of the embryo developing from this egg (including its reproductive cells) carry the newly added gene. After germ line therapy, the transferred gene is transmitted to the offspring. So far, in somatic gene therapy no methods are available to direct a gene specifically to certain organs like the liver, the lung, or the gut. However, one can imagine that in the future it might become possible to use organ-specific viruses or macromolecules like antibodies as labels to deliver a gene to a special organ inside the organism. Just recently, scientists of the University of Ann Arbor succeeded in directing a (model) gene right into the cells of a pig's arterial wall (Nabel et al. 1990).

It is anticipated that somatic gene therapy could be very helpful in a number of single gene disorders: via the addition of a functional gene to compensate for the mutant gene which has lost its function. Many people feel that there are no ethical problems with somatic gene therapy. To most people, somatic gene therapy does not seem very much different from organ transplantation.

The best candidates for somatic gene therapy are those genetic disorders in which a protein is lacking which can be transported by the blood stream to the site where it is needed. In experiments with mice and other animals, it has been shown that this method will work. However, there have been unexpected results as well. In many cases the manipulated blood cells, when transferred into the organism, ceased to synthesize the enzyme they had been producing in the test tube. The reason for this is not known, but it shows how little we understand the regulation of gene activity. Somatic gene therapy may require much more than just adding a functional gene.

SOMATIC GENE THERAPY

Among human diseases adenosine deaminase deficiency (ADA-deficiency) has been the prime candidate for somatic gene therapy for a long time. Patients with this disorder suffer from a severe immunodeficiency because their white blood cells are intoxicated by unphysiologically high concentrations of the metabolic molecule adenosine, which normally is degraded by the ADA-enzyme. Patients with an ADA-deficiency lack white blood cells and die of infections soon after birth if they are not kept in a sterile environment for the rest of their lives. Fortunately, this is a very rare genetic disorder. Worldwide there are an estimated 70 patients who suffer from an ADA-deficiency, about 20 of them seem to be amenable to gene therapy because they lack functional levels of the enzyme.

In October 1990 French Anderson and Michael Blaese of the National Institutes of Health (NIH) in Bethesda performed the very first gene therapy in humans, a somatic gene therapy in an ADA-deficient child. The physicians collected white blood cells from the four-year-old child, propagated the T-cells, a subgroup of white blood cells, in tissue culture, transferred a functional ADA-gene from a healthy person into the cells, expanded the number of the genetically corrected cells by cultivating them before infusing about a billion of the genetically engineered cells into the child's blood stream. This first example of gene therapy, however, will not cure the child. Like any other T-cells, the manipulated T-lymphocytes have only a limited life span and will produce the enzyme only for a certain period. The doctors hope that the child will need consecutive infusions of its own manipulated cells only in about six or twelve months' intervals. A permanent cure of the enzyme deficiency would only be achieved if one succeeded in manipulating stem cells which have the ability of self-renewal.

Table 3. Some milestones in the history of gene therapy

The first experiment of gene technology	1972
The first transgenic mammal (mouse)	1983
The first somatic gene therapy in man	1990

Actually, another somatic gene therapy experiment in humans is being prepared: Steven Rosenberg of the NIH plans to treat terminally ill melanoma patients with the patients' own immune cells, which he has collected from within one of their tumor sites (Culliton 1990). Rosenberg inserted the gene of a very potent substance of the human defense system, the so-called tumor necrosis factor, into these tumor-infiltrating lymphocytes (TIL). In experiments with mice it has been shown that tumor necrosis factor injected into the blood stream of the animals was able to destroy large tumors. Since tumor necrosis factor is extremely toxic to humans, patients would not survive at doses required to kill their tumors. However, white blood cells infiltrating the tumor and delivering their toxic weapon locally at the tumor site might be able to destroy the cancer cells without harming the patient.

GERM LINE GENE THERAPY

In germ line therapy a genetic defect could be corrected in the fertilized egg, resulting in an organism which carries the transferred gene in every single cell, including its reproductive cells. There have been dozens of experiments with mice and some other animal species into which a foreign gene has been stably implanted in this way. These so-called transgenic animals transmit the new genetic trait in a Mendelian way to their offspring. In a few cases scientists have even tried and succeeded in germ line gene therapies in animals. Mice have been cured of beta-thalassemia, a severe hematopoietic disorder, by transfer of a beta globin gene which provides the animal with a functioning hemoglobin for oxygen transport (Friedmann 1989). Other mice have been cured of a disease similar to multiple sclerosis, a neuromuscular disorder, by transferring the mouse gene for myelin basic protein, the insulating material of nerve cell axons, into their germ cells. Still others have been cured of dwarfism by the implantation of a growth hormone gene.

The experiments with animal models show that curing heritable disorders by germ line gene therapy is feasable in principle. However, from these experiments many caveats emerged, which would have to be observed if the same methods were applied to humans.

A few examples may illustrate this. In an attempt to transfer the gene that confers resistance to influenza virus infections from wild mice to laboratory mice, scientists discovered that a permanently active resistance gene very suprisingly interfered with the animals' embryogenesis (Arnheiter et al. 1990). Furthermore, female mice with an additional gene for growth hormone became sterile, and the transgenic males suffered from arthritis (Pursel et al. 1989). On the other hand, mice that were made deficient in the gene for an enzyme of the purine metabolism (hypoxanthine guanine phosphoribosyltransferase, HGPRT) by gene manipulation did not develop nervous system abnormalities as do humans with the Lesch-Nyhan syndrome, where the HGPRT enzyme is similarly lacking (Hooper et al. 1987). In addition, no pathological symptoms were observed in mice equipped with genes that cause sickle cell anemia in humans (Greaves et al. 1990). Thus, the genetic context seems to be very important even when adding or mutating only a single gene. Genetic imprinting, the differential chemical modification of male and female genomes, is another factor one has to cope with.

The outcome of a manipulation of germ line cells seems unpredictable as long as we do not understand the cooperation of the whole array of genes in an animal or human genome. The situation is even more complicated by the fact that the environment may decide whether a given gene behaves as a "good" or a "bad" gene. This seems to be the case with genes influencing hypertension, a condition strongly influenced by diet and other forms of life-style.

SHOULD SEVERE GENETIC DISEASES BE ERADICATED BY GERM CELL THERAPY?

It is the goal of epidemiologists to eradicate infectious diseases. This has been achieved with smallpox; polio might be next on the agenda. Will it be possible – and desirable – to eradicate genetic diseases as well? In principle, a genetic disease known in a given family could be eliminated by adding a healthy gene to germ cells carrying the genetic defect. However, one can list many arguments against germ line gene therapy:

1) Many people do not like the idea of interfering with the evolutionary forces of nature, of "playing God".
2) Since one cannot rule out the possibility of damaging the genome while manipulating it, the technique is (still) much too risky.
3) It will by no means be possible to eradicate genetic diseases completely. They are not only propagated from parents to offspring, quite a few also arise spontaneously in every generation. In some heritable diseases the rate of newly arising mutations leading to the disorder is very low as in phenylketonuria; in others, like neurofibromatosis, a severe nerve cell disorder, they may account for up to half of the disease cases. These newly arising genetic diseases theoretically could only be avoided by obligatory widespread prenatal diagnosis and subsequent abortion of any affected embryo.

Manipulating germ line cells for other than therapeutic reasons is strongly rejected by the majority of scientists in the field. Germ line gene manipulations result in the permanent alteration of the genome. Progeny generations cannot defend themselves against this. I personally think that this is one of the strongest arguments against germ cell manipulations in humans. One does not have the right to impose one's ideals on future generations. They might have different ones. Human beings should be declared inviolable in this respect.

SELECTION INSTEAD OF GERM CELL THERAPY?

The aim of any germ line gene therapy would be to allow couples with a disabled child or with a family history of a heritable disease to have healthy offspring. However, to reach this goal, germ line gene therapy may not be necessary at all. Recently, a very promising alternative has arisen: preimplantation diagnosis.

Several years ago, Anne McLaren of University College in London suggested doing prenatal diagnosis on an embryo that had not yet been implanted into the uterine wall (McLaren 1985). The so-called preimplantation embryo consists of up to a few hundred cells, and can be recovered by uterine lavage a few days after conception (Buster 1985). In mice and humans it has been shown that it is possible to take out a few cells from very early embryos, analyze their genes, select the healthy embryos, and insert only these into the uterus, while discarding those with a genetic defect. Several technical advances indicate that preimplantation diagnosis may well become feasable in humans (Coutelle et al. 1989). It even seems possible that in vitro fertilization (IVF) may not be needed to do preimplantation analysis. One can obtain several preimplantation embryos if the woman is hormone-treated before conception. A few days after conception the preimplantation embryos can be washed out of the uterus. Initial experience shows that a rather high proportion (about half) of the preimplantation embryos obtained by this means will give rise to a normal pregnancy after being inserted into a woman's uterus. A few cells can be safely removed from the preimplantation embryo for diagnostic purposes. The embryos would be frozen until the results of preimplantation diagnosis were obtained. The embryos then known to be unaffected would be thawed and reimplanted into the woman's uterus. Once a pregnancy was achieved, the couple could be fairly sure that they were expecting a healthy child. Surely most mothers would prefer to receive a "good" preimplantation embryo to having an abortion following prenatal diagnosis after several weeks or even months of pregnancy.

IS THERE A NEED FOR GENDER SELECTION?

Any method of preimplantation diagnosis – after IVF-fertilization or uterine lavage – also offers the possibility of selecting the sex of the future child.

Except for sex-chromosome-linked genetic diseases, there exists no medical indication to select a child with a specific sex. However, sex selection is being done for personal, mostly egoistic reasons and is widespread in some societies like India and China. Preimplantation diagnosis, which allows diagnosis of the sex of the child very easily, might increase the frequency of sex selection for social or individual reasons. The situation might change even more if the separation of x- and y-gametes becomes possible so that abortion and preimplantation become obsolete. One may envisage several negative consequences of sex selection:

1) An opinion poll in the United States showed that there are many more parents who would prefer to have more boys instead of more girls. Thus, sex selection might create an imbalance in the population. On the other hand, in small families it seems that many parents just want a boy as their first child and a girl as their second child, which would lead to a rather balanced situation.
2) Widespread sex selection is likely to create psychological problems. For example, if the second child is a girl, she may ask whether her parents did not want her as much as her brother because they chose her only in the second place.
3) Many societies consider it unethical to abort a child merely because it does not have the "right" sex. Being born as a male or a female should not be considered a handicap.

Some people feel that there might also be some positive consequences of sex selection:

1) Parents may enjoy the freedom to choose the sex of their child.
2) Choosing the sex of one's child may contribute to population control.

WILL THE CLONING OF HUMAN BEINGS BE POSSIBLE?

Most people are fightened by the idea of a society with subgroups of identical people like the Alphas, Betas, or Epsilons of Aldous Huxley's *Brave New World.* Thus, we have to ask whether the cloning, i.e., the production of identical copies, of human beings could become possible. In lower vertebrates like the frog, one can obtain identical replicas of the tadpole (but not of mature frogs) by splitting the early embryo into single cells and performing nuclear transplants with nuclei from these cells. In higher vertebrates this method does not work. However, Steen Willadsen of the Agriculture and

Food Research Council Institute of Animal Physiology in Cambridge has succeeded in producing identical sheep and cattle by fusing single cells taken from an early (8- or 16-cell) embryo with an enucleated unfertilized egg (Marx 1988). This method might also work in humans. But why should one clone humans? Some people argue that cloning would facilitate organ transplantations between genetically identical siblings. However, producing human beings for the purpose of spare-part donations is unethical. It would seem like a modern form of cannibalism.

SHOULD THE HUMAN SPECIES BE OPTIMIZED?

Preimplantation diagnosis – if it works safely – may render germ cell therapies, intended to cure severe genetic disorders, unnecessary. However, I do not think that this will make any further discussion on the genetic manipulation of the human genome irrelevant. Perhaps today we cannot foresee any tendency to manipulate the human genome for other reasons than for curing heritable diseases. But the experience of history teaches us that it may be better to consider even the very improbable today in order to be prepared for new and surprising developments in the future. I personally strongly dislike drawing scenarios of science fiction like Aldous Huxley's *Brave New World* to make people afraid of gene technology. However, I suggest that by taking in such extreme ideas, one can better foresee where the development should *not* go.

At first, nuclear physicists were convinced that there would not be any practical application of the atomic physics they had discovered. Yet, only a couple of years later we had the atomic bomb! Likewise, molecular biologists are convinced today that there will be no chance whatsoever in the foreseeable future of human behavior being modified by manipulating the genome. Their main argument is that practically nothing is known about how the brain works, let alone how genes and environment interact in behavioral traits. However, in the future we may get important insights into the flow and storage of information in the brain. Might it not then be possible that some people seriously propose: if by gene manipulation human beings can become better – less egoistic, less aggressive, gentler, and more affectionate – why not use it for this purpose (Weatherall and Shelley 1989)? Wouldn't we do well to prepare for this kind of question, and try to find a consensus on what the answer should be?

The different canine breedings clearly demonstrate that it is certainly possible to change behavioral traits by altering genes. Twin and family studies suggest that genes also control behavior in humans. Most behavioral traits seem to be influenced by multiple genes. Sometimes, however, even a single gene may have a very strong impact on behavior. In the Lesch-Nyhan syndrome, for instance, a molecularly minor genetical alteration, a single defect in the gene of the purine metabolism, causes the afflicted patients to severely self-mutilate their fingers and lips (besides being mentally retarded). In alcoholism genes seem to play a role as well (Weatherall and Shelley 1989).

Some scientists think that, for the time being, it is very fortunate that we do not understand how the brain works. This leaves us time to discuss where we want to go. Scientists alone should not have the right to give the direction, because like any other human beings they are prone to ambition and error in their research goal. Therefore society as a whole, including the scientists, has to decide about the application of gene technology to the human genome. Many feel that there should be strict rules for the application of the new technique to humans, and that the human genome should even be declared inviolable. Others think that specific manipulations might become justified in the future. Since ethics, which form the basis for the rules, vary according to geographic area, culture, and time, it seems very questionable to me whether we will ever reach a worldwide consensus on the question to what extent the human genome should be protected.

REFERENCES

Arnheiter, H., S. Skuntz, M Noteborn, S. Chang, and E. Meier (1990). Transgenic mice with intracellular immunity to influenza virus. *Cell* **62:** 51–61.

Buster, J.E. (1985). Embryo donation by uterine flushing and embryo transfer. *Clinical Obstetrics and Gynecology* **12:** 815–842.

Chisaka, O., and M.R. Capecchi (1991). Regionally restricted developmental defects resulting from targeted disruption of the mouse homeobox gene box-1.5. *Nature* **355:** 473–479.

Coutelle, C., C. Williams, A. Handyside, K. Hardy, and R. Winston. (1989). Genetic analysis of DNA from single human oocytes: a model for preimplantation diagnosis of cystic fibrosis. *British Medical Journal* **229:** 22–24.

Culliton, B. (1990). Gene therapy: into the home stretch. *Science* **249:** 974–976.

Friedmann, T. (1989). Progress toward human gene therapy. *Science* **244:** 1275–1281.

Greaves, D.R., P. Fraser, M.A. Vidal, M.J. Hedges, D. Ropers, L. Luzzatto, and F. Grosveld (1990). A transgenic mouse model of sickle cell disorder. *Nature* **343:** 183–185.

Hooper, M., K. Hardy, A. Handyside, S. Hunter, and M. Monk (1987). HPRT-deficient (Lesch-Nyhan) mouse embryos derived from germline colonization of cultured cells. *Nature* **326:** 292–295.

Mansour, S.L., K.R. Thomas, and M.R. Capecchi. (1988). Disruption of the proto-oncogene int-2 in mouse embryo-derived stem cells: a general strategy for targeting mutations to non-selectable genes. *Nature* **336:** 348–352.

Marx, J.L. (1988). Cloning sheep and cattle embryos. *Science* **239:** 463–464.

McLaren, A. (1985). Prenatal diagnosis before implantation: opportunities and problems. *Prenatal Diagnosis* **5:** 85–90.

Nabel, E.G., G. Plautz, and G.J. Nabel. (1990). Site-specific gene expression in vivo by direct gene transfer into the arterial wall. *Science* **249:** 1285–1288.

Pursel, V.G., C.A. Pinkert, K.F. Miller, D.J. Bolt, R.G. Campbell, R.D. Palmiter, R.L. Brinster, and R.E. Hammer. (1989). Genetic engineering of livestock. *Science* **244:** 1281–1288.

Weatherall, D., and J. Shelley. (1989). *Social Consequences of Genetic Engineering.* Amsterdam / New York / Oxford: Excerptia Medica.

Bioscience ⇌ Society
edited by D.J. Roy, B.E. Wynne, and R.W. Old, pp. 53–66
© 1991, John Wiley & Sons

Prenatal Diagnosis: Healthier, Wealthier, and Wiser?

Marcus E. Pembrey

Paediatric Genetics
Institute of Child Health
30 Guilford Street
London WC1N 1EH, U.K.

Abstract. Prenatal diagnostic tests on the fetus were developed in response to requests for help from couples facing a high risk of a child with a life-threatening or serious, handicapping disorder. Prenatal diagnosis for high-risk couples and prenatal screening for some disorders in all pregnancies have become incorporated into health services in many countries. The way such activities are promoted, delivered, and audited requires very careful consideration if they are to remain a positive contribution to society and not risk being discredited as state-inspired eugenics. It is argued that the study of the way in which families use prenatal diagnosis and selective termination provides some guidance on the objectives against which these services should be evaluated. The aim of genetic counseling is *not* to reduce the birth incidence of genetic or congenital disorders. This may be a consequence of genetic counseling, but it depends on what parents choose to do. Many couples choose prenatal diagnosis as a means of achieving their desired family of healthy children; in other words, couples use genetic services to restore or maintain family life in the face of known genetic risks.

THE TRAGEDY OF SEVERE CONGENITAL OR GENETIC DISORDERS

Diagnostic tests on the fetus were developed in response to requests for help from couples facing a high risk of a child with a life-threatening or crippling disorder. As these tests and other genetic screening procedures become widely incorporated into health services, there is a risk that they will be seen as solely "a public health measure" promoted by professionals. This can lead to stated aims that do not accord exactly with the family's point of view, and indeed it already has: a trend that is dangerous. In the area of genetics and reproductive choice, health service policies must, within the protective legal constraints determined by parliament, align with the family's needs and wishes, or risk being discredited as state-inspired eugenics. For the couple, prenatal diagnosis and its associated counseling is inextricably bound up with the threat to family life that congenital and genetic disorders represent, and their desire to resolve that threat. This paper will attempt to show that an understanding of how families use genetic services can serve as a guide in formulating policies on the promotion, delivery, and auditing of prenatal diagnosis.

For many genetic disorders, the tragedy of a child born to be seriously handicapped is compounded by the knowledge that it might well happen again within the family. The term "family ties" takes on a whole new meaning. The shock of the diagnosis of serious disease distorts, for a period of time, that set of relationships between family members that constitutes family life. This is particularly so when a longed-for baby turns out to have a congenital malformation or is soon found to have a serious genetic disease. Restoration of the family's equilibrium can be rapid when there is the prospect of a simple, permanent correction of the abnormality (such as surgery for an uncomplicated cleft confined to the lip), but in most instances it is a slow and painful business full of uncertainties and often falling short of a real "coming to terms" with the tragedy. Against this background, genetic counseling and the option of prenatal diagnosis have a major contribution to make in this resumption of a full and healthy family life.

My personal and professional experience with families makes me totally reject the notion, sometimes put forward, that society needs handicapped members because they bring out the best in people in terms of caring and personal sacrifice, that the world would be a poorer place without handicapping disorders. War engenders great heroism in people, but that is no reason for having wars. It is true that families or their friends often say that even the *severely* handicapped member has brought them joy, has caused them to count their blessings, and indeed has contributed to family life. These sentiments are more a testiment to the power of parental love to rescue something noble from the disaster of shattered dreams than a general recommendation to include such tragedies in life's rich pattern.

Williams (1990) has argued on philosophical grounds that, for a genetic disorder, the wish that "this person did not have Down's syndrome" is fundamentally different in meaning from a wish that "this person did not have a stroke". In the latter case the person without a stroke did exist before the damaging event, but not in the case of Down's syndrome or any other genetic disorder. He suggested that the real interpretation of the former wish is: "I wish you did not exist and were replaced by another person" – a subconscious desire few parents could voice, because it would be seen to be hurtful to their handicapped child whom they love and feel an obligation to protect.

In my experience, this interpretation accords well with the behavior of many parents coping with a child with a serious genetic disorder, the feelings of guilt that arise, and their requests for help in having a healthy baby. Prenatal diagnosis and selective abortion provide the parents with an opportunity to plan a family without the risk of being responsible for bringing another affected child into the world. It is a welcome opportunity for many families.

THE USE OF GENETIC AND PRENATAL DIAGNOSTIC SERVICES

I believe that families use genetic and prenatal diagnostic services as a means of maintaining or restoring family life in the face of a known genetic risk. I also believe that this view of the aims of genetic services is the only one on which the success or failure of such services should be judged. The proper focus of clinical genetics and prenatal diagnosis is in family medicine, not public health. In this perspective, prenatal diagnosis, alongside special support services for the handicapped and new therapies for genetic disease, does lead to *healthier* families.

The genetic counseling that should *always* accompany the offer of prenatal diagnosis provides the knowledge which, when combined with personal experience and beliefs, can allow a *wise* reproductive choice to be made by the couple or person concerned.

From what we know of people's choices to date, monetary considerations need not, and therefore should not, feature large in the planning and offering of prenatal diagnostic services. Indeed, with present attitudes and beliefs in Western communities, a health service can ill afford to overlook the provision of such services, for they are highly cost effective in the broadest sense of the term. Healthy children in happy families are an asset to any society. Prenatal diagnosis does not automatically mean healthier, wealthier, and wiser, but as an option for families facing a known genetic risk it can contribute to healthier and wiser families at no great cost to society in either monetary terms or in terms of undermining the fabric of a "free" society in which all its members are valued.

It should be noted that couples use prenatal diagnosis to restore or maintain family life in the face of a *known* genetic risk. This begs two important sets of questions.

1) Do the couples themselves really know the extent of genetic risk they are facing, including a true picture of the disorder, i.e. how informed will their reproductive choice be? What advances are there in determining not only that the fetus has (inherited) the disorder, but how severe it will be in a particular case? What are the prospects for successful treatment? What provision is there for the particular family with respect to medical and other help, should it be needed?
2) How did they come to know their particular risk? How far should professionals go in encouraging couples to discover any risk they might be facing, given that genetic information transfers a burden of responsibility that can extend far beyond themselves to distant relatives? What degree of (supposedly) short-lived anxiety for many is justified by the opportunity for a few to discover their high risk situation and be forewarned? This is one issue that must be addressed in genetic screening programs.

PRENATAL DIAGNOSIS OR SCREENING AND INFORMED REPRODUCTIVE CHOICE

A distinction can be made between a diagnostic test and a prenatal screening test, although in some instances the two merge. Prenatal diagnosis is the definitive test for a specific condition in the fetus, with the intention of using this information to decide whether or not to continue the pregnancy, or whether to prepare for special support or therapeutic intervention. The therapy may take place before delivery, such as surgical drainage of an obstructed fetal bladder, or soon after birth. Prenatal diagnosis is generally only offered to women who are known to be at increased risk of having a child with a specific genetic or congenital disorder. This policy is not just a matter of using limited resources for those facing the highest risk. It also stems from the fact that when the prior probability is low, the low rate of technical and human error in the test assumes relatively more importance. The proportion of "positive" test results that are false positives increases.

There are various ways in which a parent or couple can be identified as being at increased risk of having an affected child, and these are all, in a sense, screening procedures. A screening procedure is offered to a *whole population* with the intention of reassuring those with a very low risk and identifying those with a high risk, so they can be offered a definitive test which, in the context of this discussion, is a prenatal test.

From the user's point of view, of course, there is a big distinction between a screening procedure that can be carried out *before* a pregnancy and one that can only be offered *during* a pregnancy. The latter is what is referred to as prenatal screening. Taking a family history or asking a woman her age are examples of a preconceptional genetic screening test, whereas measuring maternal serum alpha feto protein (AFP) is a prenatal screening test (Table 1). Of course, sometimes it is only the existence of a pregnancy that alerts someone to offer any type of screening test. Two case histories will illustrate the extremes in terms of timing and the precision of information available to couples and therefore how variable in practice so-called informed reproductive choice can be.

Table 1. Screening procedures used to focus the offer of prenatal diagnosis on those at increased risk of having a severely handicapped child.

To identify individuals at risk
- Family history of autosomal dominant (AD) or X-linked (X-L) disorders. If positive: offer presymptomatic testing (AD) or carrier testing (X-L) using DNA/biochemical analysis, or clinical/imaging studies.
- Family history of chromosomal abnormality due to a translocation, or fragile X mental retardation. If positive: offer cytogenetic carrier testing.
- Maternal age > 35 or 38 years.
- Maternal diabetes.

To identify couples at risk
- Heterozygote screening for autosomal recessive conditions common in specified populations, e.g., hemoglobinopathies, cystic fibrosis, or Tay Sachs disease, using biochemical or DNA analysis.
- Rhesus incompatibility.

To identify a (second trimester) pregnancy at risk
- Maternal serum AFP – (risk of neural tube defects).
- Maternal serum AFP, human chorionic-gonadotrophin (hCG) and unconjugated oestriol (uE_3) (risk of Down syndrome).
- Routine anomaly ultrasound scans for gross congenital malformations.
- Exposure to rubella.

A woman whose brother and maternal uncle died of Duchenne muscular dystrophy has a 50% chance of being a carrier of a mutation in the dystrophin gene located on one of her two X chromosomes. Because of the family history she had opted for carrier testing and gene tracking. Some time ago, tests showed that she had inherited the same X chromosome from her mother as her affected brother, indicating that she is indeed a carrier. The small error rate in this prediction was explained to her and also the fact that a blood sample taken from her affected brother before he died revealed an unequivocal deletion in his dystrophin gene. Before planning a family she knows that her daughters will be unaffected, although, on average, half will be healthy carriers like herself. There is a 50% chance that any son will have the deletion and be affected, and a 50% chance he will escape the muscular dystrophy. In this situation she can be offered a definitive prenatal test to determine, first, the sex of the fetus, and if it is male, whether or not it is going to be affected. With such a family history, it is likely that the woman would not be surprised to be told that her sons might be at risk. Referral to the genetic services would be expected, and there would be ample opportunity for genetic counseling both before and during a pregnancy.

By contrast, a woman may start a pregnancy with no known increased risk of fetal malformation or genetic disease. She may or may not know that in 2–3% of all pregnancies there is an error of development that is manifest at birth or soon after (Ash et al. 1977); nor is she likely to know that 1% of people in most populations have one of the 3–4,000 simply inherited (Mendelian) disorders that occur in humans (Carter 1977). She will have become pregnant with the expectation that her baby will be healthy and develop normally. The news at 16–17 weeks gestation that her baby might be severely handicapped will come like "a bolt out of the blue". Perhaps her routine AFP blood test (which probably raised no anxiety at the time it was taken because it is a routine screening procedure offered to everyone) has shown a high risk of her baby having spina bifida. Or, more likely nowadays, the routine ultrasound at 16 weeks gestation has shown something "not quite right". Specialized "anomaly" scans will be necessary over the next two weeks, perhaps with a check of the baby's chromosomes before a conclusion can be reached. It may be a kidney maldevelopment or a heart defect. How life-threatening is it for the baby now, or when it is born? Is it surgically correctable? What is the long-term outlook – will it mean kidney or heart transplants in years to come? Despite a normal fetal chromosome complement, how certain are we that there are no associated undetectable abnormalities of the brain, for example, given that malformations tend to run together? There may be no clear-cut answers to many of these questions, and yet the couple, in conjunction with the doctors, has to make a clear-cut decision of what to do next.

These two case histories illustrate two ends of the spectrum of so-called informed reproductive choice. In the first example, the woman knew precisely what Duchenne muscular dystrophy was; she had watched her brother die of it. She and her partner had the choice of a) avoiding pregnancy and foregoing the pleasure of their own children; b) letting nature take its course and trust to luck or in the will of God; c) seeking to circumvent the genetic risk by adopting a baby or requesting ovum donation; or d) planning children of their own, but with prenatal diagnosis and abortion of males predicted to be affected by a test that is 99% reliable. By contrast, the second example is full of uncertainties, both medical and emotional. The couple is in a state of shock and unprepared for the decision they have to take. One lesson to be drawn from the above is that genuinely informed choice is more likely if the risk can be predicted prior to a pregnancy. However, to provide such an opportunity of knowing in advance can mean the loss of innocence for many more couples than would otherwise have had to face up to their risk.

A second lesson to be learned is that the widespread use of obstetric ultrasound examination currently carries with it significant risks, not of causing physical or genetic damage to the fetus, but of being misinterpreted. Faced at 17 weeks gestation with a structural abnormality in the fetus of uncertain long term clinical significance, a proportion of couples will opt for termination of pregnancy in the absence of reassurance about normal development. There is an urgent need to fund longitudinal studies with follow-up into childhood of those babies with mid-trimester ultrasound features of uncertain significance. An informed reproductive choice for many couples in the future will depend on such information.

PROSPECTIVE GENETIC TESTING AND THE LOSS OF INNOCENCE

We have seen that knowing of their risk before a pregnancy gives a couple a wider informed reproductive choice, and can allow them to take avoiding action before an affected child alerts them to the risk they were unwittingly facing all along. With the general anxiety generated by carrier testing of the healthy population (even in those who do not, in the end, turn out to be at high risk of having an affected child), some care has to be taken before rushing into screening programs. Take, for example, autosomal recessive conditions such as beta-thalassemia or cystic fibrosis. If both would-be parents are found to be healthy carriers by prospective population screening, then they face a 25% chance with each pregnancy that the child will inherit the mutation from both of them and be affected. If couples in general have only two children, then *without* prospective screening about half the couples at risk ($3/4 \times 3/4$, or 56%) will have only unaffected children and never have to know the risk they faced, with all its attendant fears and the possible decision not to risk trying for a family at all. However, for the others, an opportunity to be forewarned and take avoiding action has been denied. In the equation – "anxieties for some" vs. "help for the few who need it", the first part, anxieties for some, must be kept to a minimum by good counseling and general education. If, for example, every couple where just *one* member was found to be a carrier *wrongly* thought their children were at risk of the disease, we would have 36 couples panicking for every couple who would have actually had an affected child in the absence of prospective carrier screening. A widespread understanding of genetics and genetic risks will engender some anxiety. Accurate information and sympathetic help at the appropriate time will do much to keep this anxiety down to a level of concern that is appropriate to responsible parenthood.

What is the appropriate time to raise the matter of genetic risks and to offer screening tests that are available? This is, in part, dependent on the type of inheritance. For autosomal dominant and X-linked disorders, it is the individual who is at risk of affected offspring, and so theoretically the genetic advice could be offered at any time. Some, essentially sporadic, conditions (such as spina bifida or congenital heart defects) can only be screened for in the pregnancy itself. In autosomal recessive conditions it is the *couple* who is at risk, and so genetic screening is probably best offered when a relationship has developed to a point where having children is under consideration. For purely logistic reasons the antenatal clinic is the easiest situation in which to ascertain couples, with the pregnant woman being offered the test first and, if she is a carrier, an approach to her partner being made. However, with a pregnancy already established, this puts many couples under some pressure to consider a prenatal test with little time to think things through. On this point it is interesting to note what has happened during the evolution of population heterozygote screening for beta-thalassemia in Cyprus. Modell and Petrou (1989) describe it as follows:

> Initially, because the social implications of screening in school or prior to marriage were uncertain, and because of limited laboratory resources, screening was offered only to those at most immediate risk, i.e., to pregnant women. But when opening the new Thalassaemia Centre (in 1984), the Archbishop of Cyprus pointed out that this form of screening did not allow couples the full range of choices; he felt

that, for moral reasons, knowledge should be available prior to marriage, and he instituted an ecclesiastical ruling that every couple to be married should present a certificate that they had been screened, and suitably advised, at the laboratory. This certificate does *not* give the diagnosis, which is confidential to the couple. Since there is no civil marriage in Cyprus, this is tantamount to a law requiring premarital screening. The possibility of introducing such a civil law had been considered, and ruled unconstitutional. Since premarital screening has become a regular practice, only 2% of the couples who are both carriers have decided not to go ahead with the marriage: the majority proceed with the marriage and plan to use prenatal diagnosis in due course.

Another reason for aiming to offer preconceptional testing rather than screening during pregnancy is the fact that new advances in techniques are allowing many tests to be done earlier in pregnancy, at about ten weeks gestation. Preimplantation diagnosis in the preembryo is now a technical reality, although the overall reliability of the procedure still has to be tested (Handyside et al. 1990).

THE IMPACT OF RECENT ADVANCES IN PRENATAL DIAGNOSIS AND SCREENING

It is not the intention of this paper to describe the new technical advances in detail. In essence there are three basic approaches for prenatal diagnosis and screening: a) to examine fetal structure by ultrasound imaging techniques; b) to obtain tissue for chromosomal, DNA, or biochemical analysis that faithfully represents the constitution of the fetus; and c) to use changes in the biochemical composition of the mother's blood as a rough predictor of a specific condition of the fetus. These are summarized in Table 2, and many of the issues raised for couples, health professionals, and health service planners are discussed in the London Royal College of Physicians report on *Prenatal Diagnosis and Genetic Screening: Community and Service Implications,* published in September, 1989.

AUDITING AND HEALTH SERVICE POLICIES

There is a fear that scientists involved in molecular genetics and medical geneticists who seek to apply their discoveries in clinical practice are not giving sufficient thought to the long-term consequences of their work. The

Table 2. Present fetal sampling/imaging procedures, risks and time required to obtain a diagnosis.

Obstetric aspects			Time to diagnosis					
			Karyotyping		Biochemistry		DNA	
Sampling procedure	Weeks' gestation	Risk to pregnancy (%)	Culture	Rapid	Culture	Direct	Culture	Direct
Amniocentesis	14–17	0.5–1	2–4wk	–	2–4wk	–	5wk	> 10d[a]
Fetal blood sampling	>18	1–7	–	3d	–	2–7d	–	10d
Chorionic villus sampling	>9	2–4	2wk	2d	2wk	1d	–	10d
Ultrasound	~9	+[b]	Time to a definitive ultrasound diagnosis depends on many variables. Rapid karyotyping may be required.					

[a] New DNA methods promise to reduce the time to diagnosis to 1–2 days.
[b] The risk is of false positive diagnosis leading to abortion of a healthy fetus.

(Taken from Royal College of Physicians Report, 1989)

public perception of these long-term consequences is a Brave New World in which direct genetic manipulation or embryo selection will not be confined to avoiding severe genetic disorders, but will be used to create a preferred human race. Preferred by whom is usually not stated, but the fear would be that it would be a state preference. This matter is being addressed by multidisciplinary groups that bring together scientists, theologians, lawyers, philosophers and clinicians (e.g., Ciba Foundation Symposium 149. *Human Genetic Information: Science, Law and Ethics* 1990). What come out of these week long discussions – and rightly, in my view – are not any clear guidelines with respect to some imaginative scenario for the distant future, but rather some honest discussions about present-day practices in the field of clinical genetics, and the need for careful analysis of what the current anxieties and conflicts of interest are and how they should be resolved. I believe that a great potential danger in the area of prenatal diagnosis would be hasty implementation of medical audit based on a wrong premise about the aims of these services.

The object of genetic counseling is to provide information about risks to offspring at a time appropriate to the options available for modifying the outcome, and to put those risks into perspective. The aim of genetic counseling is *not* to reduce the birth incidence of genetic or congenital disorders. This may be, and often is, a consequence of genetic counseling, but it depends on what parents choose to do. Health service planners and public health professionals are interested to know how many handicapped children will be born and what might influence the birth incidence. They are inclined to see prenatal diagnostic services as worthwhile, because they can result in a reduction of the incidence of a disorder. Parents, as judged by their actions, do not see prenatal diagnosis in this way.

When the genetic basis of a life-threatening disorder is not recognized by families, they tend to have the same number of children as others in their community, even though several affected children might be born. The sheer burden of caring for handicapped children may have some effect on reducing the number of children planned. If the genetic risk is high, as soon as genetic counseling alone is introduced, the number of children planned is reduced. In other words, a known genetic high risk in the absence of (acceptable) prenatal diagnosis frightens parents into stopping. The birth incidence of the disorder falls, but individual family life suffers. The consequences of this stage of the development of genetic services for a disorder might be highly satisfactory from the health service planners' point of view, but is often deeply distressing to the individual couple, particularly if it is their first and only child who is affected. When acceptable prenatal diagnosis is made available to those who want it, a remarkable change in reproductive behavior can be observed. Couples who have in the past put off planning further children often have repeated pregnancies at short intervals in order to achieve their desired family size of healthy children. Their family size now increases to become comparable with those who do not face a genetic risk. This three stage effect has been documented for beta-thalassemia (Modell et al. 1980), and we are now seeing the same effect with the introduction of early prenatal diagnosis for cystic fibrosis in quite a different population. Given the 1-in-4 recurrence risk, there will be couples who are unfortunate and have one, two, or more fetuses diagnosed as affected by chorionic villus sampling in successive pregnancies. What has been remarkable in these cystic fibrosis families using our service is the very short interval between receiving a "bad news" result after first trimester prenatal diagnosis and their returning for the same prenatal test in the next pregnancy (Fig. 1). The average interval in the first nine repeat pregnancies after an earlier termination of pregnancy was only 228 days. Obviously the interval is somewhat longer after a good news result.

Some very unfortunate couples have had several affected fetuses, which they chose to abort, in successive pregnancies. Two important points are highlighted by these tragic families. First, the couple would dearly love to begin a pregnancy in the knowledge that the embryo has a winning combination of their genes. Without doubt these families were the impetus for developing preimplantation diagnosis and selective embryo transfer in England. Second, the health economists would be quite wrong in claiming that each affected fetus that was aborted represented a potential patient with cystic fibrosis whose costly health care had been avoided. These women would not have become pregnant so frequently, if at all, if early prenatal diagnosis and selective abortion was not available and acceptable to them. This interpretation is supported by the families in Figure 1 who were the first nine to have *repeat* prenatal tests by chorionic villus sampling in the four years following the introduction of the test in September, 1985. It can be seen that these families tended to avoid further pregnancies often for many years after the birth of their child with cystic fibrosis, until early prenatal diagnosis gave them the opportunity to try for unaffected children but avoiding the birth of a second affected child. Three of the nine families achieved two healthy children in the period of 4.3 years, and two have one healthy child.

In the families in Figure 1, there were two spontaneous miscarriages in the ten continuing pregnancies where prenatal diagnosis had shown the fetus to be unaffected with cystic fibrosis. In any one instance, one cannot easily judge whether an abortion in an early pregnancy was caused by chorionic villus

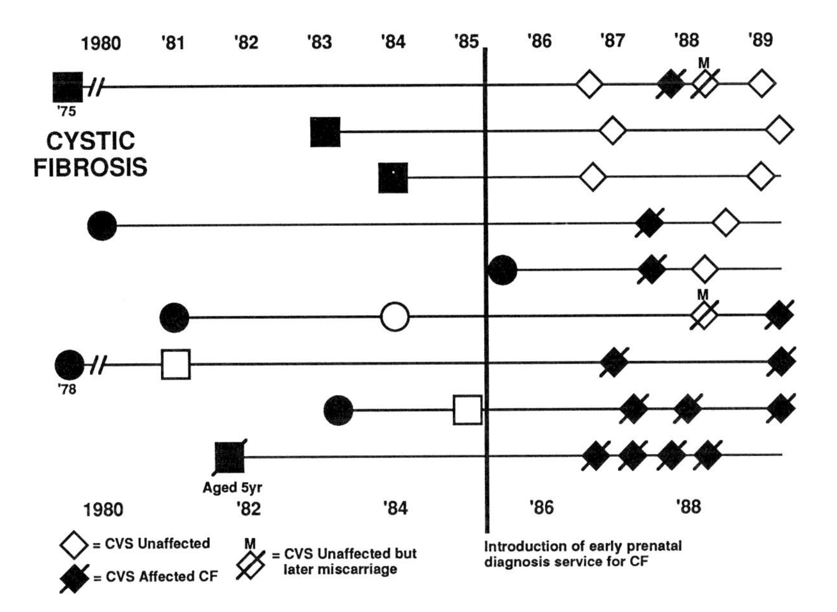

Fig. 1. The reproductive histories following the birth of a child with CF of the nine women who have had a *repeat* chorionic villus sampling (cvs) for early prenatal diagnosis of cystic fibrosis since this service was introduced at the Institute of Child Health, London in September 1985. Note the rapid succession of pregnancies after, compared with before, 1985.

sampling. International figures agree on a CVS procedure related miscarriage rate of 2–4%, over and above the background rate (Table 2). In the face of a 25% recurrence risk for a severe genetic disorder, most couples are not greatly influenced by this procedure related miscarriage rate. It is, of course, quite a different matter for low risk situations such as advanced maternal age. What does influence couples greatly is their beliefs and attitudes to selective abortion, whether they already have healthy children before their genetic risk becomes known, and no doubt many social and family factors. What first trimester prenatal diagnosis does is give the couple the option of not disclosing the pregnancy to friends and family until after the test result. This is usually not possible with amniocentesis-based tests in the second trimester.

It cannot be emphasized too strongly that what has been discussed here is the use of prenatal diagnostic services by families in a country with a National Health Service. It tells us nothing about those families who actively choose not to seek prenatal diagnosis, or those who are not aware of what can be offered, or the impact of a private health insurance approach to medical services. Experience in the Netherlands (Galjaard 1989) is fairly typical of Northern European communities in which genetic services are well established. Of women over 38 years, at least 50% have amniocentesis for the prenatal diagnosis of Down's syndrome, about 25% decline because they object to abortion on religious or moral grounds, and about 25% have other reasons for not having amniocentesis.

It is clear that a substantial proportion of the population will seek prenatal diagnosis if they are aware of a specific genetic risk for a severe disorder. We should understand in much more detail the reasons for this behavior before choosing the criteria by which the success or otherwise of these services should be judged if we are to avoid repeating some of the errors of the past.

REFERENCES

Ash, P., J. Vennart, and C.O. Carter (1977). The incidence of hereditary disease in man. *Lancet* **i:** 849–851.

Carter, C.O. (1977). Monogenic disorders. *Journal of Medical Genetics* **14:** 316–320.

Ciba Foundation Symposium 149 (1990). *Human Genetic Information: Science, Law and Ethics.* Chichester: John Wiley and Sons.

Galjaard, H. (1989). Interaction between technical and social developments in human genetics. In: *Early Influences Shaping the Individual,* ed. S. Doxiadis, vol. 160, pp. 55–66. New York: Plenum Press.

Handyside, A.H., E.H. Kontogianni, K. Hardy, and R.M.L. Winston, (1990). Pregnancies from biopsied human preimplantation embryos sexed by Y-specific DNA amplification. *Nature* **344:** 768–770.

Modell B., and M. Petrou (1989). Thalassaemia screening in ethics and practice. In: *Ethnic Factors in Health and Disease,* eds. J.K. Cruikshank and D.G. Beevers, pp. 179–86. London: Wright.

Modell, B., R.H.T. Ward, and D.V.I. Fairweather (1980). Effect of introducing antenatal diagnosis on the reproductive behaviour of families at risk for thalassemia major. *British Medical Journal* **2:** 737.

Report of the Royal College of Physicians (1989). *Prenatal Diagnosis and Genetic Screening: Community and Service Implications.* London: Royal College of Physicians.

Williams, B. (1990). Who might I have been? In: *Human Genetic Information: Science, Law and Ethics,* Ciba Foundation Symposium 149, pp. 167–179. Chichester: John Wiley and Sons.

Bioscience ⇌ Society
edited by D.J. Roy, B.E. Wynne, and R.W. Old, pp. 67–76
© 1991, John Wiley & Sons

Bioscience in Totalitarian Regimes: The Lesson to be Learned from Nazi Germany

Benno Müller-Hill

Institut für Genetik
der Universität zu Köln
Weyertal 121
D–5000 Köln 41, Germany

Abstract. The various involvements of the biosciences (biology, ethology, human genetics, anthropology, ethnology, psychology, psychiatry, and medicine in general) in the abominable reality of Nazi Germany should be carefully investigated, analyzed, and remembered. This past provides models of what should not be done. It is an illusion to believe that similar things can not occur again. Various possible strategies to stop similar developments in the future are discussed.

Politics is applied biology.
– *Nazi slogan*

What stands higher, truth or goodness? Truth! But truth may not be found in the absence of goodness.
– *Rabbinical wisdom*

INTRODUCTION

The German participants of this conference know, and the others may have heard, that all forms of genetics ("gene technology") are viewed skeptically by the German public. DNA manipulation even in *E. coli* is regarded here as dangerous. The constant talk about the geneticist's responsibilities may have exacerbated the fear that the geneticist's work is really dangerous. Responsibility signalizes danger. Where there is no danger, there is no responsibility.

But behind this is the common knowledge of the crimes of human geneticists and other bioscientists in Nazi Germany, crimes for which there was benign neglect and no punishment whatever. Until very recently U.S. historians of science could just pretend that eugenics was a more or less harmless, strictly Anglo-American affair. Daniel Kevles, (1985) in his most scholarly book on eugenics, did not even waste a line on the history of eugenics in Central Europe.

Even today many scientists think that what was performed in Nazi Germany was "pseudoscience" and therefore outside the interest of true scientists. On the other hand an increasing segment of public opinion in present-day Germany sees no difference between the human genetics of Nazi Germany and human genetics as it is done today. The breakdown of socialism and the clear victory of capitalism leaves people in East and West looking for new enemies. Scientists, and particularly geneticists, may fill this void worldwide. A careful analysis of the historical facts is thus most important for scientists, to enable them to enter the public debate in a qualified manner.

Berlin is a most suitable place to look back at the past of bioscience in general, and human genetics especially. In 1932 this city was one of the best places for science, including bioscience. Its university was certainly the best in Germany. Berlin housed the prestigious institutes of the Kaiser-Wilhelm-Gesellschaft (now Max-Planck-Gesellschaft). Aldous Huxley (1932) placed the institute which produced his *Brave New World* in Dahlem, the suburb of Berlin which housed the Kaiser Wilhelm Institutes. Twelve years later, Nyiszli, Mengele's slave worker in Auschwitz, sent the specimens of the Auschwitz victims to the Kaiser Wilhelm Dahlem Institute of Anthropology, Eugenics, and Human Heredity (Nyiszli 1966). The Notgemeinschaft Deutscher Forschung (after 1936, Deutsche Forschungsgemeinschaft, DFG), which supported Mengele's experiments, had its office in Berlin. The Schering company, the sponsor of this conference, was also located here. Its chief chemist, Dr. Johannes Goebel, collaborated with Professor Carl Clauberg in Auschwitz to find the fastest possible method for mass sterilization of females of the unwanted races (Lifton 1986). And of course, from 1933–1945 Berlin housed the offices of the Nazi government, which shaped racial policy for Europe in those years.

Before I give my brief summary, I must point out that my short text may sound exaggerated or unconvincing in the absence of details and references. I would like to ask the uninitiated reader to consult one or, even better, several of the books in English dealing with the matter (Kater 1989; Lifton 1986; Müller-Hill 1988; Proctor 1989; Weindling 1989).

HOW DID THE COALITION BETWEEN BIOSCIENTISTS AND NAZIS COME ABOUT?

In 1932 some key German bioscientists, eager to apply their ideas to society, began to view the rising Nazi party as their natural ally. A central issue was the nonwillingness of all other parties to introduce a law which would make compulsory sterilization legal. Why did top human geneticists and psychiatrists think that such a law was necessary?

Let me recall here that human genetics had shown unambiguously that certain human physical phenotypes are inherited. The human geneticists assumed furthermore that they had demonstrated unambiguously (I doubt that, but this does not matter; others will agree) that such important traits as schizophrenia, manic depressiveness, and imbecility were inherited. Population surveys seemed to indicate that such "inferior" people bred faster than the healthy "superior" people. As responsible scientists, they wanted to stop what they foresaw as a catastrophy for civilized Europe. One of the apparently minimal measures available was compulsory sterilization of schizophrenics, manic depressives, imbeciles, and others.

Most German biologists and medical scientists saw forces at work in human history similar to those in natural evolution. Darwinism was seen as the ruthless survival of the fittest. The individual had no safe place in this dog-eats-dog universe. A glimpse of this view can be seen in the reminiscences of an American zoologist who sat night after night in the German Army Headquarters in "neutral", occupied Belgium before the U.S. entered the war (Kellog 1917; see also Müller-Hill 1991).

Race mixing was thought to produce offspring that no society would want. Marriages and sexual relations between blacks and whites were penalized in some states of the U.S. and in South Africa. The German Nazi party asked for a law to stop race mixing between (German or non-German) Jews and other Germans. The above-mentioned bioscientists were aware of the high cultural achievements of assimilated German Jews, so they did not dare to propose such measures openly before the Nazis came to power. But they certainly discussed early the negative aspects of intermarriages of Germans with Polish Jews or Blacks. When the Nürnberg Laws were enacted, they were immediately heralded by those scientists as a scientific necessity.

THE ROLE OF BIOSCIENTISTS IN NAZI GERMANY

After 1933 most of the non-Jewish bioscientists became true partners of the Nazis. Some of the most assimilated German Jewish scientists, such as the human geneticist Kallmann, also tried to become partners – but fortunately they were unsuccessful in this attempt. The Nazis wanted to abolish all remnants of democracy, also in science, despite of the fact that they had come legally to power through elections. Some of the top Nazis (including Hitler) disliked science and favored some kind of "Green" enterprises. Yet the anticipated war necessitated the support of science and technology. The attempted restructuring of the population of Europe demanded racist gut feeling and science. Support increased both for applied and fundamental bioscience during this period. So the biosciences prospered materially, despite the fact that the Jews were eliminated from all positions.

Those few who disdained the racial policies of the Nazis kept silent. They had to keep silent, so as not to endanger themselves or their families. Who can blame them? Yet it must to be said that most biologists remained silent even when they could have protested, after 1945, when the arch-Nazi professors remained in their positions. None of those who remained wrote a history of the biosciences of those years.

WHAT WAS THE ROLE OF THE GERMAN BIOSCIENTISTS DURING THESE YEARS?

1. German bioscientists had to show on every possible level (in newspaper articles, books, textbooks, university courses) that the racial measures were sound and necessary from a purely scientific point of view. Here the arguments of the human geneticist, the zoologist, the ethologist, the psychiatrist, the ethnologist, and even the botanist were to be heard. Bioscientists were encouraged to do fundamental research in biology, and thus to provide the proper decorum for the whole murderous enterprise.
2. After the war began, bioscientists were encouraged to use those who were to die anyway (the inmates of psychiatric asylums and concentration camps) as substitutes for laboratory animals. In fact, German legislation gave animals, such as dogs, more rights than the Jews.
3. Bioscientists had to help in devising reproducible (i.e., scientific) procedures in the planned process of racial and genetic segregation. They had to sit in the courts which decided upon compulsory sterilizations. They had to single out the rare true Romanies from those who were products of racial

mixing. They had to single out the incurable psychiatric patients fit for the gas chamber. They had to supervise the selection process in Auschwitz which separated those still fit for the production line from those fit for the gas chamber. They had to divide the population of the conquered nations, particularly Poland and the Soviet Union, into three major groups: a) those who could be counted as Germans, as masters for the top, and b) most others as slaves for the bottom. For the third group, c) the Jews, there was no place in this universe.

And MDs and PhDs alike fulfilled all these tasks with true dedication. There was so much competition for the truly murderous jobs that those who disliked them could get out without suffering any personal disadvantage – as long as they did not publicly question Nazi policies.

NUMBER OF VICTIMS

It is important to mention the numbers of the victims to make it clear that the past cannot be discarded as insignificant.

- *1934–1939:* About 350,000 real or potential medical patients were legally sterilized in Germany, without informed consent, and with the help of the medical and legal profession.
- *1938–1942:* In anticipation of a law which never materialized, about 20,000 German Romanies were singled out for sterilization and concentration camps by a team headed by a psychiatrist. Most were sterilized, and more than 17,000 were killed in Auschwitz.
- *1940–1941:* 70,000 German psychiatric patients were diagnosed as incurable by a team of psychiatrists, among them ten professors of psychiatry. The patients were gassed by killer teams headed by psychiatrists. Prominent psychiatrists and human geneticists tried to frame a law legalizing these murders.
- *1941–1945:* About 80% of the surviving German psychiatric patients died of hunger, infections, or mistreatment in the psychiatric institutions.
- *1933–1945:* German Jews were defined as Jews by the nonexistence of an entry in a parish register of their ancesters before 1800. In the cases where a child of a Jewish father questioned his paternity, human geneticists performed paternity tests (blood group analysis etc.), and the outcome could mean life or death.
- *1940–1941:* Jewish psychiatric patients were killed by gas by the same teams which killed the non-Jewish patients. When the killing by gas ended

in Germany, the experienced killers moved to Poland and the U.S.S.R. in 1942 to set up the first death camps for the mass murder there of Jews and Romanies.

- *Medical doctors* had the exclusive right in Auschwitz to decide whether a person would be killed by gas or sent onto the production line.

If we analyze the reality created by the Nazis, we see that the populations of Germany and Europe were divided into groups with different rights. At the bottom were the Jews with no rights, in the middle were various slave groups with some rights, and at the top the Germans with most rights. The supposed differences in the genotype, which presumably made virtually every phenotype eternal in the coming generations, were used as rationalisations for this division.

THE PRESENT STATE OF AFFAIRS IN THE U.S. AND WESTERN EUROPE

In the Western democracies, like the U.S. and most European countries, we apparently find equal rights for all citizens (but not for the growing crowd of legal and illegal immigrants). One may argue that this fundamental difference already proves that the Nazi experience is so unique that nothing can be learned from it. I would like to argue that this is only part of the truth. Equal rights imply that the social status of everybody has to reach at least a certain level. The homeless population of the U.S., which now constitutes about 1% of the population in New York, and possibly elsewhere in the U.S., have a de facto status which may be compared to that of an inmate of a ghetto under German rule – but without guards or barbed wire. They are condemned to die soon.

I have been told by a U.S. scientist who is actively involved in the Human Genome Project that there will be no abuse since "the genotype is the exclusive business of its carrier". This phrase implies that the carrier is exclusively entitled to sell the knowledge of his genotype (maybe without receiving a cent – but that is his business) to his employer, who may fire him, or to the health insurance company who may not insure him. If this is indeed the future, a division in society will take place which – in contrast to the rather dilettante science of the Nazis – will become truly scientific. Those genetically inferior, catapulted to the bottom in increasing numbers, will be allowed to walk around freely in the cities, but they will find it impossible to have a home or a family. What the Nazis enforced through a plan from above could become true through a truly selective process from below, driven by the forces of the market.

THE PRESENT STATE OF AFFAIRS IN THE U.S.S.R., CHINA, AND THE THIRD WORLD

Medical practice does not allow any free choice for the patient of a genetic disease in the U.S.S.R. There the doctor decides what he believes to be the best solution, for example an abortion. I doubt very much that the local employer is not informed, for example, that a particular person is at risk for Huntington's disease. In China the situation is even worse. The strict policy to curb the birthrate gives doctors total power to decide about abortions on genetic grounds. I do not know how far this practice has now gone. Assuming little change in government and real change in the availability of genetic tests, the prospects in China are just wonderful for old-fashioned eugenicists. I do not know the state of affairs in other totalitarian or half-totalitarian regimes of the Third World.

THE ROLE OF WORDS

The words scientists use seem rather innocent to most users. They are not. I will just discuss three cases. It was, for example, common practice among all German psychiatrists before 1945 to call schizophrenics "inferior" (minder wertig). Even more strongly than its English counterpart, the German word suggests has a connotation of disgust. If a whole medical community uses such a term over many years, is it not almost inevitable that some members of this community will finally act?

The one-sentence letter signed by Hitler authorizing the murder of permanent psychiatric patients contained the word "responsibility" ("in Verantwortung"). Hitler's letter makes clear that responsibility may imply no more than the power to do legally abominable things in the face of an imagined or real threat or danger. The inflationary use of the word "responsible" after the Asilomar conference on the possible dangers of cloning of DNA in *E. coli* signalizes evil ahead.

Finally, the word "population control" awakens old memories. Medical care is clearly not directed here towards the individual but rather towards the population as a whole. This process should be resisted.

THE ROLE OF PROMISES

The eugenicists promised that they knew what measures were necessary to build a better world. Then along came the Nazis and tried some of these recipes. Following his scientific advisors, Nixon promised to conquer cancer by declaring war against it. We certainly know much more about cancer, and more cancers are curable today – but the main cancers pose problems in treatment today similar to the ones of fifteen years ago. Now scientists are promising a massive betterment in preventive medicine after the "holy grail" of the human genome has been attained, and "man will then be understood through his DNA". I doubt all that. These promises cannot be kept. The public will become discontented when it realizes that all these expensive promises are not being fulfilled. Scientists should not sell hope.

A LOOK INTO THE FUTURE

More and more genetic defects will become testable at the DNA level with simpler and simpler techniques. A small amount of saliva will suffice to quickly determine at the drugstore the allelic status of not one, but several genes. Some of these alleles will truly be grounds to prevent its carrier from entering the production line in the industrial society of the 21st century. The following arguments can already be heard, and will continue to be raised, advancing development towards genetic segregation:

- What a careful cost-benefit calculation shows to be advantageous for the community (insurance company, employer, nation) is by definition ethical.
- The medical doctor has an obligation to the patient and to the community (insurance company, employer, nation). His obligation to the community overrides his obligations to the patient.
- It is the duty of the medical doctor to give the best possible care for those who will eventually recover. Those who will never get better may be neglected, particularly when resources are severly limited.
- It is not the business of bioscientists to take a stand on these ethical matters. Their views are, after all, not more important than those of anyone else. These questions have to be decided by new experts, the bioethicists and the relevant authorities.

I do not know whether these arguments will win the day in the next century. If they do, I would dare to prophecy that first this will lead to a boom in the biosciences; later it will definitely destroy all credibility of the biosciences, and in particular of human genetics. Since I find this prospect unnecessary and undesirable, I propose that those working in the biosciences, and in particular in human genetics, should consider the past, and support human rights or civil rights movements trying to prevent the transformation of genetic injustice into an even larger social injustice. The creation of a lower race without health care and work violates equal rights. I propose here in particular that all bioscientists should support the following demands.

1) Every person has the right to know his genotype in full, in part, or, if he wishes, not at all.
2) The genotype of a person should not be determined unless the person has specifically asked for this. It should be a criminal offence to determine the genotype of a person without authorization by this person. Exceptions, such as genetic fingerprinting, should be carefully restricted.
3) Parents have the right to ask, or not to ask, for particular genotypes of their children when the onset of the disease brought about by the particular genotype is earlier than the age of eighteen. They have not the right to ask for genotypes of their children which will not disable them before the age of eighteen.
4) Every pregnant mother has the right to ask, or not to ask, for the genotypes of the unborn. The genotypes of the unborn should not be determined unless specifically authorized by the pregnant woman. Insurance companies should not have the right to ask for the genotype of an unborn child. Abortion should be decided by the pregnant woman and not by any other persons.
5) No third person, insurance company, employer, and the like, has the right to ask for the genotype of a particular person, or to determine it. Breach of this should be made a criminal offence.

So far, so good. But imagine that an allele which makes it almost impossible to survive successfully in an industrial society is found to occur at very high frequency in a certain ethnic group. This ethnic group may be identified by language or color of the skin. Everybody will know then that members of this group are likely to be "imbecile", or "schizophrenic", or whatever medical authority may call them. Present-day geneticists exclude this possibility as being part of the beliefs held by racists. But DNA testing will allow us to test other carriers of the allele which do not belong to this ethnic group, and thereby verify or negate such a proposition. Were this to happen, it would be an occasion for mourning for scientists. Then help should be provided to the afflicted, so far as possible.

REFERENCES

Huxley, A. (1932). *Brave New World.* London: Chatto and Windus.

Kater, M.H. (1989). *Doctors under Hitler.* Chapel Hill and London: The University of North Carolina Press.

Kellog, V. (1917). *Headquarters Nights. A Record of Conversations and Experiences at the Headquarters of the German Army in France and Belgium.* Boston: The Atlantic Monthly Press.

Kevles, D.J. (1985). *In the Name of Eugenics. Genetics and the Uses of Human Heredity.* New York: Alfred A. Knopf.

Lifton, R.J. (1986). *The Nazi Doctors. Medical Killing and the Psychology of Genocide.* New York: Basic Books, Inc., Publishers.

Müller-Hill, B. (1988). *Murderous Science. Elimination by Scientific Selection of Jews, Gypsies and Others, Germany 1933–1945.* Oxford: Oxford University Press.

Müller-Hill, B. (1991). Selektion. Die Wissenschaft von der biologischen Auslese des Menschen durch den Menschen. In: *Medizin und Gesundheitswesen in der NS-Zeit,* ed. N. Frei, pp. 137–155. München: Oldenburg.

Nyiszli, M. (1960). *Auschwitz. A Doctor's Eyewitness Account.* New York: Federick Fell. Translation from the Hungarian: *Dr. Mengele Boncoloorvosa Voltam AZ Auschwitz-I Krematoriumban. A cimlap Ruzisckai György munkaja* (1946).

Proctor, N. (1989). *Racial Hygiene. Medicine under the Nazis.* Cambridge, MA: Harvard University Press.

Weindling, P. (1989). *Health, Race and German Politics between National Unification and Nazism 1870–1945.* Cambridge / New York / Melbourne: Cambridge University Press.

Bioscience ⇌ Society
edited by D.J. Roy, B.E. Wynne, and R.W. Old, pp. 77–96
© 1991, John Wiley & Sons

Our Brains, Our Selves: Reflections on Neuroethical Questions

Patricia Smith Churchland

Department of Philosophy
University of California
San Diego, CA 92093, U.S.A.
and
Salk Institute
La Jolla, CA, U.S.A

Abstract. Research in neuroscience raises a variety of ethical issues, including a) whether humans ought to acquire knowledge about how their brains work, and b) assuming neuroscience does yield such knowledge, what are the ethical limitations on application of that knowledge? The issues are complex, and social policy decisions require understanding of the facts. Religion appears to be neither a consistent guide nor a source of moral understanding. The formulation of sound social policy will depend on nondogmatic and tolerant humans reasoning together. In particular, several common fallacies in reasoning are avoidable.

INTRODUCTION

It is with more trepidation than I care to admit that I have embarked upon this essay. First, because I find ethical issues, unlike their empirical brethren, exasperatingly hard to settle. I am not a utilitarian, because all extant formulations of the utilitarian principle can be counterexampled to absurdity. I do recognize, however, that utilitarian calculations often help in our moral deliberations. I am not a Kantian because the basic Kantian criterion ("moral rules are universalizable rules") either can be grittily adopted by blatantly immoral scoundrels, or the scoundrel can so fine-tune the description of the rule as to trivialize its application. Moreover, any rule I can make sense of – and apply without cheap tricks – has exceptions that are all too easily dreamt up by any moderately imaginative undergraduate.

Is God perhaps the source of what is right and wrong? Dostoyevsky thought so. But two thousand years before he worried about what is permitted, Socrates had the decisive argument against that theory. He presents it as he questions the student Euthyphro: do the gods approve of something because it is right, or is it right because they approve of it? And Euthyphro, plodding student though he is, comes to realize that the dilemma scotches the gods as the moral fountainhead. For if the first alternative were correct, then standards would be independent of the gods and so the gods cannot be the fountainhead. If the second were correct, then morality would be arbitrary, in that it would be merely a function of the gods' whim, whatever sort of character, nasty or nice, they have. this is a simple destructive dilemma, as the argument form is called in logic. Even if God were to exist, it would not help one iota with moral theory.

It might be expected that a philosopher should have a position on the question of where moral standards come from, i.e., a moral theory, and that the theory would be defended rigorously and vigorously to the hilt. Lest I disappoint, I wish to make a clean breast of things at the outset: I do not have a moral theory, or anything approximating such. I recognize that deep-seated if vague agreements on elementary dispositions concerning justice and fairness must have some basis in our evolutionary history and our genes, but how to make that connection more precise is unknown. Until much more is understood about the brain and what in the wiring is genetically prescribed, the details of the genetic component will probably remain out of reach. Despite these lacunae, we can say that *Homo sapiens* is clearly a social species, like wolves and ants, but unlike orangutans or polar bears, and it is overwhelmingly likely that our genes are an essential element in producing the social disposition characteristic of human brains.

Paul Churchland (1989) is probably correct when he suggests that specific moral concepts are acquired in much the way we acquire concepts for most other categories. That is, our genes give us a headstart with some *in situ* but modifiable neural organization, then the neural networks learn from examples. By error-correcting and/or associating to ever greater consistency with experience, a network constructs representations that cognitively resemble those of its fellows. From example, the child learns what it is to hurt and be hurt, what taking turns, sharing, and exploitation are, and assuming that he is "characterologically competent" (can empathize, can feel sorry, embarrassed, or whatever), he comes to act on his inner representations in social contexts, not always perfectly, not always without backsliding and weakness of will but often with remarkable courage and selflessness. At least, that would be the run-of-the-mill case. Moral blindess, on this view, is due to genetic differences, missocialization, or both.

If I do not have a moral theory, how can I presume to go on and talk about morality in the application of neuroscientific knowledge? I shall do it the way we all do it when we are not selfconsciously practicing moral philosophy: we reason together, with tolerance, patience, sympathy and common sense. We draw on such wisdom and empirical knowledge as we may be lucky enough to possess, and we engage in give-and-take. This is crude, I grant you, but I would rather not pretend to more.

There is a second reason why I approach this essay gingerly. The questions we are asked to consider concern social policy on matters that are emotionally inflammable. They seem tremendously important, and they are, yet they become palpable only in tandem with progress in brain science. This means that there are frequently too few facts to sustain a considered judgment. I fear, therefore, that I shall often flounder, and consequently I shall be gratified if even a small amount of light can be shed upon these questions.

What follows is my perspective on some questions concerning the possible applications of neuroscientific knowledge. I must confess, however, that it leaves out much, and I sometimes skate rather fast over the ice – especially the thin bits. What I shall do is sketch a kind of logical geography of the major features in the landscape, as I see it, in hopes that this will provide a relatively clear basis for further discussion. In this survey of the terrain, I have tried to include not only matters that exercise me, but also some that are of greater concern to others. They are not all equally compelling or profitably discussable.

A wide range of morally relevant issues concerning neuroscience can be listed, and for convenience, I divide them into two broad classes: 1) those that pertain to the just and proper application of knowledge, and 2) those that focus on the propriety of having knowledge about brain mechanisms, quite apart from any practical application the knowledge might have. In the first category, we shall consider a) *potential gains* (diagnostic, preventative and therapeutic) and b) *potential abuses* (clinical, political, judicial, and experimental). In the second category, we shall consider a) the *"Faustian potential"*, as I call it; that is, the possible loss of virtue or "humaness" that neuroscientific knowledge may visit on us, because it is "knowledge we should not have"; and b) the experimental cost.

I shall address these two classes of issue in reverse order.

IS NEUROSCIENTIFIC KNOWLEDGE WORTH HAVING?

The "Faustian potential"

Should we know how our brains work? Progress in understanding the world has often been viewed with suspicion and disgust, and even now scientists are not infrequently portrayed as unscrupulous, antisocial, and impractical, as well as having a dose of madness and a fondness for the darker arts. There is a long-standing tradition that objects to accretions in empirical knowledge, usually because in some way the discoveries are thought to entail a loss of something special, something spiritual, or something "essentially human". In this instance, the view, as I reconstruct it, is that, as we come to understand the nature of the mind-brain, it entails that we, as human individuals, are demystified, and hence that we are dehumanized, demoralized, and dissected like planaria. Falling in love is a favorite example of what, apparently, we ought to prefer unexplained, for, the argument goes, should we explain it, love would become just another physical process, and a process that might even be duplicatable in a robot. The excitement, the mystery, the abandon that is love would be as lost in the dissection, even as life is lost in a dissected laboratory specimen.

This point of view is best opposed by a liberal education, for no single argument or two will have much force against a generalized conviction that ignorance is to be preferred to knowledge. Although the idea that knowing less is understanding more may be a consolation to the ignorant, it is more lamentable than anything else. In any case, it is gratifying to note that in the hurly burly of life, the dictum may be mouthed but not lived.

Unlike my college freshman class, this audience does not need perorations on the practical value *and* aesthetic virtues of empirical knowledge. Apart from the genuine aesthetic pleasures resulting sheerly from solving the mysteries, and apart from the glaring fallacy in equating understanding to killing-by-dissection, the medical and creature-comfort benefits deriving from science are obviously stupendous. Perhaps because the heyday of torturing the insane to drive out devils is long past, because we scarcely remember smallpox, polio, or undulent fever, and because we are long past the days when the Catholic church condemned anethetics and vaccines as instruments of Satan (White 1896), and because antibiotics have demoted pneumonia from a main event to a sideshow, the technological spin-off from science is taken for granted. The debt to science as a headwater for practical benefits is very easy to neglect.

My further personal perspective on the Blakeian dogma that knowledge entails loss of a precious spiritual innocence and that the loss outweighs the gain, is that, by and large, it just does not make sense. Some chord or other must have been struck by Blake, and perhaps I am just lamentably tone deaf. But I have never seen the argument, only the poem[1]; I have never heard the justification, only the rhetoric. As a farm child struggling to help make a living in a world where practical benefits meant the difference between food and hunger, or between machines and hand-toil, I long believed the choice to be dead obvious.

Nevertheless, because the arguments that champion ignorance over scientific understanding have a sympathetic following both in so-called "New Age" groups as well as in certain segments of organized religion and the entertainment industry[2], they are disregarded at our peril. They do have an effect.

What price knowledge?

Animal experimentation. The main questions under this rubric have concerned whether humans ought to test hypotheses on animals, and if so, which ones and under what conditions. Although computer modeling is playing a growing role in neuroscience, there is no doubt that it cannot replace animal experiments. If neuroscience is to continue, animal experiments are necessary. In the history of biomedical discoveries, animal experimentation played a crucial role. Banting and Best managed to isolate insulin, but only with the aid of live animals; Salk's discovery of polio vaccine, and the development of birth control drugs, depended essentially on knowledge gleaned from experiments on animals. Antivivisection groups are sometimes opposed to all animal experimentation, on whatever kind of animal, and for whatever purpose, whether it is testing AIDS vaccine in monkeys, or contraception in mice, or locomotion in cockroaches.

In the U.S.A, the Society for Neuroscience, especially under the leadership of David Hubel from Harvard and Patricia Goldman-Rakic from Yale, has begun to organize responses to the antivivsectionist campaign. In August of 1990, The British Association for the Advancement of Science launched a

---

1 William Blake, "The Songs of Innocence and Experience". In: Keynes (1966).

2 For a wonderful if blood-souring sampler, see Martin Gardner's *The New Age: Notes of a Fringe Watcher* (Gardner 1988). This includes discussions of the views of Margaret Mead, Shirley MacLaine, Arthur Koestler, Oral Roberts, Pat Roberston, Jerry Falwell, L. Ron Hubbard, Russell Targ, Harold Puthoff, and many others.

declaration concerning animals in medical research, which is balanced, principled, and sane. It forthrightly articulates the need for animal experimentation in research and for scientists to explain this to the public at large. [3] The International Brain Research Organization (IBRO) has established a committee[4] to address the threats posed by animal liberation groups in many different countries.

These developments strike me as extremely important, since academics are more inclined to hope the problem will go away than to respond politically, and because widespread, hideous diseases such as schizophrenia and Alzheimer's disease cannot be brought under control in the absence of animal experimentation. Barring a new target to siphon off the antivivisectionist fervor, I gloomily forecast that the interference with medical research will continue unabated. Although the antivivisectionists may have done some good in requesting adherence to standards by cosmetic companies and abattoirs, they have also done great damage to medical research programs and to animal study programs that, paradoxically for the animal liberationists, provide results that precisely benefit animals and promote the humane treatment of animals. Research on pain has virtually ground to a halt, because scientists cannot undertake the risk of testing drugs on animals given experimental pain. The number of humans and animals suffering pain, caused by many forms of cancer, spinal injuries, burns, and so forth, has not, however, declined. So far as the moral principle is concerned, it is certainly not obvious to me that no

---

3 See *IBRO News*, **18(3):** 1. The text reads as follows:
In view of the threat to medical research posed by increasingly vocal and violent campaigns for the abolition of animal experimentation, we make the following declaration:
a) Experiments on animals have made an important contribution to advances in medicine and surgery, which have brought major improvements in the health of human beings and animals.
b) Much basic research on physiological, pathological and therapeutic processes still requires animal experimentation. Such research has provided and continues to provide the essential foundation for improvments in medical and veterinary knowledge, education, and practice.
c) The scientific and medical community has a duty to explain the aims and methods of its research, and to disseminate information about the benefits derived from animal experimentation.
d) The comprehensive legislation governing the use of animals in scientific procedures must be strictly adhered to. Those involved must respect animal life, using animals only when essential and as humanely as possible, and they should adopt alternative methods as soon as they are proved to be reliable.
e) Freedom of opinion and discussion on this subject must be safeguarded, but violent attacks on people and property, hostile campaigns against individual scientists, and the use of distorted, inaccurate or misleading evidence should be publicly condemned.

4 Ibid. pp. 1 and 6.

rat should be caused pain in order to relieve a child from suffering terrible oncological pain.

Are scientific research animals in fact poorly treated? This is a very broad question, but let me say that, in my experience, the answer is emphatically "no". The regulations governing animal research are actually very strict, and in California, for example, laboratories are routinely inspected, unannounced, by county, state, and federal officials. Typically, researchers care a great deal about the comfort and health of their animals, both for reasons of human decency and because sick or stressed animals do not give reliable results.

It is fairly obvious that if the antivivisection campaign were successful, much of neuroscience and other biomedical research would be impossible. Assuming that human suffering has some place in the calculation of whether this would be acceptable, I can but conclude that animal experimentation is an issue where moral fervor may fail, and fail egregiously, to coincide with moral probity and moral decency. It may also be worth noting than in San Diego, for example, the number of stray dogs put to sleep by the Humane Society is two orders of magnitude greater that the number of dogs transferred to laboratories. My guess is that these figures are not atypical of other major cities.

Human experimentation. i) *Studies using invasive techniques.* These are of course limited in the extreme, but under careful conditions, some data can be gathered in the course of surgery or chronic electrode implant. For example, George Ojemann, a neurosurgeon in Seattle, has done outstanding work this way (Ojemann 1983; Ojemann et al. 1989). Ethics committees watch this sort of research with great care, as do granting agencies and the journals to which the research papers are sent, and Ojemann himself is a clinician first, and a researcher second, by which I mean that he puts the interest of the patient ahead of research goals. With the advent of new techniques, the human subjects committees will expand their province.

ii) *Studies using noninvasive techniques.* Although these are generally preferred to invasive techniques from an ethical point of view, they too can be hazardous. Positron emission tomography (PET) uses radioactive material, albeit in tiny amounts, and while magnetic resonance imaging (MRI) appears to be safe, no one knows for sure what the subtle effects might be of putting a brain in a strong magnetic field. Electroencephalography (EEG) is probably the safest technique, but because of problems in spatial resolution, it is best used in conjunction with other techniques. So it is not a replacement for the other techniques.

iii) *Use of Foetal tissue.* Tissue grafts in damaged brains may be an important treatment, especially for Parkinson's disease, and conceivably for Alzheimer's disease. Although the research is still in its infancy, and it is not yet clear what exactly is accomplished by the implanted cells, the results so far (Bjorklund et al. 1990) are mildly encouraging. It goes without saying that the possibility of treating Alzheimer's and Parkinson's diseases has sparked widespread humanitarian interest. There is also much to be learned about neural development and developmental disorders by research on foetal tissue. Here, also, the potential benefits are immense.

The objections to research on foetal tissue derive mainly from anti-abortionists. From their point of view, abortion is murder, and foetal research simply compounds the felony. On the other side, it is argued that so long as abortion is legal – and in the U.S.A and elsewhere it is – destroying tissue which could benefit the suffering is irresponsible. Consider a slightly different case. Jehovah's Witnesses regard blood transfusions as deeply sinful, yet the medical practice is perfectly legal. Government restrictions to prevent transfusions, on grounds that the Jehova's Witnesses are convinced of its sinfulness, would generally be considered an attempt to foist a personal religious dogma on one and all. Does not the same response apply to those whose personal religious convictions deem abortion sinful and the use of foetal tissue heinous profiteering?

My own view is that it is as unjustifable to forbid early abortion of a foetus as to forbid blood transfusions, or anesthesia or consumption of pork or driving of cars by women, or cremation, or any one of a dozen other bylaws of some particular religion. One is free to follow these strictures oneself for one's own peculiar reasons, but there is no justification for imposing them on others who decline to submit to the creed. My view on the permissibility of abortion coincides with the 1973 decision of the U.S. Supreme Court in the case *Roe vs. Wade,* which upheld a woman's right to an abortion, and to a similar decision of the Canadian Supreme Court in 1989.[5] Since this is not the occasion for a searching discussion of that issue, I conclude by noting that when different and substantial constituencies of sane, decent, kind, and thoughtful humans disagree about the moral permissability of an action, then tolerance and mutual respect, not self-righteousness, coercion, and intolerance, are the virtues that ought to prevail. Moral conviction is a psychological state with a welter of epigenetic origins; it is not, as human history shows all to clearly, a singular, let alone an infallible, guide to moral recitude.

---

5 See Kristin Luker's *Abortion and the Politics of Motherhood* (Luker 1984).

Regulating use of foetal tissue is of course desirable, as is regulation of kidney, heart, and blood donations. Indeed, when it is permitted, it is regulated. To prevent abuse of donor-recipient interactions, research hospitals have typically adopted anonymity policies, which keep the two parties and their doctors quite separate, and which require consent of the pregnant woman to the abortion and the use of the tissue.

Did the patient consent? The matter of informed consent is not new and it is virtually always very thorny. It is even more thorny in the case of diseases of the brain, because the patient may be cognitively disabled by the disease. The disease itself may prevent the patient from giving his consent to treatment, as, for example, in alcoholism and other addictions. This matter has already concerned psychiatry for some time, independently of recent discoveries in neuroscience. Consequently, ethics committees are very stringent in evaluating proposals for treatment without consent, as well as for clinical trials and human epidemiological studies. So there is a foundation of policy in place on which to build. Nevertheless, the possibility of abuse remains very real, either because the patient was improperly treated, or because he was *untreated,* owing to the clinician's fear of malpractice litigation. (The question of coercive therapy will arise in part iv in the section "The Ethics of Knowledge Application")

How much will it cost? Research is expensive, and modern research in neuroscience, though cheaper than particle physics, generally requires a heavy research investment. For example, the start up cost for a PET scanner is estimated to be roughly seven or eight million dollars. Operating costs are added on thereafter. Some societies may conclude that they cannot afford high technology neuroscience, and either depend on other nations to do the research, or agree to continue at their current level of medical practice. Funds for neuroscience research in the U.S. are currently very tight as the funds for some government agencies are either cut, or fail to keep pace with increasing costs and increasing numbers of researchers. The costs of research have to be balanced against the costs of other projects society values, such as savings and loan rescues in the U.S. (at least 500 billion).

Against the presumption that neuroscience will have only biomedical benefits as the payoff on the investment, we must pause to acknowledge that pure research often yields discoveries that lead to unforseen but monumental technological breakthroughs. It is entirely probable that discoveries in neuroscience will fit the mould. Indeed, it has already been conjectured that the brain style information processing currently under study may lead to a revolution in computer technology, with profound economic sequelae (Caudill and Butler 1990; Hecht-Neilson 1990; Churchland and Sejnowski 1991). Given extant network machines, this seems quite likely, and some devices, such as a network for detecting plastique hidden in suitcases at airport checkpoints, are already functioning.

Politicians regularly benefit from a gentle instruction regarding the value of pure research and its unpredictable practical advantages. The investment in computer research in the sixties or into properties of substances at supercold temperatures might have seemed unwise – might have been "Proxmired"[6] – at the time, but with the development of chips, electronics, and superconductivity, the investment reaped a bonanza.

THE ETHICS OF KNOWLEDGE APPLICATION

Educated guesses and wild speculation

What knowledge will neuroscience bring? Could we apply it to halt the progress of Alzheimer's disease, or "rewire" a psychopath, or control thoughts and emotions by "brain bediddlers"? Although some of the speculations can be amusing, my own prefence is to concentrate on questions of knowledge application in the light of what knowledge is actually available now, and what we can reasonably expect to know in the foreseeable future. Uninformed speculation can be an awful time-waster, as, for example, in working out the details of irrigating from the canals on Mars before the properties of the planet were more clearly apprehended. There is no sharp line, of course, between educated speculation and educated-but-slightly-nutty wool gathering. For the most part, however, I prefer to think of myself as staying mostly with the first, but I allow that from time to time I may have strayed into the second.

Intervention and exploitation

One major focus of concern as knowledge of the brain progesses is the potential for abuse. Important advances in clinical interventions that treat brain disorders, both psychiatric and neurological, have made impressive progress

6 William Proxmire is a Democrat senator from the state of Wisconsin, who became well known for his "Golden Fleece Award", which sometimes included scientific projects whose motivation and significance he was unable to understand, as well as military costs that would have been hilarious but for the impoverished social programs.

since the 1950s. The medical opportunities for therapy are based on research in a number of fields, but especially neuroscience and molecular biology. Psychoactive drugs, neurosurgical procedures, and genetic analyses that reveal markers for disorders such as trisomy 21 (Down syndrome) or Huntington's chorea are prominent in the roster of remedies. Based on their current capacities and future potential, one might also speculate about further developments in pharmacology and genetics, and perhaps even wonder whether "wiring pathologies" might someday be revealed by high resolution scanning machines.

There is a plethora of moral issues here, ranging from whether antidepressants prevent some persons from confronting and changing their truly depressing situations, to whether one has a duty to abort a foetus carrying a gene for a disease certain to cause dreadful suffering. My own view is that each sort of problem has to be dealt with in the context of the facts pertaining to it – the pharmacological, genetic, diagnostic, epidemiological, etiological, and sociological data. They cannot be satisfactorily discussed all of a bunch, since answers to one problem may be tangential to another. Nevertheless, certain general principles and misconceptions can be efficiently introduced early in the conversation. I have taken the liberty of assuming that others, more qualified than I, will present various problems in their specific settings. My focus will instead be on some abstract and general questions concerning the dangers inherent in using our knowledge of how the brain works to alter human brains.

i) Not my good but *their* good. In its most general formulation, the worry is this: therapeutic techniques for treating damaged and disordered brains will be used for control and exploitation – not only of diseased patients, but of perfectly normal humans so that they become pawns or slaves of someone else. The envisaged scenario is very different from enslavement documented elsewhere in history, for here the brain might be altered to change the person's desires, motives, feelings, and the very way he reasons and plans, and even his conscience about right and wrong. The very possibility of slave revolts would be abolished from within, by making the slave brains compliant and content brains, by making the slaves happy slaves.

This is not a far-fetched danger. The heydays of lobotomies (Valenstein 1986) and valium and electroconvulsive therapy (ECT) for social malcontents have provided a harrowing glimpse into the effects of unscrupulous medical (mal)practice. What can be done to forestall the abuse of neuroscience?

My response here is not very glamorous, but it is the only one that makes sense to me: continue trying to regulate use of techniques, to educate widely, and to pressure the politicans and clinicians, so that we reduce the abuse to the minimum possible. Abuse of knowledge is of course something for which society must always be vigilant, but for which general rules – other than rules with a hollow ring, such as "do not abuse knowledge" – are not forthcoming.

Undoubtedly the early kindling of fire led to using fire as a weapon and a means of power; the discovery of anesthetics led to witholding in exchange for compliance; harassing phone calls are a menace, and so on *ad nauseam*. If we merely sermonize, "Use knowledge wisely and do not abuse it", that is not going to deter those who convince themselves they are wise and are making good use of their knowledge. More generally, we cannot expect that fair and decent rules will successfully impede determined and devious rule breakers. Not in medicine, or football, or politics, or anywhere else. We do the best we can within the social reality we start with. Specific rules, regulating where fires can be lit, who can own guns, and the penalties for phone abuse are, on the other hand, reasonable. The updating of regulations and the introduction of new ones, as new technology becomes available, is a never-ending process.

In view of the potential for harmful application of neuroscience, a "hands *completely* off" policy may be urged. According to this argument, the risk is such that we should collectively abjure any sort of medical intervention in human brains. The implications of this recommendation bear long reflection. About 1% of the population is schizophrenic, 12% of men and 18% of women have a depressive disorder, 2% have epilepsy, 20% of people over 65 have Parkinson's disease, 12% over 60 have Alzheimer's disease, 10% of American are alcoholics, and so on. What of tumors, gunshot wounds, encephalitis, meningitis, spinal cord paralysis, hydrocephalus, and so forth?

Presumably the "hands off" argument rests on some sort of cost-benefit analysis, but in the absence of having it to hand to study, one can only suspect that a decimal point must have strayed far from where it should be. In any case, it is sometimes forgotten that failing to act can be every bit as immoral as an overt act. Sins of commission may be more colorful, but they are not necessarily more heinous, than sins of omission. Throwing an infant overboard is morally unacceptable; so is failing to pull a drowning infant out of a puddle because one does not want to muddy one's shoes. Dosing a rambunctious child with ritalin, because we would rather watch TV, is morally unacceptable; letting a hyperactive child become a social pariah, because someone else might use ritalin selfishly, is also unacceptable. We cannot keep our hands clean and our consciences clear by omitting altogether to apply neuroscientific knowledge. What do we do then? The homespun wisdom here is that we muck on as decently well we can, avoiding extremes, and relying on those with proven character, breadth of experience, and depth of understanding to lead the way. In a more literary vein, one might say that between the Scylla of abuse and the Charybdis of abolition is the painstaking and thoughtful working out of social policy.

ii) The fallacy of the beard. Sometimes in the debate over social policy, it will be noted that clean-cut divisions, between what should count as within the rule and what not, are hard to make. When legislation lays on a division even so, it may seem abritrary and unjustified. Reflecting on when it is permissable to abort a foetus, some might suggest that drawing the line at five months is arbitrary – after all, there are minute differences between a foetus at 150 days and at 151 or 149 days. And there are many more arguments cut to the same pattern. Why, it might be argued, should the freeway speed limit be set at 55 mph, rather than 56 or 59? Why should garden watering be permitted in San Diego between 10 pm and 8 am, rather than 9:30 pm and 8:30 am?

Observing that there are continua is commonplace; the fallacy consists in concluding that therefore no meaningful distinctions can be drawn. The fallacy of the beard takes its name from the case where it is noticed that there is an uninterrupted continuum between the face of a man who is smooth shaven, the same man at 5 o'clock later that day with the legendary "shadow", the same man next morning with stubble, and finally, the same man a month later with a full beard. To deny that we can draw any distinction between being clean shaven and having a beard, on grounds that any two adjacent points on the continuum are scarcely distinguishable, runs foul of common sense. Large rivers are easily distinguished from small creeks, though some creeks at flood are larger than some small rivers in high summer. But so what? The fuzzy intermediate cases do not negate the contrasts in distant points: hills and mountains, blue and aquamarine, juvenile and adult, manslaughter and negligence, aggressive marketing and price gouging, education and propaganda, conceptus and infant.

In a continuum, the end points typically contrast vividly. A neonate is radically different from a conceptus; a submachine gun is very different from a slingshot; a manic depressive from a blue and lonely first year student. Consequently, when practical life demands it, we can make distinctions on the continuum, recognizing that adjacent points may be virtually indistinguishable. How do we do that? There is no generally applicable answer, save that in moral matters it is wisest to err on the side of caution. Thus, legislating five months rather than seven months as the limiting date for abortion errs, if it errs at all, on the side of caution. So far as distinctions relevant to applications of neuroscience are concerned, here, too, we shall be guided by available information, moral willingness, and the rule of thumb to err, if at all, on the side of caution.

iii) The Rifkin fallacy. Another temptation in this arena consists in 1) imagining future possibilities based on freely speculative and grossly under-defined technology, 2) discovering that no clear answer about social policy is reachable in these conditions, and then 3) generalizing to conclude that moral matters arising from currently planned and quite well-defined technology are also beyond our collective wisdom. The corollary is that if we do not know how to solve the moral problems served up by a technology, we have no right to use the technology.

"Suppose", a student will ask, "that neuroscience allows us to produce a device for rewiring sociopaths so that they became socially responsible, but the side-effect is that their mature height is three inches shorter than it would otherwise be, or that they become musically insensitive. Would that be morally acceptable?" The instructor heaves a sigh. When, as in this example, the conditions are woefully underdescribed, who on earth knows *what* to say? To generalize from a freshly hatched "what-if" story to a fledgling medical application is evidently not justified. Factually rounded-out examples are very different from fictional line drawings, and, in particular the richness of the available data in the former case enables us to sink our collective teeth in and figure out a fair-minded course of action. By way of contrast to the fanciful rewiring story, consider more palpable questions such as the use of foetal grafts in Parkinson's patients, or whether to abort a foetus carrying a genetic marker indicating it is probably a psychopath, or whether, since an hour of clinician's time is worth about two months of fluoxetine capsules, psychotherapy should be paid for from the patient's purse, not the purse of the insurance company or the government.

I should perhaps end this section by explaining that I refer to the inference from a fanciful case to realistic cases as "The Rifkin Fallacy" because at least some of Jeremy Rifkin's objections to scientific research and biomedical innovation have that character. Some, though probably not all, of his acolytes too have acquired a reputation for sliding into the fallacy in the name of public good.

A crude graph with a "trait undesirability" continuum on the X axis and a "biological knowledge availability" continuum on the Y axis, can be constructed, thereby plotting out various genetic counselling issues as a function of what we know and how seriously debilitating is the gene. (Fig. 1) Thus cystic fibrosis and trisomy 21 and trisomy 18 are located approximately in the upper-left, whereas gender is located in the upper-right quadrant. But traits such as knock knees and IQ are somewhere in the lower-right quadrant. Now, what I am calling fallacious is the idea that arguments pertaining to the acceptability of genetic choice of traits in the lower-right applies equally to genetic choice of traits in the upper-left. The Rifkin fallacy in this arena consists in exactly that. Thus it may seem convincing that one is not justifed in aborting a fetus that merely has an average IQ, in hopes that the next fertilization will produce a specimen with a higher IQ. But even should the argument be convincing for traits in the lower-right quadrant, it cannot be neatly rolled over into an argument against aborting a fetus with traits in the upper-left quadrant, such as trisomy 21.

iv) Responsibility: Full, diminished, and absent. Determining social policy for violations of laws concerning public safety is often anguished; it is, however, a domain where practical decisions are unavoidable. In the forefront of unacceptable behavior are violence (including murder, rape, assault and manslaughter), theft, fraud, kidnapping, bribery and tax evasion. Listed among the issues concerning punishment and blame are these: 1) protection of others, 2) deterrence of potential violators, 3) prevention of recurrence by the same person, 4) compensation to the victim, and possibly 5) retribution. A crucial concern in deciding how to treat the person whose behavior is problematic is whether he or she was responsible for what was done. There is a contrast between someone who, during an epileptic seizure, strangled a seatmate, and a young man who plans and carries out his wife's death, leaving him rich and free to marry his mistress.

When is someone responsible for what he does? As more is understood about behavior and its causes, the more complex become the answers. My approach here will be to outline a framework for discussion by focusing on 3) above, namely, prevention of recurrence by the same person. One view of punishment is that *inter alia* it serves as negative reinforcement, and hence deters recurrence, but only if the agent was genuinely responsible for what he did. Moreover, it is generally agreed that it would not be just to punish someone for something that was not, strictly speaking, his fault. While we can distinguish cases of sheer accident from cases of plotting by a sane man, there are many cases where the issue is less clear.

Sometimes the brain has properties which entail that the person has diminished responsibility for the action. For example, the person is paranoid and believes resolutely that the postman is Himmler, and, in the name of the public good, does him in. There are cases where the mechanisms for motor paralysis during REM sleep are defective, and the sleeper acts out his dream of strangling a mad dog, killing his wife as he does so. In these sorts of cases,

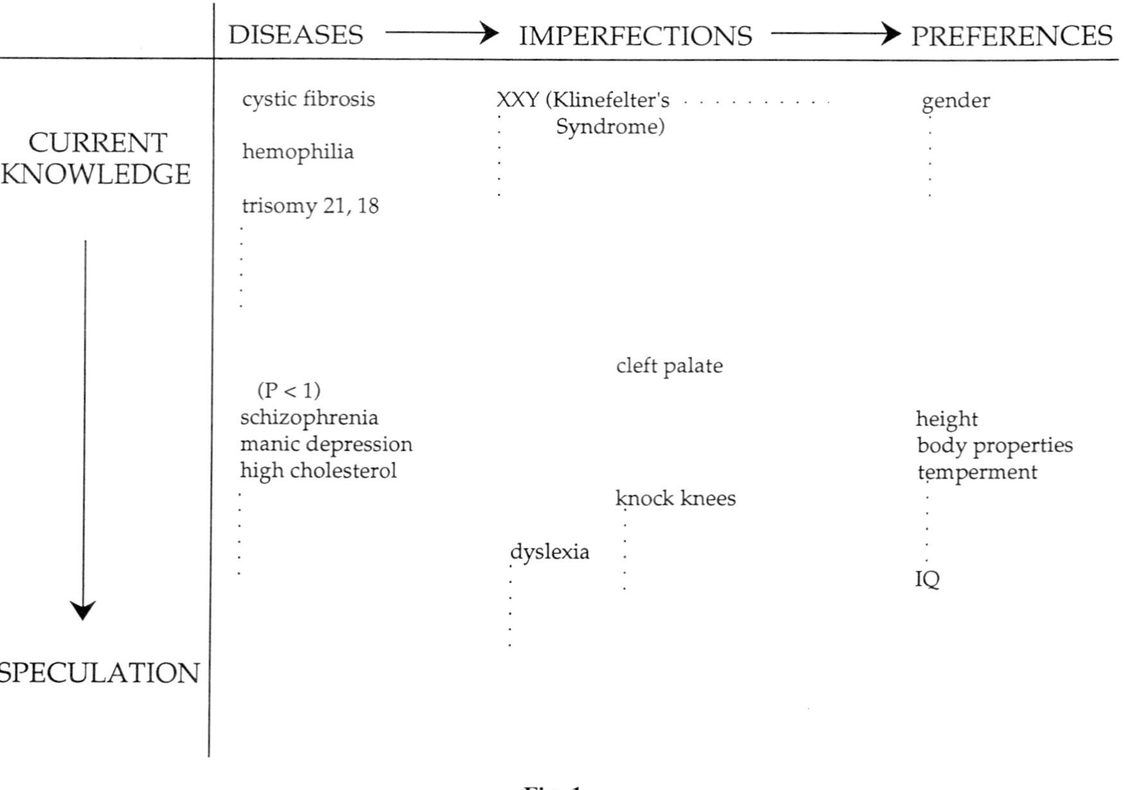
DISEASES ⟶ IMPERFECTIONS ⟶ PREFERENCES
CURRENT
KNOWLEDGE
SPECULATION
cystic fibrosis
hemophilia
trisomy 21, 18
(P < 1)
schizophrenia
manic depression
high cholesterol
XXY (Klinefelter's
Syndrome)
cleft palate
knock knees
dyslexia
gender
height
body properties
temperment
IQ

Fig. 1.

punishment may be inefficacious and inhumane. Instead, it may be preferable to intervene in the brain directly. That is easy enough to do in the case of the dreamer, since REM blockers can be used with great success. The person is no longer a threat, but he also no longer dreams: a fair exchange, one would assume.

The paranoid schizophrenic is also treatable with drugs, as is the epileptic. Protective custody may still be in order, but a conventional prison term would probably not. Obsessive-compulsive disorder is now successfully treated with drugs in many cases, and "uncontrollable rage", manic depression, and erratic acts associated with premenstrual syndrome (PMS) are also argued in court as more appropriately treated with biomedical intervention such as surgery and drugs than with conventional punishment. Genetic markers for some disorders such as schizophrenia and sociopathy may be found, raising the further question whether preventative treatment would be justifiable. If we had known Ted Bundy was a homicidal sociopath before he killed several scores of people, would it have been appropriate to treat the condition with drugs or surgery? How much different would life in Uganda have been with Idi Amin on haldol?

So far we have considered mainly the needs of society, given the deviant behavior of an individual, but we must balance these against the needs of the individual. Privacy, diversity of personality types, eccentricity, and so forth are not features we want to crush on behalf of public protection. Fanaticism about the latter can lead to injustice on a massive scale.

In character, the problem is really one of constraint satisfaction: there are many important constraints – protection of society, and freedom of the individual, to mention only two – where some constraints may be weighted more or less than others. These are tough problems, not to be solved by the mechanical application of set of a hide-bound rules. Constraint satisfaction problems can be elegantly solved by connectionist (neural net) models, given feedback corrections by a "teacher" who knows the answer.[7] The human condition is such that we do not have anything really analogous to the connectionist "teacher". The real world with its uncompromising causal reactions is a kind of teacher, but typically an unforgiving one. Nevertheless, we do have each other, and we do have our own neural networks in our heads. Perhaps, reasoning together, continually reassessing and revamping, using gentle persuasion, is the way to approach our constraint satisfaction problem with

7 For an introductory account of the computing style of neural nets, see Paul M. Churchland's *The Neurocomputational Perspective* (Churchland 1989) and *Matter and Consciousness*(Churchland 1984); or Patricia Churchland and Terrence Sejnowski's *The Net Effect: A Computational Neuroscience Primer* (Churchland and Sejnowski 1991).

neural networks. As always, it is a mistake to suppose that the issues can be settled on "purely moral grounds alone" and in the absence of relevant scientific data. Such data provide crucial additional constraints.

The adversarial system of courtroom justice (prosecution, defence judge and/or jury) appears to work better than anything else, but sometimes, as we all know, justice is not done, and there is no "Court of Angels" to swoop down and do the job properly. Sometimes we are overwhelmed by the problem of constraint satisfaction, sometimes, as with such problems generally, there may be no one right answer, only a set of "local minima", and sometimes, because we are ignorant or frightened, we get stuck in a local minimum,[7] and it takes a shock to bounce us out.

v) Will neuroscience eventually show that no one is responsible for anything? My considered view is that, as we come to understand in greater detail how the brain works, we will deepen and broaden both our conception of agency and of the fair means to intervene when there are defects of agency. But a wholesale dumping of the notions of agency and responsibility is not something forseen from where we are in 1991. At least because it may turn out that "normal" individuals are at their best when they conceive of themselves as responsible agents, and because education and nurturing may proceed most effectively if children are taught to conduct themselves as rational, responsible agents. On the other hand, it is also conceivable that new concepts – neuroscientifically informed concepts – may function even better in these roles, and that, with developments in neuroscience, they will evolve and come to replace the orthodox conceptions of agency and responsibility. What such notions could look like is not predictable at this stage of science, at least not by me. But it is worth allowing for the possibility that we can do better than we are.[8]

CONCLUDING COMMENTS

The possibilities before us for understanding and treating human brains are both inspiring and sobering. While there is a clear potential for the abuse of knowledge – as there is in every field of human affairs – vigilance, goodwill, and common sense can do a great deal to steer us around the hazards. I forsee the greater dangers to lie not in new knowledge and its regulated

8 See my *Neurophilosophy* (Churchland 1986) or Paul M. Churchland's *The Neurocomputational Perspective* (Churchland 1989).

application, but in fanaticism, superstition, and ignorance, both moral and empirical. I am probably more afraid of the moral fanatic, the self-righteous inquisitor, and the hot-blooded do-gooder, than I am of the self-interested hedonist, who often can be counted upon at least to be reasonable, if not altruistic. I expect we have more to fear from overpopulation than from neural transplants, from holy wars than from genetic engineering. Despots who control us from afar by cognitive/affective "brain bediddling machines" are a science fiction that divert us from the living, breathing despots who control by mob hysteria, advertising, chauvinism, sexism, and religion, none of which require sophisticated neuroscience. This is not to detract from the real neuroethical questions that need to be addressed, but only to suggest that we put them in perspective within the wider range of moral issues that involve changing the brain.

Aristotle believed in moral progress. In his view, as we search and reason about social life and its perils, as we experience life, and reflect on its complexities and surprises, we come to a finer appreciation of what is decent and fair. We learn from each other, and from those whose lives exemplify the human virtues. We learn from errors of the past; our own, and those in the history of the species. This is not a flashy theory of the archetypal Good, nor is it is the sort of theory to whip up moral zeal. Nevertheless, it is a reasonable and sensible approach to achieving some measure of human good, without succumbing either to irrational invocations of the supernatural or to self-destructive scepticism, but with grace, dignity, and realism. Although I am not generally an optimist, I think Aristotle is right about moral progress. On the other hand, if the ecologists are right, population growth may sunder human progress in all its manifestations, moral progress included.

Acknowledgements. Thanks especially to Paul Churchland, Anne Churchland, Sarah Bellam, William Peabody, Steve Quartz, and David Sharp. This work was supported by a UC President's Humanities Fellowship, 1990–91.

REFERENCES

Bjorklund, A., O.G. Nilsson, and P. Kalen (1990). Reafferentation of the subcortically denervated hippocampus as a model for transplant-induced functional recovery in the CNS. *Progress in Brain Research* **83:** 411–426.

Caudill, Maureen and Charles Butler (1990). *Naturally Intelligent Systems.* Cambridge, MA: MIT Press.

Churchland, Patricia Smith (1986). *Neurophilosophy: Towards a Unified Science of the Mind-Brain.* Cambridge, MA: MIT Press.

Churchland, Patricia Smith and Terrence J. Sejnowski (1991). *The Net Effect: Models and Methods on the Frontiers of Computational Neuroscience.* Cambridge, MA: MIT Press.

Churchland, Paul M. (1984). *Matter and Consciousness.* Cambridge, MA: MIT Press.

Churchland, Paul M. (1989). *A Neurocomputational Perspective.* Cambridge, MA: MIT Press.

Gardner, Martin (1988). *The New Age: Notes of a Fringe-Watcher.* Buffalo, NY: Prometheus Books.

Hecht-Neilson, Robert. (1990). *Neurocomputing.* Reading, MA: Addison-Wesley.

Keynes, G., ed. (1966). *Blake, Complete Writings, with Variant Readings.* Oxford: Oxford University Press.

Luker, Kristin (1984). *Abortion and the Politics of Motherhood.* Berkeley and Los Angeles, CA: University of California Press.

Ojemann, George. (1983). Brain organization for language from the perspective of electrical stimulation mapping. *Behavioral and Brain Sciences* **6:** 189–230.

Ojemann, Goerge, J. Ojemann, E. Lettich, and M. Berger. (1989). Cortical language localization in left, dominant hemisphere. An electrical stimulation mapping investigation in 117 patients. *Journal of Neurosurgery* **71:** 316–326.

White, Andrew Dickson (1896). *History of the Warfare of Science with Theology in Christendom,* vols. 1 and 2. Reprinted (1978). Gloucester, MA: Peter Smith.

Bioscience ⇌ Society
edited by D.J. Roy, B.E. Wynne, and R.W. Old, pp. 97–108
© 1991, John Wiley & Sons

Where Bioscience Could Take Us in the Future

Rolf Andreas Zell

Journalistenbüro KLARTEXT
Osianderstr. 13
D–7000 Stuttgart 1, Germany

Abstract. The general public has virtually no understanding of modern biotechnology. That is one reason why this technology is not generally accepted by a large proportion of the population in many countries. Another reason is that many recombinant products developed thus far lack direct benefits for the consumer. A more open-minded, informal, and early product assessment process is needed to improve public acceptance of gene technology. If a "symmetrical" dialogue between industry, research institutions, and the public – including the critics – cannot be achieved, there is little chance of modern biotechnolgy establishing itself as an option for the future.

INTRODUCTION

Scenarios of the future pertaining to science research and its applications should be left to journalists. This was an answer given recently by Nobel laureate Georges Köhler in an interview. His argument was that scientists should be involved primarily in the progress of knowledge rather than in the evaluation of the possible applications of their discoveries.

Armed with this sort of legitimacy, I would like to outline my vision of the future of modern biotechnology within our societies. I am, however, not interested in painting a naive picture of a world with new products and a new technology able to solve all – or, at least, all important – global problems. Nor do I wish to conjure up the apocalyptic vision of worldwide catastrophes due to the unforeseen consequences of new technology. Instead, I would like to outline some of the possible routes that biotechnology could take to establish itself in the long run as an accepted part of future technological development.

1) Scientists have been using recombinant technology for some 20 years now. This new scientific tool box can now be found in almost all biomedical laboratories throughout the world. Even such "old-fashioned" biological disciplines as taxonomy utilize recombinant technology, e.g., gene cloning, DNA-fingerprinting, or PCR techniques, to gain new insights.

This widespread diffusion of the technology contrasts with an almost complete lack of public knowledge concerning the methods, goals, possibilities, or risks involved in gene technology. My first hypothesis is therefore that the great majority of people outside of science have no notion whatever of what recombinant technology really is, what ends it serves, and what sorts of risks might be associated with it.

A piece of evidence for this hypothesis derives from one of the few empirical studies available on this topic. A poll conducted by the research institute of the Dutch consumer organization SWOKA recently revealed (Scholten, personal communication) that only slightly more than half of the Dutch population (57%) claims to have an inkling of what the term "biotechnology" means. If the people who fall within this 57% are then asked to classify individual products or industrial processes in terms of whether they are recombinant ones or not, it becomes evident that only 64% of these 57% have a reasonable notion of what kind of products or processes fall into the category in question. In other words: 63% of the Dutch population does not have even a vague idea of the topic we are discussing. And I am quite sure that this finding would differ only slightly in any other country.

2) The fact that a majority of people do not understand what recombinant technology really means does not at all imply that most people have no understanding of gene technology at all. On the contrary, only in rare cases does insufficient knowledge or a lack of knowledge engender indifference. It much more frequently leads to spontaneous rejection. To some extent, a reaction of this sort is a protective mechanism, a natural and sensible response. Do we not all, in the last analysis, react in this manner? Overtaking another car in a blind curve is a game for gamblers only.

3) This having been said, it is important to be aware that it is not possible to draw the opposite conclusion either. More knowledge does not automatically produce more acceptance of a technology among the public. Evidence for this comes again from the Dutch research institute SWOKA. A couple of years

ago this institute conducted a study on food irradiation (Feenstra 1988). The most interesting finding of the study was that there is no positive correlation between depth of knowledge on food irradiation and acceptance of this method of food preservation. Those who believe that public education can provide the solution to the acceptance problem should take this consideration into account.

4) A further false conclusion can be inferred from what has just been stated: one could be misled into failing to seek a dialogue with the public. However, a democratic understanding of science and technology necessarily presupposes a readiness for such a dialogue. Moreover, the people "in the know" have, as it were, an obligation to share their knowledge unreservedly. Only if this is the case can a public discussion emerge in which all of the participants – industry, scientific institutions, and the public – are able to contribute on an equal footing.

5) My next hypothesis is that in Europe – and, as far as I can see, the same is true for the United States – this situation is far from having been being successfully achieved. This leads me to the following conclusion: if politicians, scientists, and/or industrialists do not succeed in properly educating the public, in order to render such a dialogue possible, then it is very doubtful whether there will be a future for modern biotechnology in democratic societies at all.

THE PROBLEM OF ACCEPTANCE

Recombinant technology – together with military weapons development – belongs to the very few technological fields for which public opinion was negative from the very outset. This is true for all countries in which empirical data are available on the acceptance of various technological sectors. Moreover, recent polls from different countries have revealed that the low acceptance of biotechnology within the public continues to decrease. These are the findings of a comparative secondary analysis of national demoscopic data that was performed by the International Institute of International Socioeconomic Studies (INIFES) in Augsburg on behalf of the (German) Federal Ministry of Science and Technology (Jaufmann 1989).

If one accepts the validity of representative polls at all, this data does not bode well for a bright future for biotechnology. Although I am well aware that molecular biologists are not fond of comparisons between gene technology and the utilization of nuclear energy, it might be instructive to look at this technological field to shore up one point: nonacceptance of a technology can result in a failure of this technology to become established in democratic societies.

This appears to be the case with nuclear power plants. Although to my knowledge there is not a single country in favor of an immediate shutdown of nuclear power plants, in many countries no new plants are planned for construction. In other words, in the long run many industrial nations will decrease their share of nuclear energy production down to zero. This is thus a technology which has been unable to be established on a lasting basis.

Attributing the failure of this technology to become established to the activities of the various antinuclear movements would be too simple an explanation. Even in Germany, where there is a conservative government in power, no further nuclear power stations are to be built, even though this technology is explicitly protected under German nuclear energy legislation.

This is all the more surprising as energy production is of crucial importance for an industrial nation. The benefit of the availability of energy is self-evident. I would like to discuss this point a bit later with regard to recombinant technology, because I believe that in this respect – the assessment of benefits – gene technology really differs from nuclear energy.

The long-term failure of nuclear energy to become established as a technology in many countries has much more to do with risk assessment. Whether the arguments are scientifically sound or not, in the eyes of the public the risks of nuclear power plants are too high to be accepted. This is an important point: technology assessment among the public does not refer only to "real" risks – "real" being defined as the risk that can be calculated from theory or that can be established empirically – but also to risk as it is perceived by the public.

"Acceptance" is a widely-used but rather poorly defined term. It implies at least two important factors – risk and benefit –, but for the sake of this discussion it is even more important to look at the way in which these two factors are "measured" among the public.

I wish to illustrate the point with the aid of an example. In Germany more than 14,000 people die annually as a result of automobile accidents. Notwithstanding, only very few people in this country are willing to abolish this technology or render it more safe. Only a small minority of Germans are in favor of very simple risk-reduction measures that have proved effective in a variety of countries, e.g., speed limits on highways.

This example does not necessarily stand in contradiction to the fact that the acceptance of a technology is "measured" in terms of risk-benefit categories, but it does demonstrate that this sort of "quasi-quantification" does not rely on the mathematical terms of both parameters. Instead, subjective and emotional parameters play an important role in such assessments.

If this is true, it means for the discussion of gene technology that acceptance cannot be achieved by means of legal regulations governing the application of a technology on a purely rational basis. But, here again, one must be careful to avoid drawing the opposite conclusion. Of course it would be fatal to believe that we could abandon legal regulations because they fail to increase acceptance. However, legal regulations alone will not help to solve the problem of acceptance.

The reader might at this point get the impression that my view on biotechnology is ruled by defeatism. To counteract this impression, I wish to turn to some factors that increase acceptance of technology. To illustrate the point, I would like to present two examples from electronics and communication technology: CD-players and fax machines have really won a significant share of the world market within only a few years. Many people – including myself – cannot even imagine how they could have survived so long without these fabulous inventions.

The factor that has facilitated their real breakthrough in public acceptance is the direct usefulness experienced by even the casual user of these machines. The second point is that CD-players are even cheaper than comparable analogue equipment. However, buyers of digital discs even accept higher prices because they can hear – or at least they believe they can hear – the superior sound quality of digital over analogue discs.

The perception of a direct benefit is in my mind one of the surest strategies to create acceptance. Of course I have to admit that CD-players do not appear to involve any particular risks that might hamper the technology's acceptance. (This is a bit different with fax machines, where at least some people see disadvantages). However, if we return to the automobile, we will find ourselves face to face with a technology which has proved more dangerous by far (in terms of deaths) than the production of nuclear energy. But the important point about automobiles is that the benefits by far outweigh the lethal risks (automobile accidents not being the only risks).

If it is true that the perception of benefits is crucial to technology acceptance, this has consequences for gene technology. It is very important for the establishment of a technology that the products which first leave the new production line offer direct benefits to consumers.

The propagators of gene technology have, however, neglected this point. Instead, recombinant products have been presented that meet the benefit criterion only in very rare cases. (Of course there are exceptions in the field of therapeutic agents and vaccines.) I would like to illustrate my criticism with two fields of application in which recombinant products have already become available or are just around the corner.

The problem of acceptance in the pharmaceutical industry

Therapeutic agents are forced to struggle against a rather strange subtype of the acceptance problem: the administration of a drug does not appear to augment the patient's well-being. At best, the drug is able to restore the patient's original state of health. In doing so, the drug continuously reminds the patient of his health problem. This might explain why so few patients gratefully welcome the availability of a drug.

Taking this into account, it is not surprising that the side-effects of a particular substance are much more often on the agenda of public discussion than is the case with the intended and useful effects of drugs. And I can hardly imagine that this state of affairs will change dramatically in the future.

But this particular situation does have an enormous impact on acceptance. If – in the future – there are cases of recombinant drugs that produce serious side-effects, the pharmaceutical industry will face even greater difficulties in gaining public acceptance. The recent discussion on drugs containing the recombinant version of the amino acid L-tryptophan points in precisely this direction.

What can be done to improve the current situation? I am not convinced that a "breakthrough" of public acceptance within the pharmaceutical sector will be based on single recombinant drugs which prove to be useful. Let me illustrate this point with the aid of an example. Recombinant tissue plasminogen activator (rt-PA) appeared to offer a therapeutical advantage over all other fibrinolytic agents.

In the meantime things have changed to some extent. Although rt-PA convincingly turned out to be effective against life-threatening diseases, a controversial discussion still continues on whether its effectiveness can be translated into a greater survival advantage for patients treated with rt-PA compared to those that are treated with another type of fibrinolytic substance. And in the case of rt-PA one has to consider that, with this drug in mind, even prominent critics of gene technology – at least in Germany – have conceded some sort of usefulness of recombinant technology in medicine.

In speaking of recombinant products, the pharmaceutical industry has to be even more careful not to overestimate the usefulness of given drugs. When claims that have been made prove unsound, not only the individual product but the technology as a whole is at stake.

The danger of overestimation is an important issue not only for existing drugs, but for future drugs as well. Some proponents of gene technology endeavor to convince the public with the argument that all sorts of major disease categories – from cancer and Alzheimer's to AIDS – can be tackled only by recombinant technology. This is to my mind a dangerous gamble, because these people are presenting to the public a bad check, hoping that the public will wait before cashing it until sufficient funds have accumulated in the account. The big promises and expectations concerning interferon as the "magic bullet" against cancer – made even by scientists involved in the field – should serve as a warning.

To my mind, other strategies could be useful. Recombinant technology has not only made possible the production of a safe and effective hepatitis B vaccine, the new technology also permits the vaccine to be produced much more cheaply. The costs per vaccination have sunk during the past few years from around $100 to, at present, nearly $1 (U.S. currency).

It has now for the first time become possible to envision vaccination not only in the rich and industrialized countries. A vaccination program against hepatitis B in the Third World countries in which this viral disease still causes millions of deaths is now within reach. A concerted action of the vaccine-producing companies, the World Health Organization, and the health administration of the various nations affected could lead to a situation in which hepatitis B will lose its horror, as was the case with smallpox.

And, to put it somewhat sarcastically, how long has the World Health Organization been living off its image and profile of having achieved the total eradication of smallpox from the earth? A hepatitis B vaccination program on the basis of a recombinant product could be a milestone in medicine, a case in which everyone could perceive and understand at once the power and usefulness of this technology.

And this strategy can be developed even further: many tropical diseases can be pushed back economically with the aid of vaccination programs. The availability of a vaccine is a necessary precondition, but not the only one. Millions of children and adults die every year of measles and polio, although cheap and effective vaccines are on the shelves. For many tropical diseases, however, no vaccines exist yet at all. The more research that has been done in malaria or schistosomiasis vaccine projects, the more it has become clear that traditional vaccines will probably not work to immunize against parasitic diseases. Most scientists involved in such projects believe that recombinant subunit or vector vaccines might be more promising.

The pharmaceutical companies involved in such vaccine projects stress the fact that even in the case of success the products of their research will hardly reach the break-even point. If this should turn out to be true, it can well be envisioned that companies which otherwise compete with one another will cooperate in projects such as these, which appear risky in economic as well as in scientific terms. From this sort of cooperation it would be only a minor step towards an international fund in which all pharmaceutical companies contribute to funding such "philanthropic" and – in the global meaning of the word – "orphan" drug and vaccine projects.

And to put the finishing touches to my biotech Utopia: the commitment of the pharmaceutical industry should not stop at funding or managing the research and development of such products. It would be necessary to initiate concerted actions together with the World Health Organization and the tropical countries to make sure that available vaccines will actually reach the people in the field. Many companies have better logistics in Third World countries than, for instance, the World Health Organization. Such commitment to research and development and to vaccination campaigns could illustrate in a very simple fashion how important recombinant technology is for major world health problems.

The problem of acceptance in the agricultural sector

According to many experts in the field, gene technology will have its second "debut" (medicine having been the first) in the fields of agriculture and food processing. To my mind, it is exactly this area in which proponents of this new technology have made the most striking mistakes in improving public acceptance.

Three lines of recombinant products dominate the public discussion on agbiotech:

1) transgenic animals with additional genes coded for growth hormones;
2) recombinant growth hormones to increase milk production or to accelerate meat production; and
3) herbicide-resistant, transgenic plants.

Each line of products currently faces opposition by many politicians, by most direct users, and by almost all consumers. This rejection front was perfectly foreseeable. For many years now the European Community has had to struggle against a costly overproduction of various food products such as milk, wheat, and meat. As long as I can think back, the political discussion on European agriculture has been formostly dominated by the problem of reducing surplus production. In recent years the discussion on how to structure a sustainable agricultural system has turned out to be the second most important topic in Brussels.

In a situation like this, it is little short of intellectual suicide to propagate a new technology with products that – at least in the eyes of the public – augment the problem of surplus production and environmental damage caused by agriculture. Even the most positively inclined person will be touched with pity at the sight of a suffering creature like the transgenic pig, which has a hard time even standing on its own feet. From there it is only a small step towards hate of, and opposition to, the technology that has allowed scientists to create such an animal.

Of course I am well aware that the first recombinant agbiotech products that are now trying to enter the market were chosen by industry because these genes were the only ones that were available when the projects started in the early 1980s. But is this a sufficient justification for the choice? Market research does not have to ask, "What is available?" in the first instance, but rather, "What is needed most?"

What can be done? If industry does not succeed in developing agricultural products which have direct benefits for the consumers, and which allow sufficient food production with less harm to the environment, or which can reduce the suffering of animals, I see no chance of the public accepting the use of recombinant technology in this field.

Milk from cows treated with bovine somatotropin (bST) is by no means superior to "normal" milk as far as product safety and quality is concerned. At best, milk from animals treated with bST is comparable to normal milk. However, discussions are ongoing on the amino acid composition of milk from cows treated with bST as well as on safety aspects. I cannot see how it would be possible to improve public acceptance of recombinant products in the food industry, even if it should prove to be the case that bST offers some economic advantage in terms of productivity and production costs for farmers.

Acceptance could be achieved much more easily if food products which contain transgenic plants or animals or which have been processed by recombinant food additives were to offer direct and tangible benefits for the consumer. Plants containing a "healthier" fatty acid combination or a higher share of essential amino acids can be envisioned. And the development of pest-resistant crop plants requiring fewer or even no pesticides would be gratefully welcomed by the public. But this is certainly not the case with herbicide-resistant plants, for which it is quite complicated to demonstrate such a positive effect at all, e.g., by switching from environmentally harmful and long-life pesticides to other compounds which do not accumulate in the soil and contaminate ground water.

I believe that it is essential for the agricultural industry to reevaluate the validity and reliability of its market research. Even some producers of bST acknowledge today that they failed to evaluate the consumer acceptance of bST, looking only for market potentials and benefits for the direct user of the product. But it is the behavior of the consumer that is crucial for the failure or the success of bST. Even a weak consumer boycott, leading to a decrease in milk consumption on the order of a small percentage, could turn greater cost effectiveness in milk production into a net loss of income.

CONCLUSIONS AND OPEN QUESTIONS

In the discussion on regulating recombinant technology in the European Community, the "fourth hurdle" argument was recently brought forward. Not only safety, efficacy, and quality should be the criteria for evaluating new products, but social and economic needs must be included as the fourth criterion.

The arguments and examples which I have presented in this paper may appear to support the fourth hurdle argument. But this is not at all the case. There are at least two reasons for not including this fourth hurdle in the regulation process.

1) Not only can a new product fail to gain legal authorization to enter the market place; it is also possible for products already on the market to fail because they offer no benefits for the buyer or because their benefits do not justify their price. In other words, the fourth hurdle is already functioning properly. Every product has to surmount this hurdle, and the behavior of the consumers will decide whether the product will fail to do so or not.
2) If the fourth hurdle is added to the other three criteria in the regulation process, a considerable degree of arbitrariness could be the price. It would be quite easy for one country to ban a new foreign product from its markets on the basis of the fourth hurdle only in order to protect its own economy. The fourth hurdle could very soon develop into a powerful weapon in the hands of economic protectionists. And I cannot reject completely the U.S. viewpoint stating that this is actually the case with bST in the European Community.

However, even as far as public acceptance is concerned, the fourth hurdle does not provide any improvement of the situation. On the contrary, the implementation of this criterion would deprive the consumer of his right to decide on his own whether to use or reject safe, efficient, and high-quality products. This cannot and must not be the task of government and/or public administration.

It is much more important to strengthen the competence of the public with regard to the assessment of new technology products. Knowledge is a key factor in this process. I therefore believe that it is necessary for industry and research institutions to inform the public earlier and in a more open-minded fashion than they have done in the past. The "glasnost principle" could be useful in industry as well. The better informed the public is, the more precisely companies will be able to evaluate future market opportunities for their products.

Of course, There is one problem linked to the early disclosure of relevant data. Many companies fear that by doing so they will be forced to share their particular know-how with competitors. On the other hand, there does exist a very sophisticated tool that offers economic protection of know-how – the patent laws. A patent application is a sort of deal between the inventor and the public. A patent is granted only when the inventor makes certain particular knowledge available to the public. He in turn receives the right to use this knowledge exclusively, or to licence it.

I am well aware of the fact that patents in the field of biotechnology are themselves an issue of public concern. And of course it will not be possible to safeguard all of the information disclosed under current patent legislation. This is why my reference to patent protection can be seen only as an illustrative example rather than as the solution to the conflict between the public need for more and early information and an effective protection of know-how.

HYPOTHESIS AND OPEN QUESTIONS

1) The worldwide distribution of recombinant technologies contrasts with the widespread lack of public knowledge concerning the methods, goals and possible risks of gene technology.
2) Lack of knowledge or insufficient knowledge creates not indifference but fear and rejection.
3) A better-informed public does not automatically lead to a broader acceptance of a technology among the public.
4) Acceptance can be "created" only in an "equal status dialogue".
5) If politicians, scientists and/or industrialists do not succeed in adequately educating the public and rendering such a dialogue possible, there is con-

siderable reason to doubt whether there will be a future for modern biotechnology at all.

6) Technology assessment within the public does not only refer to "real" risks, but also to risks as they are anticipated by the public.
7) Legal regulation of gene technology does not necessarily improve public acceptance.
8) Direct benefits for the consumer are a key factor in improving acceptance, even in the case of "risky" technologies.

Open question: How can we solve the conflict between the need of the public to have access to all relevant information at an early stage of the assessment process and the need for protection of know-how?

REFERENCES

Feenstra, M. et al. (1988), Irradiation – A long-life method? SWOKA Working Paper No. 7, Den Haag, The Netherlands.

Jaufmann, D. et al. (1989). *Jugend und Technik.* Frankfurt a.M.: Campus.

Scholten A.H., SWOKA Institute of Consumer Research, Den Haag, personal communication.

Standing, left to right: Benno Müller-Hill, Barbara Hobom, David Sharp, Maurice McGregor, Eric-Olof Backlund, Rolf Zell
Seated, left to right: Patricia Churchland, Diana Brahams, Robert Old, Marcus Pembrey, David Roy

Bioscience ⇄ Society
edited by D.J. Roy, B.E. Wynne, and R.W. Old, pp. 111–118
© 1991, John Wiley & Sons

Group Report
Does Bioscience Threaten Human Integrity?

Rapporteur: D.W. Sharp
E.-O. Backlund
D.J. Brahams
P.S. Churchland
B. Hobom
M. McGregor
B. Muller-Hill
R.W. Old
M.E. Pembrey
D.J. Roy
R.A. Zell

INTRODUCTION

For the purpose of these discussions it was agreed that "covenant with society" would be taken to mean both "understanding with society" and "undertaking to society". Scientists are part of society. Why then is a covenant necessary? It is necessary not only because scientists use public money, but also because the rapid expansion in knowledge carries a much increased potential for good and for harm. The real wish is for science to acknowledge the ethical norms of society. In the multicultural societies of today there are no longer any authoritative and easy-to-follow directives. Ethical norms must be worked out on each scientific issue as it arises. Thus, any covenant must include the affirmation that science will inform, discuss, and listen – and do so frequently and repeatedly.

A manifestation of this new "democratization" of ethical decision-making is the creation of the ethics committees whose agreement is already considered essential before any experimentation involving human beings is undertaken. The present assembly of scientists, ethicists, philosophers, and writers is another. What follows is an account of the engagement of this group in topics they consider important. The intention was not to pronounce on every issue. Neither was it to reach a consensus. But the process of exploration of common ground and the identification of differences of opinion may be helpful to others.

The group, having been invited to prepare an agenda of between six and sixty questions, wielded Occam's razor and finished with just five. The topics were: 1) genetic counseling; 2) the human genome project; 3) fetal tissue; 4) embryo research, and 5) overpopulation. The precise wordings of these questions introduce the summaries of the five sessions of discussion that follow.

GENETIC COUNSELING: CAN THERE BE A CONFLICT BETWEEN THE GOOD OF SOCIETY AND PARENTAL CHOICE?

In posing this question, the group may have thought that it was inviting controversy. In the event, there was little dispute with M.E. Pembrey's theme – namely, that it is parental choice which should determine where the genetic counselor goes and how far. If informed, couples tend to make sensible choices. With beta-thalassemia, for example, Cypriot communities have gone a long way down the road of encouraging carrier testing. The church has acquiesced in the requirement of a certificate of testing before a marriage ceremony. That does not prevent risk marriages taking place. The couple may get married anyway, opting for secondary avoidance of affected births via prenatal diagnosis and termination. So there is a sort of accord between families and the health objectives of Cyprus; this has been achieved with only a limited element of compulsion.

Sex determination, however, led to more controversy. Today (1990) sex determination really means selective abortion, or, possibly, selective transfer. Not possible yet is selection at conception by separating X- from Y-bearing sperm. Parental use of sperm selection techniques would, it was thought by most of the group, be consistent with the good of the family, but may or may not accord with the good of society. A range of possible future scenarios had to be debated, from the situation where a strict one-child-per-family policy operated, to one where an altered sex ratio would almost certainly ensue. A shift in the sex ratio at birth from roughly 50/50 to say 65/35 male to female could have impacts other than one of a population decline. What would the nature of that society be if the population were more predominantly male? More aggression, more rape, for example? If fetal sex selection is to be included among the reproductive choices parents have a right to make, the outcome would need close monitoring.

Without badly denting the consensus, other doubts about whether what parents do is necessarily for the best were heard. Children with cystic fibrosis (CF) are living longer, even into adulthood. Such lives are usually of poor health quality. With advances in CF genetics, would the existence of families with more than one CF child be deemed responsible? The trouble is that there is no single CF mutation: the most common one (the F508 deletion) is indeed severe clinically, but others may not be. Even if parents did seem in some sense to have got it wrong, no court, in a civilized society anyway, would impose on them an action such as abortion. The right to make bad choices should not be taken away. What about insurance companies in countries where health care provision was largely insurance-based? Suppose that parents, fully informed at counseling, decided to go ahead with a pregnancy known to be affected. The group did not agree on whether an insurance company has the right to say "if you go ahead, the medical costs of caring for the affected child will not be met by us." And what about the reporting of findings incidental to a procedure such as antenatal ultrasound or karyotyping – not just fetal sex (XX or XY) but also chromosome abnormalities, the example discussed being XYY? Where the specificity (taken here to mean the certainty of the medical consequences) was low, as it is in XYY, routine disclosure might be undesirable.

On the other side of the fetal sex argument, there is a hint that women who know the sex of their child via chorionic villus sampling may bond better to the child than those who learn a few weeks later via amniocentesis. For the present, at least, it seems that multifactorial genetic diseases, where the specificity would probably be low (e.g., apolipoprotein genes and coronary heart disease), do not enter the equation. However, more and more research findings are being translated into genetic counseling. Society cannot afford everything; it may well be that some of the tests without which parental choice is meaningless cannot always be afforded, at least in the public sector.

The group concluded that the guiding principle in the regulation and practice of genetic counseling and prenatal diagnosis was protection of informed parental choice. This places an obligation on society to try to provide a) information on both genetic risks and the impact of a given genotype on health and development, and b) a range of reproductive options, from contraception to selective termination of pregnancy, or support for families in caring for a handicapped child, regardless of the health-care system in place. This conclusion was based on the observation that, with respect to family life, there was such accord between what families and societies want that the actions of a fully informed couple would not undermine the moral and social fabric of their society. This presumption in favor of parental choice means that in those exceptional circumstances where survival seems threatened, such as disturbance of the sex ratio, the onus of proof is on those who argue for temporary restriction on parental choice.

INFORMATION ABOUT THE HUMAN GENOME, WHOSE IS IT, AND WHAT MAY BE DONE WITH IT?

Until recently there has been no great push behind progress in genetics. Over the next ten years or so, this will change with the search for the chromosome map of the whole human genome, a project backed largely by U.S. federal money to the tune of $110 million, and supported by France, the U.K., and Japan, but with little activity elsewhere. The focus will be on a few small chromosomes or a small one such as no. 21, avoiding "man-on-the-moon" type targets. The complete DNA sequencing of the genome is much further off. The likely pace of this advance is in dispute, but as important genes and alleles come to be identified, so that an individual can be told his or her genotype in respect of those genes, real problems could emerge. It seems inevitable that insurance companies might seek this information; employers might, too. B. Müller-Hill, introducing this session, predicted that "if this [i.e., insurance, employment, and state activities with cost efficiency as goals] goes on and on unchecked, it will create a proportion of citizens – in the U.S. for example – who are labelled 'genetically unfit'. That is potentially explosive. At that point science will lose the respect of the population." Already some 37 million Americans lack health insurance. It is scientists who should make a stand.

Some 3% of the U.S. National Institutes of Health and U.S. Department of Energy human genome budgets was to be allocated to research into the ethical, legal, and social aspects, but thus far there has been disappointment with the quality of grant applications.

What is the difference, in terms of disclosure, between phenotype and genotype? A consensus again emerged that disclosure of the DNA sequence has to be accompanied by especially careful interpretation. There could be another side to the coin. If there is a genetic element in, say, mania-depression it will be a frequent one, and that can only be explained if the DNA sequence carries some advantage, now or in the past, for the population, too. This is not to say that there is no situation where an employer's knowledge of genotype might not seem mutually helpful. An illustration of the sort of controversy that might arise here is alpha-antitrypsin deficiency, which carries a risk of emphysema, an effect exacerbated by smoking. An at-risk employee might be in danger in a factory where air pollution was a problem, raising the temptation to screen employees in such a situation. This option, as opposed to the alternative of cleaning up the factory, raises a serious ethical issue.

SHOULD THERE BE LIMITATIONS ON THE USE AND PRODUCTION OF FETAL TISSUE FOR CLINICAL PRACTICE AND RESEARCH?

This issue originates from tentative studies in recent years on transplantation of dopamine (DA) producing cells to the brain of patients with Parkinson's disease. To overcome ethical and immunological problems, autologous grafts were used initially, but with limited success. The use of DA-producing cells from the brain of aborted fetuses was assumed to be a more promising alternative.

E.-O. Backlund, introducing this discussion, noted that a debate on fetal tissue could not be separated from the debate on abortion. This did not mean that attitudes to the two issues ran on exactly parallel lines. Survey respondents might be negative about abortion on demand, but be able to rationalize the research use of fetal tissue; or they might be positive on abortion,, but still say "no" to the use of fetal tissue. Ethics and public attitudes apart, there are scientific, practical, and psychological aspects to be taken into account. Cells from an 8-10 week embryo are very potent, so in a new environment such as the brain, there may be unpredictable problems, associated with uncontrolled growth or autoimmunity. The early expectations of progress with the transplantation of fetal tissue (brain or islet cells for Parkinson's disease or diabetes, respectively) have not been fulfilled so far. If they were, issues such as commercialism and the problem raised by conception with the sole aim of producing graft tissue would be raised; and the logistics of supply (an obstetrician) and demand (a transplant surgeon) would have to be tackled. As an alternative to the use of fetal cells, a future use of cells from "domesticated" cell lines does not seem unrealistic. Another alternative, originating from a current clinical study, may be the administration of growth-promoting factors (NGF, for example) into the patient's brain, together with an autologous DA-producing graft.

Precautions against the misuse of fetal tissue include donor/recipient anonymity, no payment, and no specifying by the donor of who receives the tissue. The pregnant donor should be asked about donation only after she has decided on termination. From conservative, strict rules such as these it may be possible to liberalize later. In the United States a panel of the National

Institutes of Health has looked into the desirability of research on human fetal tissue (if federal funds such as NIH grants are involved) and voted 15 to 3 in favor, but the Secretary for Health and Human Services came down against it. There remained concerns about the pressures under which women might come if there was another member of the family who might benefit from a fetal tissue graft.

SHOULD HUMAN EMBRYO RESEARCH BE RESTRICTED?

Introducing this question, D.J. Roy noted that in the U.K. the Royal Society had set out three categories of research that might benefit from research on the human embryo: in vitro fertilization methodology, basic embryology, and tumorigenesis and chromosome abnormalities. The acceptability of research on the predefinitive embryo, when the embryo cells are still totipotential and the primitive streak has not yet developed, has been widely held. Less clear is why it has been accepted or why the reverse – experiments on older embryos (even though they were not, on current experience, sustainable in vitro for longer than even ten days) – was being rejected. There was some unease with two arguments used in favor of a 14-day guideline, based on the possibility of twinning before this stage and on the view that until neurulation began there could be no sensation of pain. The argument that nonhuman animal embryo research would suffice could be dismissed by the example of Lesch-Nyhan syndrome, where what had been learned of the enzyme defect in mice turned out not to be transferable to humans.

The time restriction (14-day rule) is becoming generally accepted in countries in which embryo research is allowed. More controversial was the creation of embryos for the sole purpose of research. This is allowed by statute in the U.K.; the U.S. Congress has taken no stand via legislation, but federal money cannot be used for embryo research, and state by state practice varies. In Germany, legislation was headed towards a total ban on all embryo research. In Sweden, the legislation will probably follow that of the U.K., whereas in Norway any "surplus" fertilized eggs/embryos obtained at IVF have to be implanted in the uterus. Denmark has so far practiced a moratorium on all embryo research, and a very restrictive attitude has been expressed by a national ethics committee. In Canada (via the Medical Research Council), only embryos surplus to a clinical IVF program can be used in this way. There were dangers, even life-threatening ones, to the woman when superovulation is induced, which adds a new dimension to the argument on oocyte production for purposes other than the treatment of infertility.

Two discussants – one a gynecologist looking at surrogacy, the other a lawyer reading up on embryo research – had seen their initial, antagonistic views of those procedures reverse totally. Surely the "new covenant" that we had been asked to consider was a question of promoting dialog, protecting the channels of communication newly opened up, and giving the public (and legislators, where appropriate) time to respond and to adjust.

HOW SHOULD THE "DEMOGRAPHIC TRAP" BE TAKEN INTO ACCOUNT IN MEDICAL INTERVENTIONS AFFECTING LARGE POPULATIONS?

For this discussion (introduced by D.W. Sharp) the group took as its stimulus a provocative paper by Maurice King and an editorial in the Sept. 15, 1990 issue of *The Lancet.* The following quotes illustrate, but do imperfect justice to, the argument:

From King's "viewpoint" article on the demographic trap:

> The view that, if the child death rate declines sufficiently, the birth rate must decline also, and that there is a causal relation between them, is untenable . . . the birth rate is unlikely to be lowered by measures designed to reduce the child death rate that are imposed on people by vertical programmes – e.g., those for mass immunisation and for oral rehydration for diarrhoea . . .
>
> The ultimate in unhappy choices is to be faced on the one hand with not doing all that is possible in public health, and on the other with increasing ecological deterioration, leading eventually to starvation and to the destruction of the very population it is intended to benefit . . .
>
> Family planning programmes must be promoted with renewed vigour, and the objective of sustainability used as the impetus to drive them.

And from the editorial:

> . . . those who have beaten their chest the hardest about the morality of abortion – some US Presidents spring to mind – have cut back on the resources going into family planning and have defunded effective organisations such as the United Nations Population Fund and the International Planned Parenthood Federation. Undoubtedly they have thereby increased the number of abortions taking place world wide . . .
>
> Let the last word come from Bangladesh, where the social marketing programme stops births by distributing 130 million condoms a year and stops infant deaths by selling oral rehydration sachets. Certainly, Bangladesh is also in the jaws of the "demographic trap" and current successes in family planning are coming

tragically late. The population is almost bound to double at least one more time before it stabilises.

King, following a gloomy recipe, has thrown into the cooking pot several ingredients – population, preventive and curative health measures, resource consumption, birth control, the North/South divide – and comes up with a thought-provoking meal. The ethical dilemma our discussion group had to tackle was that set out by the British physiologist A.V. Hill almost thirty years ago. He asked, "If ethical principles deny our right to do evil in order that good may come, are we justified in doing good when the foreseeable consequence is evil?" That family limitation as a prompt consequence of lower child mortality can no longer be relied upon to prevent ecological disaster in much of the developing world leads King to ask whether we should think the unthinkable and seriously consider the notion that elements of the GOBI (Growth charts, Oral rehydration, Breastfeeding, Immunization) vertical programs of the world's aid agencies should be withheld. All the same, the planned withholding of health care measures was totally unacceptable to the members of our discussion group. In taking this view the group was as concerned about the survival of human integrity as it was about the survival of mankind.

King's claim that U.N. agencies had been neglecting family planning was challenged. Historically, in the 1970s, in international, multilateral, and bilateral support, population control had priority over health care; in the 1980s that priority was reversed. In the 1990s both get priority, and it must not be forgotten that population control can be a health measure ("safe motherhood"). However, the decline in interest in birth control delivery, the reduction in research funding by the U.S. government, and the opposition to birth control by some world religious leaders were unacceptable to the group. Representatives of Group 2 (Gordon et al., this volume) endorsed King's warning of demographic breakdown with projections of catastrophe in 10–20 years on a world scale. The only valid doubt was when and where. Could we prove that these catastrophes would happen? Maybe we cannot prove it with certainty, but to "wait and see" would be too late.

If the appalling ethical dilemma that Maurice King conjures up is to be avoided, bioscientists with special interests in birth control research and delivery would need to join with demographers, ecologists, and others. Together they should appeal directly to an informed public, hoping that governments and churches will then alter their priorities so that international agencies can follow suit.

Bioscience ⇌ Society
edited by D.J. Roy, B.E. Wynne, and R.W. Old, pp. 119–131
© 1991, John Wiley & Sons

Are Recombinant DNA Techniques Tinkering with Evolution?

Horst Nöthel

Institut für Genetik
Freie Universität Berlin
Arnimallee 7
D–1000 Berlin 33, Germany

Abstract. Evolution does not take place in individuals but in reproductive communities, or populations. The genetic composition of populations has its own laws. They enable causal factors of evolution to be effective. These factors are "change of genetic information", "isolation", "random drift", and "selection". They have fine-tuned interactions. Changes in genetic information contribute genetic variability. Variability is altered by isolation and random drift. It fuels selection. Selection enables adaptation to ever-changing environments.

Sources for change in information are mutation, recombination, and movable DNA. Recombinant DNA is movable DNA that carries humanly selected genes. It will principally have the same effects in evolution as mutations have. It is explained that even extremely high mutation rates do not decrease the fitness of a population in the long run, but increase fitness by improving selection. Recombinant DNA will underly the above mechanisms of evolution, and will not dominate a natural genome.

INTRODUCTION

Man is part of the living world. Any artificial change in this world merits his concern. Concern has to be considerable in the case of hereditary changes. In the 1930s, attention was focused on this concern in a masterly way by Huxley in his *Brave New World.* These were times of flourishing racism, at least in Germany. Besides political stubbornness, this racism was fueled by very limited recognition of contemporary science. Only a few years later, Lyssenko, ignoring established knowledge of genetics, led Soviet agriculture into disaster. And a few years earlier, Muller had started his campaign against the spread of ionizing radiations with their mutagenic potential. It took decades to raise public interest and to gain adequate legislative restrictions to artificial radiation sources which we have today. These long debates explain the overreaction in present-day politics, which is based more on emotion than on scientific evidence. However, these debates alerted the public conscience, and when "recombinant DNA techniques" became the subject of enthusiasm for their possible benefits to agriculture, industry, and medicine, "gene manipulation" aroused tremendous fears in the general public about invisible dangers comparable to those from radiation. I am sure that in spite of all safety precautions, it is simply a question of statistics that dangerous failures will occur. It is up to society to weigh possible benefits and risks. As with mutations, these failures will affect not only individuals today, but the life of future generations. In this paper, I will deal with evolution and with evolutionary consequences possibly brought about by recombinant DNA techniques, but not with individual fates. I have to emphasize that this distinction is fundamental to all discussion of environmental effects on living systems. Therefore, I will give a short account of populations and their structure, i.e., the entities in which evolution occurs. I will then turn to the genetic basis of evolution. A closer look will be given to changes of genetic information, including artificially recombined DNA. I will conclude with some remarks about how such DNA fits into the framework of evolution.

1. THE GENE POOL

1.1. *Mendelian populations* are best suited to illustrate mechanisms of evolution. They are composed of sexually reproducing, cross-fertilizing individuals that share a common gene pool. A gene pool is the total of all genetic information from all fertile individuals of that population. Since here, as always, the total is more than the sum of its parts, the structure of a gene pool underlies its own laws. They allow for evolutionary forces to become effective.

- Mendelian populations are most often multicellular organisms. Their natural transfer of genetic material is restricted to germ cells. In general, germ lines differentiate early in individual development from the bulk of other tissues, the soma. Germ line and soma remain safely separated thereafter.

Therefore, evolutionary relevant transmission of genetic information, natural or manipulated, always requires its presence in the germ line.

- Individuals of Mendelian populations are most often haplo-diploid, i.e., their life cycle alternates between haploid stages with one set ($1n$) of information (most often only mature germ cells) and diploid cells ($2n$). Mendel's laws of heredity follow from the meiotic reduction from $2n$ to $1n$, and subsequent cross-fertilization to $2n$. Sets of information are chromosome sets. Chromosomes are equally distributed to daughter cells in mitosis. In the linear DNA of any chromosome a large number of entities of information, genes are linked. A gene consists of a stretch of DNA with several hundred base pairs. Differences in any of these bases yield a different allele of that gene. It may or may not exhibit a change in function. Several alleles from any gene exist in a gene pool. If genes are symbolized by letters and their alleles by numbers given in the exponent, this reads as a^1, a^2, a^3, a^4, a^5, etc. Diploid genotypes from two alleles are the homozygous a^1/a^1 and a^2/a^2, and the heterozygous a^1/a^2. It is easy to see from a combination square that with five alleles there will be $(5^2+5)/2$, or 15, different genotypes. This accounts for basic variability in a gene pool. Variability is very much higher if several genes are considered. Thus, free recombination between 23 genes, with 5 alleles each, produces 15^{23} or 10^{27} genotypes, i.e., all individuals will be different.

1.2. *Laws* underlying the composition of alleles in a gene pool are deduced from ideal populations. These are of infinite size, with all individuals being potentially able to intermingle, undergoing no changes in information, and having the same fitness, i.e., they all contribute equally to the gene pool of the next generation. In such an ideal population, frequencies of alleles will remain constant, and those of the genotypes will be in equilibrium. If p is the frequency of allele a^1 and q that of a^2 $(p+q=1)$, then genotype frequencies will be p^2 for a^1/a^1, q^2 for a^2/a^2, and $2pq$ for the heterozygotes (Hardy-Weinberg law). Likewise, there will be a linkage equilibrium between alleles of different genes that only depends on the relative frequencies of any of these alleles and the resulting combination frequencies. Equilibria, or constant frequencies, mean that there is no evolution.

2. GENETIC BASIS OF EVOLUTION

"Organic evolution is a series of partial or complete and irreversible transformations of the genetic composition of populations, based principally upon altered interactions with their environment" (Dobzhansky et al. 1977). Evolution is only possible if any one of the preconditions for an ideal population is not valid. The synthetic theory therefore considers 4 causal factors of evolution: 1) change of genetic information; 2) isolation that prevents panmixis; 3) random drift in populations of limited size; and 4) selection because of unequal fitness.

2.1. *Changes of genetic information* are due to mutation, recombination and movable DNA (including recombinant DNA). They will be dealt with in more detail in the section "Changes of Genetic Information".

2.2. *Isolation* prevents panmixis and accelerates divergence. Organisms of different species exhibit, by definition, total reproductive isolation. It is achieved by separation either in space or in time of reproductive maturity, either in incompatibility of reproductive rituals, organs, or germ cells, or in lethality or sterility of hybrids. Geographic isolation is most often the first step in isolation between populations, whereas, in animals, differences in courtship or social behavior are the most effective, finetuned mechanisms which are eventually achieved. For example, human populations in Europe have linguistic boundaries between languages and even dialects that correlate with clear-cut differences in allele composition (Barbujani and Sokal 1990).

2.3. *Random drift* consists of chance effects due to small numbers. This may be illustrated by families with 6 children each. With a 1:1 sex ratio, there is an average of 3 girls and 3 boys. But according to binomial statistics, this outcome will occur in less than 1/3 of such families, whereas about 1/30 of them will have only girls or boys, respectively. The same holds for different alleles of a gene. It likewise holds for population sizes of several hundred individuals, since a drastic push in one direction cannot easily be switched later on. This means that of two alleles, one becomes extinct, the other fixed by chance. This gives a direction to evolution that is not adaptive. Moreover, small populations unavoidably lose considerable amounts of genetic variability, so that further adaptations become impossible. This contributes to the present extinction edge of several vertebrates like the cheetah (O'Brien et al. 1987). Another aspect of random drift is that it may sweep newly arisen alleles all over a gene pool of even a large population. The "neutral theory of molecular evolution" (Kimura) describes this as a major force in evolution, especially for molecular variants that do not contribute clear-cut adaptive changes.

2.4. *Selection* is the most effective directional factor in evolution. It brings about adaptations to environmental changes. Per se, any genetic variant (allele, new mutant) is without any value. Only its concerted action with other genes and alleles decides whether its adaptive value is "positive" or "negative" in a given environment. Since the adaptive value of any variant depends on its internal and external environment, it will change in time and space. An adaptive maximum can be set by a breeder. Therefore, theories of artificial selection are the basis for traditional breeding techniques. Darwin based his theory of evolution on natural selection. This is directed towards maximum fitness, i.e., to maximum contribution to the gene pool of the next generation. There have since been many, and in part very sophisticated, explorations and experimental studies of various types of selection. Selection is not only directional. Balancing selection, for example, retains a steady state. This is mainly achieved by heterosis, i.e., superiority of heterozygotes above both homozygotes (frequency of $a^1/a^2 > 2pq$). It can also be due to frequency dependent selection, as in the case of mating preferences of rare types. Of special importance are recessive lethals. Lethals are genetic factors that "kill" their carriers. Recessive lethals are only effective if homozygous or hemizygous, i.e., in a single dose without a homologue. Homozygotes have a frequency of q^2, if q is the frequency of the lethal. In a gene pool, any lethal will be kept in a mutation selection equilibrium of approximately $q^2 = u$, where u is the mutation rate to the lethal condition. Lethals will persist infinitely in heterozygotes, and will give lethal offspring again and again. This is what Muller (1950) has termed "genetic load". The vision of an enormous increase of this load, with increasing mutation rates by ionizing radiations, was his main argument against artifical radiations.

The causal factors of evolution are joined in a delicate concert of interactions. This is illustrated in Figure 1. Included is the role of movable DNA in inducing mutation and recombination, and in breaking isolation even across the borderlines of species.

3. CHANGES OF GENETIC INFORMATION

Genetic information is changed by mutation (3.1) and recombination (3.2), including movable (3.3) and recombinant DNA (3.4). Effects of drastic and continuous increases in mutagen exposures on populations in a long run of generations (3.5) illustrate, by analogy, what recombinant DNA is expected to do in evolution.

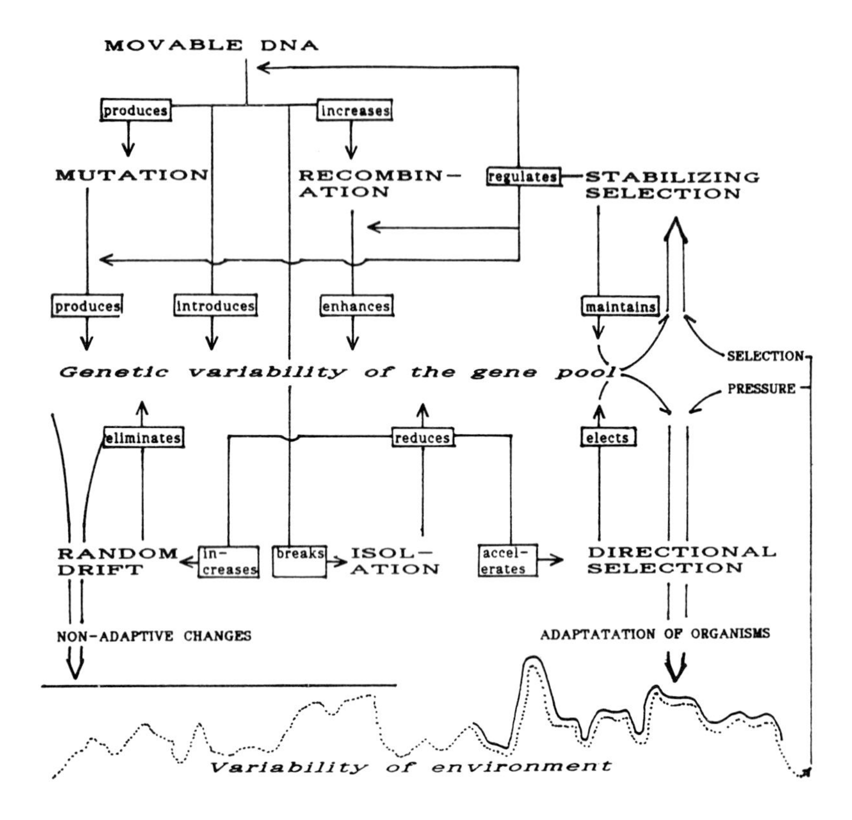

Fig. 1. Simplified model of network between factors causing evolution. Domains of variability are in italics. Factors are in bold letters. Their effects are indicated in rectangles within the pathways drawn. Main final evolutionary consequences are symbolized by double arrows.

3.1. *Mutation* of genetic information is the ultimate cause of genetic diversity. Mutations are changes of distinct genes, and structural or numerical aberrations of chromosomes. They occur naturally, but can also be induced by environmental mutagens, by artificial sources like radiations (UV, ionizing) and chemicals (alkylating agents etc), and by movable DNA.

- Numerical aberrations change the amount of DNA. The number of chromosome sets is altered in euploidy mutations like those from diploidy to triploidy ($2n$ to $3n$). The number of single chromosomes is increased $(2n+1)$ or decreased $(2n-1)$ in aneuploidy mutations. Numerical aberrations are most often due to missegregation of homologuous chromosomes during meiosis. They increase with the age of division-arrested meiotic stages, as in human females. Hypoploidies may also result from deletions, hyperploidies from translocations.

- Structural aberrations are within one chromosome (genes symbolized by A B C D), or between chromosomes (A B C D and K L M N). Intrachromosomal deletions (A D), duplications (A B C B C D), and insertions (A B z C D) change the amount of DNA; inversions (A C B D) and transpositions (A D B C) alter linkage distances between genes, i.e., they concern intrachromosomal recombination (see 3.2). Interchromosomal translocations (A B M N und K L C D) change linkage group order of genes and, thus, interchromosomal recombination (3.2).
- Gene mutations alter the base sequence in a coding DNA. Bases are either exchanged, substituted by analogues, or shifts occur within reading frames by insertions, duplications, or deletions of bases. These mutations change genetic information, i.e., they increase genetic variability of a gene pool. They are most important for further evolution when the mutated gene is high in the hierarchical order of control and regulation of gene activity.

The vast majority of errors in DNA base composition, or in the DNA backbone (single and double strand breaks), are completely repaired by a variety of general or specialized repair mechanisms that are genetically controlled. Some of these mechanisms, however, produce errors, i.e., they are mutagenic. The mixture of error-proof and error-prone mechanisms of DNA repair opens a variety of pathways for evolutionary control of mutation rates. Thereby, mutation rates are adjusted to evolutionary requirements (Nöthel 1987, 1990).

3.2. *Recombination* is interchromosomal between genes that are not linked together in one DNA molecule or chromosome. This is restricted to eukaryotes with several chromosomes that segregate independently during anaphase of the first meiotic division. It is changed by translocations that alter linkage groups.

Intrachromosomal recombination takes place in eukaryotes between homologous DNA sequences during prophase of meiosis I. It recombines alleles of linked genes. It depends in frequency on the distance between linked genes. It requires close synapsis between homologous chromosomes and is therefore only rarely observed in mitoses. This general recombination further requires a set of enzymes for various cuts and ligations of DNA in making and resolving the underlying molecular structures. These are similar in eukaryotes (between

homologues brought together by fertilization) and prokaryotes (DNAs combined by conjugation or after uptake of cell-free DNA in transformation).
3.3. *Movable DNA* is involved in bacterial conjugation, and mediates other forms of recombination.

- Movable genetic elements in conjugation are fertility- or F-plasmids. Plasmids of bacteria are small rings of extra-DNA that can become integrated into the main DNA ring. F-plasmids move from F^+ to F^- cells. If integrated, they can transport main DNA from F^+ to F^-. This leads to general recombination. Plasmids may carry additional genes as is the case with drug resistance (resistance, or R-plasmids). Via R-plasmids, resistance to antibiotics spreads much faster than is possible with Mendelian gene pools.
- Other transportation vehicles are bacterial viruses (phages). These can be integrated in the bacterial DNA at specific attachment sites by site-specific recombination. Excision of this prophage-DNA is not always precise. Therefore, the resulting free phage may carry some of the bacterial DNA. This can be incorporated along with the phage-DNA into another bacterial DNA so that bacterial information is transduced.
- Another category of movable DNA is jumping genes. These are insertional sequences and transposons. They are characterized by specific base sequences at both ends, that are either direct or inverted repeats of each other. With these repeats, the movable DNA is inserted into the host DNA. The elements carry the necessary information for the enzymes needed (transposases). Transposons are able to transport additional genes. Insertion in prokaryotes is apparently initiated by an extra replication cycle, i.e., a newly synthesized element moves. In eukaryotes, transposons may also be excized and subsequently inserted into another part of the DNA.
- Retroviruses and retroposons seem to be common in eukaryotes (see Charlesworth und Langley 1989). They use RNA instead of DNA to move. RNA mediates the information pass from DNA base sequence to protein amino acid sequence. In doing so, a RNA messenger (m-RNA) is transcribed from a DNA template. There is an ancient pathway, however, which leads the other way round, i.e., RNA is the template for DNA in reverse transcription. Retroviruses are RNA viruses that are integrated into the DNA of a host after reverse transcription. Similarly, retroposons are jumping genes that apparently are transcribed to m-RNA which is then reversely transcribed to insert in other locations of the DNA.

Movement of DNA may be restricted in multicellular organisms to specific cells, as is the case with P-transposons in the fruit fly *Drosophila melano-*

gaster (Rio et al. 1986). Insertion is frequently (as with transduction) site-specific, but seems to be possible at a great variety of sites for transposons and similar elements in eukaryotic genomes. Movement is in any case under strict genetic control, which is apparently rather quickly achieved after the movable element invades a genome (see Charlesworth und Langley 1989). Movable DNA may either stick to the genome of a species, or may use other organisms (viruses, phages) as transportation vehicles. It may thus cross the species border and produce interspecific recombination, or invade a genome with foreign DNA (see Syvanen 1984).

3.4. *Recombinant DNA techniques* make use of the above described transportation vehicles. Vehicles have been synthesized from parts of the former to introduce manipulated DNA into target cells. In principle, vehicles are loaded with foreign DNA by enzymatic cuts within specific base sequences, mixture and subsequent ligation of fragments (from vehicle and foreign DNA), and final selection of intact vehicles with desired DNA. Vehicles are then introduced into host cells. Transferred DNA is integrated into the host genome via the respective mechanism of vehicle. The main difference from movable DNA is, therefore, that with recombinant DNA, humanly selected or even humanly manufactured DNA is introduced into an organism. This may not necessarily mean that it can be introduced into the germ line of that organism. But since this is possible in many plants, and since – at least in *Drosophila* and in the mouse – vehicles are known that invade germ line cells, it is assumed that such vehicles will soon be available for many organisms. Even with these perspectives, recombinant DNA will probably not direct evolution:

- Recombinant DNAs will mainly be well-analyzed structural genes, whereas evolution is much more decisively directed by regulators whose nature is still obscure (for example, differences between humans and chimpanzees are minor in structural genes).
- Regulation of activity of transvected (foreign) genes within the only superficially understood concert of a genome will scarcely achieve the necessary degree of fine-tuning.
- Insertion will most often be at random, because targeted homologous recombination can only be of limited application (foreign DNA finds its homologue in the cell's nucleus and replaces it).
- The only certain evolutionary effect will be a change in genetic information. Like any mutant, this will be weighed by selection for its adaptive value. Only this will decide on its future in a population, if its initial frequency is high enough to avoid elimination by random drift.

3.5 *Evolutionary consequences* of recombinant DNA cannot yet be estimated from direct observations. Analoguous situations are populations under extremely high mutation pressure. In Figure 2, results are summarized from experimental populations of *Drosophila melanogaster* that have been exposed to high levels of ionizing radiations in every generation for a total of > 800 generations (Nöthel 1987, 1990). Two points are evident. 1) Any increase in radiation exposure (22 to 220 Sv) resulted in an immediate decrease in progeny number. (This was mainly due to detrimental dominant mutations.) It was always followed by adaptation that brought progeny numbers back to carrying capacity of the (artificial) ecological system (Fig. 2A). 2) Adaptation was due to both a reduction in mutability and an increase in general viability (Fig. 2B). Only in early irradiation history, variants (*rar-1*) were selected that already preexisted in the gene pool. Later adaptations were exclusively based on radiation-induced new genetic variability. It is interesting, and this may underscore the very importance of movable DNA, that at least one factor of relative radiorestance (*rar-3*) is a transposon. The point to be stressed is that the agent which produced selection pressure likewise increased genetic variability and thereby enabled evolutionary adaptation. The other important point is that, in spite of the tremendous amount of recessive lethals induced with any exposure, expressed genetic load became low, since mechanisms of lethal suppression evolved. These findings very clearly support the view of Wallace (1970) on the relative unimportance of genetic load in evolutionary terms. To make this point quite clear: there will be severe hazards to radiation-exposed individuals, and even to their immediate progeny, but there will be scarcely any harm to a population in the long run. On the contrary, because of increased genetic variability, such a population has an increased chance to adapt to an environment which produces, among other things, such an increase in mutagens. In my opinion, it would be wrong to prevent just the one parameter that is known, i.e., the input of mutations. This would prevent adaptive responses on the biological scale and would require even more interventions by human beings.

CONCLUSION

Evolution needs genetic variability. It does so even more in fast-changing environments such as are produced today by human beings. Only then can adaptive selection achieve survival of a species. Survival has its price on the individual level because adaptive peaks shift: what had been best declines with the upcoming of the better. The human being is still subject to

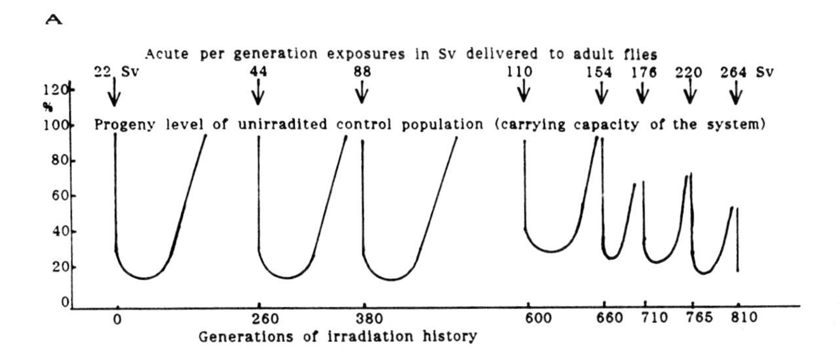

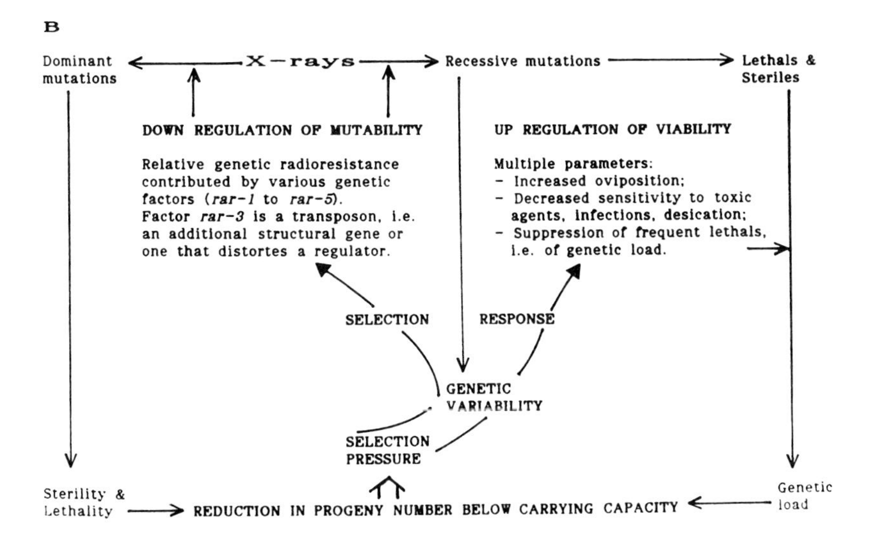

Fig. 2. Evolutionary consequences of long irradiation histories of adult *Drosophila melanogaster*.
A. Viability of irradiated flies is given in percentage of contemporaneous controls (identical origin and maintenance except for irradiations). Viability is measured by progeny numbers. These are at carrying capacity of the sytem in the control line. They are drastically reduced with any increase (arrow) in per generation exposure (from 22 to 220 Sv), but they always recover to carrying capacity.
B. Adaptation to the mutagenic action of X-rays is based on induction of genetic variants. They enable selective responses that follow two pathways, namely, reduction of mutagen sensitivity and increase in general viability

natural evolution. Even in the advanced western world, 15% of all recognized conceptions lead to spontaneous abortions, and half of these are due to chromosome aberrations – expressions of mutation and selection. Man cannot control the living world at will. One example is pests. From among the > 1 million insect species (more than all other animals together) human beings would like to eliminate several. Mosquitos, tsetse flies, screw-worm flies, and Mediterranian fruit flies provide recent examples of the fact that insecticides and other measures of eradication can fail. Likewise, microorganisms that were fought with antibiotics have always evolved mechanisms of resistance. In both groups of organisms, considerable genetic variability within large populations with short generation times and high numbers of progeny enabled adaptation. On the other side of the scale, there are organisms that we can hardly help to survive. These follow evolutionary strategies towards increased body size because of improved energy balance. Giants require more space, resources, and developmental time; thus, population size and progeny number decrease. Limited population numbers reduce genetic variability, elongated generation times handicap selective responses. This is why dinosaurs became extinct when the earth surface changed (insects and the then small mammals survived). Since the human species is in precisely the same situation today, with natural resources vanishing because of population sizes above natural carrying capacity, we need to know as much as possible about cultural evolution and use it as effectively as possible to compensate for this situation. Therefore, recombinant DNA techniques should be employed. This is further justified by the fact that evolution in general will not be harmed by the release of organisms engineered with recombinant DNA. This DNA will only be a source of genetic variability. Specific variants may, in some cases, be obtained more quickly than with random mutation, but that will not involve tinkering with evolution. Evolution itself will decide if these variants are "positive" or "negative", and its scale of values will certainly not coincide with human priorities.

REFERENCES

Ayala, F.J. and J.A. Kiger (1984). *Modern Genetics*. Menlo Park, CA: Benjamin/Cummings.

Barbujani, G. and R.R. Sokal (1990). Zones of sharp genetic change in Europe are also linguistic boundaries. *Proceedings of the National Academy of Science U.S.A.* **87:** 1816–1819.

Charlesworth, B. and C.H. Langley (1989). The population genetics of *Drosophila* transposable elements. *Annual Review of Genetics* **23:** 251–287

Dobzhansky, Th., F.J. Ayala, G.L. Stebbins, and J.V. Valentine (1977). *Evolution.* San Francisco, CA: Freeman.

Nöthel, H. (1987). Adaptation of *Drosophila melanogaster* populations to high mutation pressure: Evolutionary adjustment of mutation rates. *Proceedings of the National Acadamy of Science U.S.A.* **84:** 1045–1049.

Nöthel, H. (1990). Mutagen-mutation equilibria in evolution. *Advances in Mutagenesis Research* **1:** 70–88.

O'Brien, S.J., D.E. Wildt, M. Bush, T.M. Caro, I.A. Fitzgibbon, and R.E. Leaky (1987). East African cheetahs: Evidence for two population bottlenecks? *Proceedings of the National Academy of Science U.S.A.* **84:** 508–511.

Rio, D.C., F.A. Laski, and G.M. Rubin (1986). Identification and immunochemical analysis of biologically active *Drosophila* P element transposase. *Cell* **44:** 21–32.

Syvanen, M. (1984). The evolutionary implications of mobile genetic elements. *Annual Review of Genetics* **18:** 271–293.

Vogel, F. and A.G. Motulsky (1986). *Human Genetics.* Berlin: Springer.

Wallace, B. (1970). *Genetic Load.* Englewood Cliffs, NJ: Prentice-Hall.

Watson, J.D., J. Tooze, and D.T. Kurtz (1985). *Rekombinierte DNA.* Heidelberg: Spektrum. [*Recombinant DNA.* New York: Scientific American Books.]

Bioscience ⇌ Society
edited by D.J. Roy, B.E. Wynne, and R.W. Old, pp. 133–150
© 1991, John Wiley & Sons

Predicting the Ecological Implications of Biotechnology

Michael J. Crawley

Centre for Population Biology
Department of Biology
Imperial College
Silwood Park
Ascot, Berkshire SL5 7PY, U.K.

Abstract. Release of genetically engineered crop plants does not pose substantial *new* threats to the environment. Release of any nonnative organism is potentially risky. The evidence clearly suggests, however, that releasing engineered crop plants is likely to be safe, especially if current practice for the release of potentially dangerous substances is reinforced (Johnson 1982; Urban and Cook 1986). Nevertheless, this optimistic prognosis has to be reconciled with widespread public concern about the environment. A conclusive, experimental demonstration of the absolute safety of transgenic organisms is impossible. However, if detailed, experimental assessments of the population biology of transgenic plants are carried out before permission for commercial introduction is granted, then potentially problematical genotypes should be eliminated prior to release. At this stage, however, it is not clear how a simplified, inexpensive experimental protocol for ecological risk assessment will look. We shall need the experience gained during the kind of field trials described in this paper before any recommendations on protocol are possible.

Finally, Murphy's (Safety of Genetic Releases) Law should be borne in mind. This states that the biggest problems with genetically engineered organisms will come from those that look to be the "safest" (i.e., from those cases where we perceive no risk at all). This is because, in cases where the organisms were perceived to be dangerous from the outset, such great care would be taken that problems would be spotted early and eradicated. The unforeseen rare event that creates a novel kind of problem from a benign-looking crop will be difficult to control, because, once released, it is not practical to eliminate plants from all the habitats and seed banks into which they have found their way.

INTRODUCTION

The aim of this paper is to assess the degree to which ecologists can predict the environmental consequences of certain recent advances in biotechnology. Ecologists tend to be modest about their predictive ability. Sceptics would say they have a lot to be modest about. It is quite true that there have been some conspicuously bad failures of prediction in such fields as environmental impact analysis, especially in the ability to anticipate second-order, or knock-on effects (viz. large civil engineering projects like dam construction, river diversions, estuarine barrages, desert irrigation projects, nonselective pesticide use, and so on). Many ecological predictions are thwarted simply by the unpredictability of the weather. Until meteorologists can predict next year's weather, it is unrealistic to suppose that ecologists will be able to predict next year's ecological conditions. There are, however, a good many cases where predictive ability is reasonably good, but these have received little exposure in comparison with the failures.

In contrast to the rather gloomy picture of ecosystem-level predictions, we can be relatively optimistic about our predictive ability in relation to the issues raised by biotechnological advance, because these pose questions of a rather different kind. Here, the issue is not the replacement of entire ecosystems (or at least we hope it isn't). Rather, the issue is whether the employment of a new industrial and agricultural technology will have adverse environmental consequences. To keep matters clearly focussed, we can take as an example the issues that surround the commercial development of genetically engineered crop plants. The conjectural risks associated with transgenic crops are seen as falling into the following four categories:

1) the engineered crops will become a nuisance in agriculture (e.g., they become more persistent volunteer weeds);
2) the engineered crop plants will become invasive of natural habitats (i.e., they escape from the confines of arable agriculture and become a nuisance in other places);
3) the engineered genes escape the confines of the crop plant in which they were introduced, and become a nuisance in some other organism (e.g., the genes move via cross pollination to weedy relatives growing nearby, making the weeds more pernicious than they were before);

4) the transgenic crops will have undesirable properties as foods because of their altered constitution.

I shall say no more about the last category of risks because existing legislation on novel foods appears to be adequate to address these concerns.

ECOLOGICAL BACKGROUND

There is a view that the ecology of genetically engineered organisms is somehow different from the ecology of conventional organisms, and that the intentional release of genetically engineered organisms poses a greater threat to the balance of nature than the introduction of other kinds of organisms bred by humans. This view is mistaken. While there are risks associated with the introduction of any novel organisms into a habitat, the ecology of genetically engineered organisms is exactly the same as the ecology of any other living thing. The rules are precisely the same, no matter how the genotype is put together, and ecologists are virtually unanimous in their agreement that it is *the product not the process* that is important in risk assessment. Of course, it is possible to dream up imaginary organisms that would pose an intolerable threat to the environment. But the pressing issues concern the degree to which ecologists can predict the environmental impact of existing genetically engineered organisms. This paper addresses the question in the context of a specific example: oilseed rape *Brassica napus* engineered to express resistance to a particular kind of herbicide.

The first point to emphasise is that the ecological rules are the same for transgenic plants as for nontransgenic plants: populations must have an intrinsic rate of increase greater than zero in order to persist (see below); they must be supplied with essential resources at a sufficient rate to allow this multiplication rate to be expressed; they must resist the onslaught of predators, parasites, diseases, and competitors; and they must disperse their propagules effectively in order to survive. They may require other species (mutualists) for help with resource gathering, reproduction, defense, or dispersal. They must deal with the vagaries of changing weather and heterogeneous substrate in the same way as any other organisms.

Most of the evidence from studies carried out before the advent of genetic engineering suggests that the release of genetically engineered crop plants will be safe (Bradshaw 1984; Davies and Snaydon 1976; Hedrick 1986; NAS 1987). This conventional wisdom is based on three notions: 1) almost all genetic changes reduce fitness; 2) wild type genotypes are almost always competitively superior to introduced genotypes of the same species; 3) there is no free lunch in evolution, and all advantages accruing from the engineered genes must be paid for. This will tend to lead to reduced fitness in the wild where, in the absence of selection in favor of the trait, the engineered gene is more likely to be a liability than a benefit.

It is plausible, however, that some engineered traits (e.g., increased tolerance of extreme environmental conditions, or increased resistance to insect pests or fungal pathogens) may be advantageous under natural field conditions, and these traits are the most likely to prosper in conditions removed from the feather-bedding of arable agriculture.

GENOTYPE AND PHENOTYPE

One of the fundamental difficulties in ecology is that we have no way of predicting the behavior of the phenotype from a knowledge of the genotype. Even with a mass of data on the behavior of a given genotype under one set of environmental conditions, there is no guarantee that a change in environmental conditions will not lead to unpredictable changes in the behavior of the phenotype.

This is why the question of the safety of genetically engineered organisms is so difficult. There is no way, at present, of using theoretical arguments to establish the hazard associated with a given transgenic construct. We have absolutely no way of predicting its behavior under any given set of environmental conditions. We are always open to the accusation that under an unforeseen different set of conditions the genotype might produce a phenotype that behaved in unpredicted, undesirable ways.

It is probably safe to say that nontransgenic introduced species pose more of a hazard to the environment than do engineered varieties of native plants or familiar crop plants, and that the "exotic species model" substantially overestimates the risks associated with the introduction of genetically engineered crops (Crawley 1986, 1987). An important reason is that many of the problematic exotic plants were introduced without any thought of screening or of assessing their environmental impact. On the other hand, a large number of crop plants have become serious weeds, especially in tropical environments (Holm et al. 1977).

ECOLOGY WITHOUT TRADE-OFFS?

If the transgenic plant is to prosper, there must either be selection in favor of the engineered trait (or of some interaction effect it produces), or the engineered plants must get something for nothing: a broader niche with no loss of efficiency; a higher rate of increase with no loss of competitive ability; increased population density with no increase in susceptibility to natural enemies; and so on. A useful approach is therefore to ask explicitly: "Is it possible that genetic engineering can operate without trade-offs?" Can certain elements of phenotypic performance be improved without concomitant reduction in others? The evidence is meagre, but it all points in one direction.

Experience with traditional breeding methods has shown repeatedly that trade-offs are to be expected. For example, the spectacular increases in cereal yields achieved in the last two decades have come about not by increasing total shoot weight, but by increasing the proportion of shoot weight made up by seeds (Riggs et al. 1981). Note, also, that a large fraction of the yield increase has been the result of increased fertilizer and pesticide use, coupled with earlier planting, and has rather little to do with breeding, except in so far as varieties have been selected that respond best to these cultural practices (Snaydon 1984). Set against this, however, is the universal observation that genetic variability in natural populations is high (often much higher than anticipated). Thus, "suboptimal" genes are not always eliminated quickly from real populations. This should caution us against too quick an acceptance of conventional wisdom, and especially against the generic assumption that all manipulated genes will quickly go extinct under field conditions.

RISK ASSESSMENT EXPERIMENTS

As a result of legislative interest in the questions surrounding the intentional introduction of genetically modified organisms (GMOs), previously esoteric branches of ecological research are being called upon to provide information that can be used in risk assessment. Thus, studies of pollen flow within and between populations have become vital sources of information for attempts to predict the likely speed and distance of movement of transgenic constructs away from points of introduction of genetically engineered plants. Studies of the history of biological invasions have been dredged for information on the attributes of plants and animals that are associated with increased risks of invasiveness. The ecology of invasion-prone and invasion-resistant communities have been compared (Crawley 1986, 1988).

While it appears that our combined experience does permit a large number of broad predictions about crop plants that are almost certainly robust, the detailed issues that arise in prerelease risk assessment can at present only be answered by tailor-made field experiments. In due course, sufficient experimental material will have accumulated that large classes of construct/organism combinations can be exempted from detailed prerelease testing. There is no doubt, however, that novel construct/organism combinations will be subject to detailed, small-scale ecological testing before permission is given for wide scale release (including precommercial large scale testing) for many years to come.

THE PROSAMO EXPERIMENTS

Until regulatory authorities and the public are convinced that genetically engineered organisms behave in essentially the same way as their untransformed counterparts, the risks associated with genetic manipulation will continue to demand attention. Thus, for the foreseeable future, all workers proposing to conduct experiments outside full containment will have to address the potential risk to the environment posed by the intentional introductions of GMOs (Mooney and Bernardi 1990).

The quantification of risks posed by GMOs is a pressing environmental problem, with important implications for biotechnology industries through its effect on policy makers and public perceptions (Tiedje et al. 1989; Royal Commission on Environmental Pollution 1989). The PROSAMO initiative was undertaken to provide data on the comparative ecology of transgenic and nontransgenic varieties of crop plants, with particular attention to questions of gene transfer (e.g., through pollen flow, or via soil microorganisms) and invasivness or weediness of the engineered plants themselves. The research is jointly funded by the Department of Trade and Industry, the Agricultural and Food Research Council, and a consortium of multinational companies.

THE DEMOGRAPHY OF INVASIONS

There is a considerable literature on the dynamics of biological invasions (reviewed in: Mooney and Drake 1986; Drake et al. 1988). For convenience, we may consider the process of invasion as comprising three phases: a) colonization, b) establishment, and c) spread.

a) Colonization

For a crop in widespread use we can assume that colonization is bound to occur, because many millions of seeds will be transported around the country resulting in spillages into roadside habitats. The crop will also be "introduced" in large numbers in the arable environments where it is sown.

b) Establishment

Given that seeds of the crop are present in a particular habitat, we need to determine whether the population will persist. We formalize this by asking: "If small numbers of seed are introduced into a particular place, will the population size be greater in the next generation?" This requirement can be framed in demographic terms. Three parameters are commonly used to describe population growth (see the current standard ecology text: Begon et al. 1990); each of the parameters gives a precise condition for a population to establish.

a) Basic reproductive rate (R_0). This is the average number of seeds produced by a single seed in the absence of density dependent constraints. In formal terms

$$R_0 = \sum_x l_x m_x,$$

where l_x is the probability that an individual survives to age x and m_x is the age-specific fecundity.

b) Finite growth rate (λ). This parameter is simply the multiplicative constant at which a population, with a stable age distribution, increases over a fixed time interval (t) when freed from density dependent constraints. For a population with discrete generations, λ is defined by

$$\lambda = \frac{S_{t+1}}{S_t},$$

where S is the number of seeds and the suffixes denote successive generations.

c) Intrinsic rate of increase (r). This is the instantaneous rate at which a population increases in size when freed from density-dependent constraints. In a population with age-dependent survivorship and fecundity, once a stable age distribution is achieved the total population size ($n(t)$) increases exponentially at a rate r (i.e., $n(t) = n(0)\exp(rt)$). The intrinsic rate of increase is related to the finite rate of increase by $r = \ln(\lambda)/t$, and to the basic reproductive rate by $r \approx \ln(R_0)/T$ where T is the average age at reproduction.

All the parameters are defined when population size is small and so intraspecific competitive interactions between individuals are unimportant. For establishment, R_0 needs to be greater than 1 so the population will increase when rare. This implies that the finite rate (λ) of increase is greater than 1 and the intrinsic rate of increase (r) is greater than 0. On the other hand if R_0 is less than 1, then each seed replaces itself with less than one seed and the population will decline to extinction.

c) Spread

Given that establishment has occurred, will the population spread from the site of introduction? In order to assess the risk associated with the release of an engineered organism, we need to be able to predict the rate at which it might spread to other areas. This topic has been the basis of a large body of theoretical research which has given a number of useful results (Manasse and Kareiva, in preparation). A useful summary statistic is the asymptotic rate of spread (ARS), measured in units of distance per unit time, which is the rate at which an organism's aerial range increases in radius. The ARS provides an upper bound for the velocity of range expansion (Lubina and Levin 1988). In a homogeneous environment the ARS is given by

$$\text{ARS} = 2\sqrt{rD}$$

where r is the intrinsic rate of increase and D is the diffusion coefficient. A similar expression can be obtained for a heterogeneous environment where

$$\text{ARS} = 2\sqrt{\overline{r}\langle D \rangle},$$

in which $\overline{r}$ is the arithmetic mean intrinsic rate of increase and $\langle D \rangle$ is the harmonic mean of the spatial diffusion coefficients (Shigesada et al. 1986). In both cases the intrinsic rate of increase plays a fundamental role in determining the rate of spread of an introduced crop.

THEORETICAL BACKGROUND

I shall expand a model of invasion (Crawley 1986) for a clearer understanding of the circumstances under which a genetically engineered crop plant might become more weedy or more invasive of natural habitats than its non-transformed counterpart. I shall refer to the intentionally released, genetically engineered crop as the invader, and the resident plants as the wild type. The wild type may be genotypes of the same species if the plant is already common in the environment, or they may be unrelated species of the same guild in cases where the released species is new to the environment (for details see Crawley 1987, 1988).

The model for invasiveness is:

the rate of increase of the transgenic plant in a given habitat = plant development rate

\+ its seed production (timing and duration)

\+ survival of vegetative parts (discounted by their mortality rate)

− the effects of competition with other plants of the same kind

− the effects of competition with other plant species

− the effects of herbivores (molluscan, insect and vertebrate)

− the effects of fungi and other plant diseases

− the effects of mutualists (if they are in short supply; pollinators, seed dispersal agents, mycorrhizal fungi, etc)

\+ immigration of transgenic seed from other sites

\+ establishment of transgenic plants from dormant seeds in the soil (seed bank)

The parameters of the model determine whether the introduced genotype will increase or decrease. In terms of risk assessment, we are interested in the circumstances under which the rate of increase is positive. If the rate is positive, then we need to ask what level of abundance the plant would achieve, and what environmental consequences would follow from the plants persisting at this density.

FIELD STUDIES OF TRANSGENIC *BRASSICA NAPUS*

The experiments are designed to allow the invasive ability of transgenic oilseed rape to be evaluated. An important component of this evaluation is the use of historical information on the invasive ability of the untransformed crop. By studying local floras we were able to determine the habitats occupied by oilseed rape outside cultivation (Table 1). It is clear from Table 1 that oilseed rape plants are restricted to disturbed habitats that are open and which allow successful recruitment from seed. It is also clear that in many such habitats the plant is a "casual species" in which constant immigration of seed is necessary for persistence (e.g., seeds literally falling off the back of a lorry).

Table 1. The status of oilseed rape *Brassica napus* in some local floras. Most of the reports are of an essentially casual plant, found only in disturbed habitats where there is repeated introduction of rape seed from outside the system. There are no reports of oilseed rape from mature perennial vegetation.

County	Habitat
Anglesey	Common in corn fields and arable land.
Berkshire	An occasional casual in arable fields and on rubbish tips.
Buckinghamshire	Local on cultivated ground and waste places
Cheshire	An alien of cultivated ground and by streamsides.
Derbyshire	A rare escape from cultivation found on field borders, waysides, banks of ditches or streams and waste places.
Durham	Introduced and casual on arable, disturbed and waste ground.
Essex	Introduced relic of cultivation; not persisting. Arable land, waste places and rubbish tips.
Hampshire	Occasional as an escape from cultivation in fields and waste ground

By performing detailed comparative experiments on the demography of transformed and untransformed plants we intend to determine whether genetic engineering produces genotypes that are more, less, or equally invasive compared with the untransformed crop. We suspect that the parameters (i.e., λ, r, R_0) that determine the likelihood of successful invasion are not only genotype-specific but also habitat-specific, and so it was necessary to carry out the experiments in as wide a range of habitats as possible. Habitat specificity results not only from variation in abiotic conditions, but also from differences in herbivore pressure and the intensity of plant competition.

Experimental design

We performed the same small-scale release experiments in three different sites in Great Britain. The sites were chosen to represent a benign climate with an early start to the growing season (Cornwall, in the extreme southwest of England), an intermediate, more continental climate (Berkshire, in southeastern England) and a more hostile climate with a late start to the growing season (Sutherland, in northeastern Scotland). At each site we chose four habitats, which were selected to cover a wide range of soil and light conditions; in each site we attempted to provide wet and dry soils, and sunny and shaded conditions.

In each of the four habitats there were four experimental plots, each measuring $25\,m \times 25\,m$, positioned at random within the experimental area. Each plot was then divided into four subplots, and these were either fenced to exclude vertebrate herbivores or cultivated to remove perennial vegetation in factorial combinations (i.e., fenced/cultivated, fenced/uncultivated, unfenced/cultivated and unfenced/uncultivated). Within each of the subplots we sowed seed of three genotypes of oilseed rape:

1) Untransformed – control
2) Transformed – with kanamycin marker
3) Transformed – with kanamycin marker and resistance to Basta herbicide

The seeds were sown into quadrats which received factorial combinations of insecticide and fungicide.

In addition to the seed sowing experiment described above, we also buried nylon mesh bags of seeds at three different depths in order to determine the carry-over of seeds from one year to the next.

Throughout the growing season we monitored the quadrats to determine how many seeds had germinated, the proportion that survived to reproduce, and the seed production of these plants. These data indicate enormous site-to-site and also habitat-to-habitat variation in seed production, and dramatically underline the importance of performing this kind of risk assessment field trial in a wide range of different habitats, and over a protracted span of years.

Within each habitat we manipulated the intensity of vertebrate grazing pressure and plant competition. Both these processes had a profound effect on total seed production. A total of 33,408 seeds were sown into uncultivated areas, and these seeds failed to produce a single reproductive plant. Of course it would be wrong to assume that all the seeds sown into uncultivated areas are no longer viable, but it is clear that the finite rate of increase could not be greater than one. This result indicates that recruitment failure is the reason that oilseed rape does not occur outside disturbed habitats.

The dramatic effect of herbivore exclusion is illustrated by the fact that fencing resulted in a twofold increase in the number of surviving plants recorded at maturity (averaged over all three sites). However, the degree of impact of vertebrate herbivore exclusion was site specific: herbivory resulted in a substantial reduction in the number of plants recorded at the Berkshire and Scotland sites but had no effect on plant density in Cornwall, perhaps because the weather conditions there permitted regrowth of damaged plants. Note, however, that even in Cornwall, grazing led to a substantial reduction in the numbers of seeds produced by each surviving plant.

There was great variation in per capita fecundity within a habitat; most seeds that germinated failed to produce any seed at all. Of particular note from the point of view of risk assessment is the extreme skewness of the statistical distribution of seeds per plant, which has a long tail to the right. It is vital, therefore, that there is sufficient replication to allow the shape of the tail of this distribution to be estimated with reasonable precision.

An example analysis

Using the 1990 PROSAMO field data we can produce crude estimates of the finite rate of increase in the different habitats. As an example of the type of calculation, consider the following data. The average number of seeds produced per seed that germinated was 17.6, while the probability of seed germination was 0.026. The probability of seed mortality is at present unknown, but will be estimated from the seed burial experiment in due course. If we were to assume that all those seeds that failed to germinate have died, then the finite rate of increase is

$$\lambda = g\overline{F} = 0.026 * 17.6 = 0.4576.$$

Alternatively, if we assume that all the seeds that failed to germinate are still viable, then the finite rate of increase is much greater, namely:

$$\lambda = (1 - g) + g\overline{F} = 0.974 + 0.026 * 17.6 = 1.43$$

The true value of λ therefore lies between these two extremes. Since under one set of assumptions, $\lambda < 1$ while under another set $\lambda > 1$, it is clear that unless there is considerable carry-over of seeds then the population cannot persist. It is also clear that unless we have data for the complete life cycle it will be impossible to assess the invasive ability of introduced crops.

ECOLOGICAL EXPERIMENTS

It is a fact of ecological life that ecologists do not repeat one another's experiments. In physics or in medicine, every lab in the world tries to repeat each exciting new experimental result. In ecology, on the other hand, a new experimental result tends to be met with a shrug of the shoulders. It is assumed that experimental results are so context-specific that it is not worth even thinking about repeating them because there is so little chance of obtaining a similar result. In truth, ecologists are capable or making a great many well-informed predictions. Several of these predictions are relevant to issues of biotechnology and the safety implications of that technology.

How predictable was the PROSAMO result in the first field season? Most parts of the experiment were highly predictable: for example, the prediction that the factors resisting invasion could be ranked from most to least important was published before the experiment was begun (Crawley 1990); interspecific plant competition was the most important factor, then vertebrate herbivory, then fungal pathogens and invertebrate herbivores. The unpredictable part of the experiment was the difference, if any, that would appear between the transgenic and nontransgenic plants. The null hypothesis was that genetic engineering would make no difference to ecological performance, and to the extent that this is what we observed, then our predictions were borne out here as well. In fact, our prediction is not as good as this suggests. If the transgenics did less well than the controls, perhaps because of the "genetic baggage hypothesis", then we predicted that the plants with Basta and kanamycin resistance would do least well, followed by the plants that were transgenic for kanamycin resistance alone. What we observed is that the plants with kanamycin resistance alone produced the fewest seeds, followed by the Basta-resistant plants, with the nontransgenic controls producing the largest average seed crops. We shall need to see whether this result is consistent over the three years of the experiment before we read too much into this one result.

POLITICAL OPTIONS

On really big issues like carbon dioxide production, global climate change, or human population control, the solutions are obvious, but the ability and commitment of political institutions to tackle them is conspicuously absent. At present, for example, it would be political suicide for an American political party to ration each citizen to 500 gallons of gasoline per year, but that is what it would take to make a significant reduction in transport-generated CO_2 emissions. The scope for black market corruption in any such scheme also beggars description. Again, there appears to be no prospect that the Catholic Church will speak in favor of birth control, and human population grows inexorably by thousands every day.

We shall have to content ourselves, therefore, with predictions about relatively minor issues that do fall within the political scope (e.g., release of genetically engineered organisms), and with post hoc rationalizations about the consequences of the bigger issues (e.g., the impact of climate change on terrestrial temperate ecosystems). It is not at all clear that scientific information will play a major role in formulating policy decisions, but the discovery of the ozone hole and the banning of CFCs is an interesting case. This example succeeded, one supposes, because the perceived solution (stop using CFCs as refrigerants and aerosol propellants, and use something else instead) was so simple and so cheap. The image of a hole in our atmosphere was also enormously evocative (like being told you have a hole in the heart). It is hard to see what the ecological equivalent of the discovery of the hole in the ozone layer would be. It is even more difficult to imagine that the image it would conjure up would be sufficiently compelling as to stir the political community into action (the destruction of tropical forests is perhaps as close as we can get, and although this has been moderately successful, the degree to which ecological thinking has been involved is slight indeed).

CONCLUSIONS

This paper has concentrated on the ecological issues associated with the introduction of GMOs, and it has been tacitly assumed that evolutionary processes can be ignored. However, as Bradshaw (1984) states, "Evolution is likely to be most rapid in those situations in which selection has not already acted, because in relation to the particular selection pressure there can be a store of hidden unselected variability." This is exactly the situation in crop plants which have been artificially selected for agricultural characters, such as high yield and

Table 2. The invasibility of plant communities expressed in terms of the invasion criterion λ. Invasion and persistence will occur when $\lambda > 1$, while any beach-head population will quickly disappear if $\lambda << 1$. Short term persistence is possible following very large initial introductions if λ is close to 1 or if there is persistent seed dormancy (plants from the initial introduction can appear over a number of years, suggestive of there having been repeated reproduction whereas, in fact, it was simply delayed recruitment of the introduced material). The table shows the severity of environmental conditions increasing from top to bottom, and severity of biological conditions increasing from left to right. We have substantial predictive ability about the outcome of this experiment; the only quibbling is over the precise values of the parameters. For example, the shape of the results table is wholly predictable, with λ declining from the top left of the table to the bottom right. For a ruderal plant like oilseed rape, we would predict no recruitment at all in undisturbed, intact perennial vegetation (i.e., in the right-most column). This, indeed, is what we observe. We would argue that the precise parameter values will always be context-specific so that, without some local data input, a precise quantitative prediction will always be impossible. We would need to be able to predict the weather, and then some more besides. But we can predict the important qualitative patterns, and this allows us to generalize about the likely environmental impact of transgenic crop plants.

Biotic	Complete protection	No pathogens	No herbivores	No plant competition	Intact vegetation
Perfect	**64**	**16**	**4**	**2**	*0*
Benign	**12**	**4**	**2**	**1.2**	*0*
Moderate	**2**	**1.5**	*0.8*	*0.3*	*0*
Harsh	*0*	*0*	*0*	*0*	*0*

low seed dormancy, and subsequently are inadvertently released into natural habitats. In natural habitats these crops might evolve rapidly away from their domesticated ancestors. Therefore our attempts at understanding the ecology of transgenic crops are aimed at a moving target. However, the precise nature of genetic engineering means that transgenic plants are unlikely to have any more or less hidden variability than untransformed conventional crops, and so the scope for rapid evolution should be no different in transformed than in conventional crop plants.

It is possible, but as yet undocumented, that transformation will produce unexpected changes in the ecological behavior of the genotype (Campbell 1990). This strongly reinforces the importance of empirical studies of the complete life-cycle, so that subtle changes can be detected and their effects assessed by comparison with conventional genotypes. These comparisons are best achieved using the demographic techniques described earlier.

The preliminary results of the PROSAMO experiments suggest that transgenic rape plants are unlikely to pose more of a threat to the environment than their conventional counterparts. The experiments also demonstrate the importance of multisite release experiments, and results obtained in any one place are strongly dependent on the local conditions during the particular year of the experiment. It is worth stressing that yield trials in arable habitats are not adequate for ecological risk assessment, because important aspects of the life cycle are ignored, and the range of environmental conditions to which the crop is exposed is so limited.

REFERENCES

Begon, M., J.L. Harper, and C. Townsend (1990). *Ecology: Individuals, Populations and Communities,* 2nd Edition. Oxford: Blackwell Scientific Publications.

Bradshaw, A.D. (1984). Ecological significance of genetic variation between populations. In: *Perspectives on Plant Population Ecology*, eds. R. Dirzo and J. Sarukhan, pp. 213–228. Sunderland, MA: Sinauer.

Campbell, A. (1990). Epistatic and pleiotropic effects on genetic manipulation. In: *Introduction of Genetically Modified Organisms into the Environment*, eds. H.A. Mooney and G. Bernardi. Scope 44. Chichester: John Wiley and Sons.

Crawley, M.J. (1986). The population biology of invaders. *Philosophical Transactions of the Royal Society of London Series B* **314:** 711–731.

Crawley, M.J. (1987). What makes a community invasible? In: *Colonization, Succession and Stability*, eds. A.J. Gray, M.J. Crawley and P.J. Edwards, pp. 429–453. Oxford: Blackwell Scientific Publications.

Crawley, M.J. (1988). Chance and timing in biological invasions. In: *Biological Invasions*, ed. J.A. Drake, pp. 407–423. New York: John Wiley.

Crawley, M.J. (1990). The population dynamics of plants. *Philosophical Transactions of the Royal Society of London B* **330:** 125–140.

Davies, M.S., and R.W. Snaydon (1976). Rapid population differentiation in a mosaic environment. III. Measures of selection pressures. *Heredity* **36:** 59–66.

Drake, J.A., F. di Castri, R. Groves, F. Kruger, H. Mooney, A. Rejmanek and M. Williamson (1988). *Biological Invasions: A Global Perspectives*. Chichester: John Wiley and Sons.

Hedrick, P.W. (1986). Genetic polymorphism in heterogeneous environments: A decade later. *Annual Review of Ecology and Systematics* **17:** 535–566.

Holm, L.G., D.L. Pluncknett, J.V. Pancho, J.V., and J.P. Herberger (1977). *The World's Worst Weeds*. Honolulu: University of Hawaii.

Johnson, E.L. (1982). Risk assessment in an administrative agency. *The American Statistician* **36:** 232–239.

Lubina, J.A., and S.A. Levin (1988). The spread of a reinvading species: Range expansion in the Californian sea otter. *American Naturalist* **131:** 526–543.

Manasse, R., and P. Kareiva. Quantitative approaches to questions about the spread of recombinant genes or organisms. To appear.

Mooney, H.A., and G. Bernardi, eds. (1990). *Introduction of Genetically Modified Organisms into the Environment.* Scope 44. Chichester: John Wiley and Sons.

Mooney, H.A., and J.A. Drake, eds. (1986). *Ecology of Biological Invasions of North America and Hawaii.* New York: Springer-Verlag.

NAS. (1987). *Introduction of Recombinant DNA-engineered Organisms into the Environment: Key Issues.* Washington, D.C.: National Academy Press.

Riggs, T.J., P.R. Hanson, N.D. Start, D.M. Miles, C.L. Morgan, and M.A. Ford (1981). Comparison of spring barley varieties grown in England and Wales between 1880 and 1980. *Journal of Agricultural Science* **97:** 599–610.

Royal Commission on Environmental Pollution (1989). *The Release of Genetically Engineered Organisms to the Environment.* Thirteenth report. London: Her Majesty's Stationary Office.

Shigesada, N., K. Kawasaka, and E. Teramoto (1986). Travelling periodic waves in heterogeneous environments. *Theoretical Population Biology* **30:** 143–160.

Snaydon, R.W. (1984). Plant demography in an agricultural context. In: *Perspectives on Plant Ecology*, edited by R. Dirzo and J. Sarukhan, pp. 389–407. Massachusetts: Sinauer.

Tiedje, J.M., R.K. Colwell, Y.L. Grossman, R.E. Hodson, R.E. Lenski, R.N. Mack, and P.J. Regal (1989). The planned introduction of genetically engineered organisms: Ecological considerations and recommendations. *Ecology* **70:** 298–315.

Urban, D.J., and N.J. Cook (1986). Hazard evaluation division. Standard evaluation procedure. Ecological risk assessment. EPA 540/9-85-001. Washington, D.C.: US Environmental Protection Agency.

Bioscience ⇌ Society
edited by D.J. Roy, B.E. Wynne, and R.W. Old, pp. 151–162
© 1991, John Wiley & Sons

Ecological Restoration and the Reintegration of Ecological Systems

William R. Jordan III

University of Wisconsin-Madison Arboretum
1207 Seminole Highway
Madison, WI 53711, U.S.A.

Abstract. Any meaningful consideration of threats to ecosystems will depend on an evaluation of the cost and likelihood of their achieving reintegration, either through natural recovery, or through recovery assisted by active efforts at restoration. The science and art of ecological restoration has recently emerged as one of several disciplines committed to this task. The purpose of this paper is to provide an overview of the state of the art in this area, considering both the ability of restorationists to recreate authentic replicas of naturally occurring ecosystems, and also the value of restoration as a way of achieving reintegration of human beings and the natural landscape.

Though it is generally agreed that restoration is in the early stages of its development as a discipline and an environmental technology, restorationists have had considerable success in bringing some kinds of ecosystems back to a more or less "natural" condition following their disruption or even destruction as a result of human activities. In general, the likelihood of success and the authenticity of the resulting ecosystems vary not only with the type of ecosystem, but from project to project, and of course they depend in part on unpredictable influences such as the weather, over which the restorationist has little control.

At the same time, attention is directed to the act of restoration itself as a means of achieving a positive relationship between the restorationist, the landscape being restored, and nature in general. The act of restoration offers opportunities for establishing such a relationship through an exploration of the nature-culture relationship in both the spatial (ecological or landscape) dimension and the temporal dimension (history and cultural evolution). Since conservation depends on the quality of this relationship – at least in any landscape influenced by human activities – this may prove to be the most valuable aspect of the process of restoration.

INTRODUCTION

The theme of this paper is our ability to achieve the reintegration of ecosystems through natural processes of recovery that may be aided, speeded up, and perhaps guided by active efforts at restoration. It seems to me that this perspective is necessary if we are either to assess the gravity of the risks of ecosystem disintegration, or to conceive coherent strategies for dealing with such problems.

In recent years a number of new initiatives has emerged in the area of environmental health care, represented by organizations such as the Society for Conservation Biology, the International Association of Landscape Ecologists, various societies which are concerned with the reclamation of lands degraded by mining and, most recently, the Society for Ecological Restoration. Since, however, it is restoration that represents this commitment in its most explicit and perhaps most ambitious form, and also because I am most familiar with this area, I will confine my remarks to it. I define ecological restoration as any attempt to return a whole ecological system (whether conceived as an ecosystem, an ecological community, or a landscape) to an earlier condition following a period of alteration, disturbance, or even destruction as a result of human activities. The degree of degradation may be severe (as in strip mining) or subtle (as may result from a period of protection from fire). An example would be the attempt to upgrade and expand a remnant prairie in Illinois on a site originally occupied by prairie but more recently used as a cornfield.

In this paper I will describe as best I can the state of the art in this area, considering first of all the ability of restorationists to achieve the reintegration of natural ecological systems. I will then offer some reflections on the importance of restoration as a way of achieving the reintegration of ecosystems at a higher level, by providing the basis for their reentry and reinhabitation by human beings.

FROM PROCESS TO PRODUCT – THE REINTEGRATION OF ECOSYSTEMS

Since restoration is defined in terms of its goal or product – an artificial ecosystem intended to resemble its natural counterpart in every particular – it makes sense to begin with the question of how successful restorationists have been in actually achieving this goal. Since there is as yet very little published information on this subject, I will draw as best I can on the rich and, in some cases, provocative oral tradition that has developed within the restoration community.

Probably the most systematic efforts at evaluation of the quality of the resulting systems have been made in connection with the restoration of various kinds of wetlands, ranging from offshore seagrass beds to isolated inland wetlands. The reason for this is that wetlands have been prime candidates for a practice known as "mitigation", in which a developer agrees to compensate for the destruction of a natural community by restoring or creating comparable habitat at another, usually less useful or less expensive, site. This has led to the development of a minor industry devoted to the restoration and creation of wetlands, and a growing concern among environmentalists regarding the quality of the resulting systems.

As a result, documentation of the quality of restored wetlands, though still fragmentary, is probably the best available for any community type. In a recent survey of the state of the art in this area, Jon Kusler and Mary Kentula conclude that total duplication of any wetland community is impossible, but that some kinds of wetlands can at least be approximated under favorable conditions, the quality of the resulting systems varying widely, both for various kinds of wetlands and with respect to different parameters of authenticity.

They report that the best results overall have been obtained for wetlands along shorelines where the supply of water is likely to remain substantially unaltered – especially in the lower elevations along coastlines, where the water supply varies with the mathematical regularity of the tides. By the same token, the most difficult systems are isolated inland marshes that depend on supplies of surface or groundwater which are likely to be difficult or impossible to control (Kusler and Kentula 1989).

In all cases, of course, the degree of success achieved depends on the objectives – those features of the system the restorationist has attempted to restore. Here, not surprisingly, it is various engineering and "agricultural" features that are easiest to reproduce. Thus, it is fairly easy to design an artificial wetland that resembles a natural wetland in its ability to retain and release runoff water, or in its value for waterfowl. It is much more difficult to restore subtler functions or biological processes. Recently, for example, Joy Zedler and her colleagues at San Diego State University have been studying a five-year-old restored cordgrass (*Spartina*) wetland in San Diego Bay, comparing it with a nearby natural wetland with respect to attributes such as organic matter accumulation, nitrogen concentrations in soil, water, and vegetation, plant height and biomass, and the number and densities of invertebrate species. They found that overall values for the restored wetland averaged only 57% of those for the reference wetland. They also noted that this wetland has so far failed to meet its original objective of providing habitat for the light-footed clapper rail (Zedler and Langis 1990).

One element that is virtually impossible to replace quickly, Zedler noted, is organic matter, which ordinarily accumulates gradually as plants die and form layers of peat. This material plays a key role in many wetland ecosystems, not only as a critical component of the habitat of many plants and animals, but as a buffer against seasonal variations in water supply. Yet its replacement is usually impossible on a large scale, and even if a supply of appropriate material were available, recent research by Cal DeWitt indicates that it is difficult, if not impossible, to move it without disrupting the fine structure on which its water-holding and transporting properties depends. (DeWitt 1991).

PRAIRIES

While the quality of restored prairies has not yet been studied as systematically as that of various restored wetlands, prairies – especially the tallgrass prairies of the American Midwest – are of special interest in this respect, because these systems were among the first to be subject to systematic restoration efforts, and because prairie restoration has come to represent for many a paradigm for restoration in its purest sense, guided exclusively by environmental, historical, and aesthetic interests uncompromised by other concerns. In addition, and partly for this reason, prairie restorationists have accumulated a wealth of experience and an intimacy with the system that, though not yet systematized, clearly represents major progress toward the evaluation of authenticity for this system.

Can prairie restorationists really restore prairies? A recent evaluation of the restored prairies at the University of Wisconsin-Madison Arboretum, which at 55 and 45 years are the oldest restored prairies anywhere, notes that the number and distribution of species on these prairies compare favorably with those remnants of natural prairies on similar sites, but notes persistent problems with a few exotic weeds such as Kentucky bluegrass (*Poa pratensis*), sweet clover (*Melitotus spp*) and leafy spurge (*Euphorbia esula*) (Kline and Cottam 1989). One experienced restorationist, however, questions the fine-scale distribution of species in restored prairies, noting that the mingling of species is generally coarser than she finds in natural remnants, the effect being patchy – a course plaid rather than the fine-grained tweed of the natural prairie tapestry (Powers 1987).

In an attempt to quantify the progress of a prairie restoration, Floyd Swink and Gerould Wilhelm have devised a system that involves assigning to species values that reflect both their rarity and the extent to which they are confined to the prairie community, then evaluating the restored community for the presence and abundance of these species (Swink and Wilhelm 1979). Interestingly, and encouragingly, Steve Packard of the Illinois Nature Conservancy has found that this "Swink-Wilhelm Index" increases gradually for both prairies and oak openings subject to restorative management, principally with fire and hand weeding of certain especially troublesome species (Packard 1990).

In general, attempts to evaluate restored prairies have concentrated on vegetation. Studies of insects in restored prairies have barely begun, but suggest that a large fraction of species native to the prairies still exist in fence rows and old fields and readily find their way back into restored prairies (Panzer 1990). An additional fraction, however – possibly in the neighborhood of 20% – seems to be confined to prairie remnants, which are now rare and widely scattered. These do not generally reappear spontaneously in restored prairies, and only a few attempts have been made to reintroduce them. Since even this small fraction probably amounts to somewhere in the neighborhood of 2,000 species, this could be a formidable task. This being the case, Panzer advocates restoring prairies around remnant natural prairies whenever possible. Even very small remnants, he says, can serve as sources of insects and perhaps other organisms that would almost surely be overlooked in the restoration process.

In another series of studies, mainly on a large (250 ha) restored prairie under restoration at Fermi National Laboratory near Chicago, Michael Miller and Julie Jastrow, scientists working for the U.S. Department of Energy, have been studying the underground component of the prairie ecosystem. What they are finding is that a key to recovery of the soil community is the reestablishment of the symbiotic relationship between plants and mycorrhizal fungi, which inhabit the roots and play a key role in the uptake of water and nutrients, and in the recovery of the crumb structure of the soil. Interestingly, this process has reached 95% of full recovery on parts of the Fermilab prairie within about a decade. Miller and Jastrow believe this is a crucial step toward the recovery of the biological community in the soil and the recovery of soil properties such as organic carbon, which has risen from the 3.6% characteristic of degraded farmland to 4.5% at Fermilab – that is, about half the level believed to be characteristic of an undisturbed tallgrass prairie (Miller and Jastrow 1990).

It would be possible to provide similar outlines of what is known about the restoration of a number of other community types. None, however, would reveal more than is known about wetlands and prairies, which may be taken, therefore, to represent the state of the art with respect to this aspect of restoration.

Clearly, much more needs to be learned before environmental planners can count on restoration as a strategy for the long-term conservation of high-quality natural areas. At the same time, I would like to conclude this part of my discussion by noting that this very situation reveals two of the great values of restoration for conservation. The first of these is precisely that restoration makes a promise actually to conserve something. This is certainly one reason why it has been more vulnerable to criticism than preservation, which, at least in principle, promises only to leave the system alone. This can actually lead to the loss of entire biota as a result of succession under altered, postsettlement conditions, as our experience with the prairies and oak openings of the upper Midwest of the U.S. clearly shows. Here conservation depends on an active, ongoing program of restoration, and, while this situation may be unusual with respect to the speed of the response, it may be taken as paradigmatic: ecological communities *inevitably* change in response to changes in the behavior of the human community. Thus, while preservation certainly must be a *component* of conservation, it must be, so to speak, surrounded by a process of more or less continual restoration.

This points toward the second great strength of restoration, which is that rather than ignoring or denying human influence on the landscape, and making a radical distinction between change brought about by human beings and "natural" change, restoration acknowledges human influence on the landscape, and rejects a radical distinction between human beings and nature. What restoration is, in effect, is an attempt to compensate for human influence in a precise and effective manner. In this way restoration (in this strict, highly conservative sense) not only represents the best hope for conservation of ecosystems, it also becomes an exploration of the continually changing terms of our relationship with nature. It thus becomes a way of defining in ecological terms exactly who we are, and so it becomes the key to resolving the dilemma of our place in nature.

FROM PROCESS TO PERFORMANCE: THE REINTEGRATION OF NATURE AND CULTURE

At this point I would like to turn from a discussion of restoration as a way of producing a certain kind of product – a way of creating or repairing ecosystems conceived as objects in the landscape – to a discussion of its implications as an act or performance. I think that restoration, in this sense, offers a way of transcending the idea of the natural landscape as an "environment" and presents the possibility of making it once again a habitat for human beings. Note that with this idea we have by no means ended our consideration of the reintegration of nature, but we have come to the very core of this issue, since it is the divergence of human beings from the rest of nature in the course of cultural evolution that underlies everything we now identify as an "environmental" problem.

Reintegration, then, must begin with the reintegration of our own species into nature. Environmental writers generally acknowledge this. Some years ago, for example, Loren Eiseley wrote of our need to return to the "sunflower forest" of the earliest human experience – without, however, abandoning the knowledge gained "on the pathway to the moon" (Eiseley 1970). Eiseley, however, offered no clear idea of how this is to be achieved. What I want to explore here is the idea that the act of ecological restoration enacts a mutually beneficial relationship with nature, and represents nothing less than a technique for achieving a healthy relationship between human beings and the natural or archaic landscape (Jordan 1986).

To realize this, it is necessary to turn from a consideration of the products of restoration to a consideration of restoration as a process, and even as a significant human act – a performance – regarding it not merely as a way of changing the landscape, but as a way of carrying on a dialogue with it.

To do this we must attempt to "read" or interpret the act of restoration much as we would a book or a play. When I began doing this thirteen years ago, the first thing that struck me about restoration was that it was a way of learning about the ecosystems that were being restored. A dramatic example of this was the "discovery" of prairie fire, which was a direct result of the early attempts at prairie restoration at the Arboretum (Curtis and Partch 1948). This not only gave us a tool that has proved crucial to the conservation of the prairie, it deepened our understanding both of the prairies and of our relationship with them (Jordan et al. 1987).

The notion of restoration as a technique for basic research and a kind of acid test of ecological understanding has now been explored in some detail and even given a name: "restoration ecology" (Jordan et al. 1987). At the same time restoration is more than a way of doing research. It is also, perhaps most obviously, a form of agriculture; and it entails the processes of hunting and gathering as well, in the gathering of seed, for example, or in hunting to acquire animals for restoration, or to substitute for the role extirpated predators play in the control of prey species.

What this adds up to is the realization that, viewed as an act or performance (rather than merely as a process or means of production), restoration is a form of reenactment. What it reenacts is the history of our relationship with nature and with a particular landscape. And it does this at several levels, or on several time-scales: first, the level of the individual and his culture, as the restorationist seeks to undo the consequences of its influence on the landscape; and second, the level of the species itself, through a recapitulation of all the major phases of cultural evolution and their characteristic ways of perceiving and interacting with nature.

This provides a second clue to the value of restoration as a way of negotiating a satisfactory relationship between ourselves and the natural (other-than-human) landscape: it gives us a way of dealing with the problem of history and cultural change as a component of that relationship. This is a component of the relationship with nature that is peculiarly human, yet it is crucial if our relationship is to be anything other than superficial and sentimental – if, that is, it is to be complete, natural, and ecological.

With this in mind, let us consider for a moment how restoration works as the basis for a healthy relationship between human beings and nature. To begin with, to have a relationship with anything you need the thing itself, in this case the classic ecosystems of the natural or archaic landscape. Restoration offers this in the form of the restored system, conceived now not so much as the product of a process, but almost as a by-product, the souvenir of a journey in time. Second, we need a working relationship with that landscape that exercises all our abilities, but that is practical and ecological – that is, that entails an exchange of goods and services. Restoration provides this by admitting us into the ecosystem as participants, actual members of the land community with business to carry out there, business that is distinctly and fully human, but that tends to the continual "improvement" (or renaturalizing) of the ecosystem. It also provides a way of integrating productive and progressive human technologies such as agriculture into the natural economy, literally joining economy with ecology, by providing ways to recycle back and forth between the two. In this way Illinois could become once again, say, 20% prairie in a continual rotation with corn and soybeans, and the farmer-restorationist could become once again an honorable member of the land community.

Third, our relationship with nature must be historical and progressive. It simply must take account of both past and ongoing changes in our relationship with the natural landscape. As noted above, restoration, precisely by attempting to reverse the effects of such change, offers a powerful way of exploring it and becoming aware of its implications. It does this not by attempting to freeze-frame the past, but precisely by doing what needs to be done actively to sustain the system against the pressure of changing conditions.

This idea rests on the assumption that conservation of natural landscape in a world inhabited by human beings will ultimately depend on human understanding of that landscape. It is this that brings us to the essence of what restoration has to offer conservation: a way of bringing the classical ecosystems both into human understanding and within human regard, so that nature and culture can ultimately co-exist.

The key here is essentially the transcription of the wisdom of the landscape from chemical (DNA) form to the electronic form that mediates cultural evolution. This makes the two compatible, opening up at least the possibility of their co-existence. Restoration, moreover, contributes to this process in two ways: first, by contributing to our understanding of the classic ecosystems (transcription); and second, by serving as a technique for propagating them into the future (translation).

Finally, as a self-conscious species, dependent on the use of language, we need ways to articulate, and even celebrate, our evolving relationship with nature. This is where we come to what I believe will prove to be the most important aspect of ecological restoration: its development as a performing art, a framework for the development of rituals and liturgies concerned with our reinhabitation of wild nature.

It is precisely here that we reach the critical problem that has prevented modern environmentalism not only from recognizing the potential of restoration as a technique for the reintegration of human beings and nature, but even from envisioning a relationship of any kind other than the abstemious, self-effacing, and attenuated relationship of the nonparticipating observer. The problem is precisely that this relationship cannot be achieved outside the ritual dimension, and so it is essentially inaccessible to a culture like ours with its diminished sense of the efficacy of ritual.

This, it may well be, is the real root of our environmental problems – not just the increase in the distance between ourselves and archaic nature that took place as a result of the scientific and technological revolutions of the past four centuries, but the fact that, just as these revolutions occurred, Western civilization began systematically discrediting the very technique (ritual) that had always provided the means for negotiating a satisfactory relationship with nature (Roszak 1978).

In my view, the act of restoration has at least three distinctive advantages as a scaffolding for the development of a tradition of rituals for this purpose. The first is that it is practically as well as ritually effective – that is, it actually brings about an improvement in the landscape and establishes the basis for a reciprocal relationship with it. The second is that it provides for the exercise of the full range of human abilities, including those dependent on culture.

The third is that it provides a means of entry into the ritual dimension based on activities such as gardening and hunting that are already conventional, and even popular, in our society.

What it does, in fact, is integrate more or less traditional activities into a larger, transcendent activity, the purpose of which is not merely recreation, entertainment, or even personal fulfillment, but the healing and reinhabitation of the natural landscape.

This, finally, is the ultimate lesson of a restorative event such as the prescribed burning of a prairie. It announces what Frederick Turner has called our emancipation from the role of mere observers of nature (Turner 1988). It dramatizes the fact that dependence really is mutual, and participation both possible and necessary. Hence the excitement – even the joy – that surrounds the burning of the prairies in the Midwest each spring, with its bright promise for the future of the natural world.

REFERENCES

Curtis, J.T. and M.L. Partch (1948). Effect of fire on the competition between bluegrass and certain prairie plants. *American Midland Naturalist* **39(2):** 437–443.

DeWitt (1991). The restorability of ecosystems. *Restoration and Management Notes* (in press).

Eiseley, L. (1970) The last magician. In.: *The Invisible Pyramid.* New York: Charles Scribners Sons.

Jordan, W.R. III, M.E. Gilpin, and J.D. Aber (1987), Restoration ecology: Ecological restoration as a technique for basic research. In: *Restoration Ecology: A Synthetic Approach to Ecological Research,* eds. W.R. Jordan III, M.E. Gilpin, and J.D. Aber. Cambridge: Cambridge University Press

Jordan, W.R. III (1986) Restoration and the reentry of nature. *Restoration and Management Notes* **4(1):** 2; and other editorials in recent issues of *Restoration and Management Notes.*

Kline, V.M. and G. Cottam (1989). The University of Wisconsin Arboretum: Restoring biotic communities. *Wisconsin Academy Review,* December 1989.

Kusler and Kentula (1990). *Wetland Creation and Restoration: The Status of the Science. Executive Summary.* Washington, D.C. and Covelo, CA: Island Press.

Miller, R.M. and J.D. Jastrow (1990), Hierarchy of restoration and mycorrhizal fungi interactions with soil aggregates. *Soil Biology and Biochemistry* **22:** 579–584.

Packard, S. (1990). S. Packard, The Nature Conservancy of Illinois, 79 W. Monroe Street, Chicago, IL 60603. Personal communication.

Panzer, R. (1988). Managing prairie remnants for insect conservation. *Natural Areas Journal* **8(2):** 83–90.

Powers, J. (1987). Restoration practice raises questions. In: *Restoration Ecology: A Synthetic Approach to Ecological Research,* eds. W.R. Jordan III, M.E. Gilpin, and J.D. Aber, pp 85–87. Cambridge: Cambridge University Press.

Roszak, T. (1978). The sin of idolatry. In: *Where the Wasteland Ends,* chapter 4. Garden City, NY: Doubleday & Co., Inc.

Swink, F. and G. Wilhelm (1979). *Plants of the Chicago Region.* Lisle, IL: The Morton Arboretum.

Turner, F. (1988). A field guide to the synthetic landscape. *Harper's Magazine,* April 1988: 49–55.

Zedler and Langis (1990), Urban wetland restoration: A San Diego Bay example. In: *Country in the City Symposium,* ed. M. Houck. Portland, OR: Audubon Society of Portland (Oregon) (in press).

COMMENTS AND QUESTIONS RELATED TO ECOLOGICAL RESTORATION AND ITS IMPLICATIONS FOR THE THEMES OF THIS CONFERENCE:

1. As a challenge that transcends the boundaries of traditional disciplines, restoration offers an arena of activity in which to develop more effective and more fruitful relationships between disciplines, between theorists and practitioners and between professionals and amateurs. Indeed the business of restoration has generally been pioneered by dedicated amateurs, and some of the best work in this area is still done by amateurs.

2. Biologist Peter Medawar has criticized the impersonal style of reporting scientific research that has become conventional during the past few generations as anti-historical and even deliberately misleading. At the very least this style serves to obscure the process by which science is conducted and the origins of its ideas. In addition, its compulsive use of the passive voice reflects a conception of science as an impersonal, objective activity that is no longer accepted by historians and philosophers of science.

The idea of "restoration ecology" or restoration as a technique for basic research suggests a return to a more personal, narrative style. Since it suggests that research is basically the unfolding of a kind of story – the story of an attempt to puzzle a complex system back together again – then the best and most honest way to report it is in a personal, narrative manner.

3. This, in turn has obvious implications for the relationship between scientists and society, since a more straightforward, honest and historically accurate style would inevitably lead to better communication with nonspecialists.

4. In much the same way, restoration suggests the basis for a more egalitarian form of environmentalism – one as concerned with environmental junk-picking as with the preservation of selected environmental "crown jewels" that can be practiced in inner cities as well as in remote, pristine natural areas, and that suggests ways of using nature intensively without using it up.

5. The heuristic value of restoration, evident in restoration ecology, also has important implications for education. Restoration is proving a powerful way to teach people about the environment and their relationship with it. This is the basis for "Earthkeeping", a program being developed by the Society for Ecological Restoration and the University of Wisconsin-Madison Arboretum to provide opportunities for people to participate in restoration projects as a way of learning about nature and their realtionship with it.

Bioscience ⇌ Society
edited by D.J. Roy, B.E. Wynne, and R.W. Old, pp. 163–182
© 1991, John Wiley & Sons

Human Population Growth and Ecological Integrity

Stuart L. Pimm

Department of Zoology and
Graduate Program of Ecology
The University of Tennessee
Knoxville, TN 37996, U.S.A.
and
Centre for Population Biology
Imperial College at Silwood Park
Ascot, Berkshire, SL5 7PY, U.K.

Abstract. Scientific advances are at least partly responsible for the historically faster than exponential growth of the human population. Such growth cannot be sustained, and the evidence points to substantial ecological damage in the near future and a consequent rapid slowing of the human population. A major shift for science must be to address how it can mitigate this ecological damage. I review ideas on how the environment responds to the damages we inflict upon it. Ecologists are used to the idea of finite, connected systems. Changes to the population dynamics of one population generally have consequences for the dynamics of other populations, sometimes ones that are remote in space and time. For example, by greatly reducing infant mortality in developing countries, one would expect to have far-reaching impacts on adult human mortality in those countries and world-wide changes in the transmission of human diseases. Those who do not expect ecological consequences from ecological actions appear to be merely badly informed, rather than following a plausible hypothesis of expected no effects. Yet, equally, the environment is not so fragile that all human impacts are devastating, nor do all effects magnify as they spread through the ecosystem. We must understand what we are doing to the ecosystem, predict what the effects will be, and establish priorities for minimizing impacts. A major human impact is to simplify nature; there is no simple balance to nature that leads to restoration of the original complexity. Rather, simplification has a wide range of complicated, though not incomprehensible, effects on the dynamics of ecological communities and their constituent species.

THE CHANGING COVENANT

For an ecologist, the question of whether a profound change is imminent in the relationship between science and society has an almost trivial answer. The worldwide rate of human population growth, at least until very recently, has been faster than exponential. It has been estimated that the human population grew at an equivalent rate of a doubling per three thousand years during the millenium ending in 500 A.D., was doubling every 700 years in the millenium ending 1500 A.D., in this century was doubling every 56 years, and is now doubling every 35 years (Baxter 1988). Science has made it possible for humans to live longer and for children to have much better chances of surviving into adulthood. Our population has grown to over five billion, and it is certain to continue to grow within our lifetimes. Yet the population cannot continue to grow exponentially. Estimates, often very hopeful estimates, of final equilibrium numbers are generally in the order of ten to twenty billion people (Ehrlich and Ehrlich 1990).

The qualitatively new relationship between science and society that emerges from these arguments is that while science has in the past promoted exponential population growth, in the future it must tackle the consequences of this growth and the limits that will be placed upon it. Science has taken the admonition of Genesis I,28 very seriously: we have been fruitful and multiplied, have subdued the earth, and have dominion over every living thing. In the future we shall have to pay particular attention to the missing advice, that is, to replenish the earth.

There is compelling evidence that the limits to growth will act relatively quickly. The five billion or so humans on this planet are already felling large parts of (particularly rain-) forest, and very widely fragmenting natural ecosystems. We are losing the "trees" – the individual species, some of economic importance (tropical plants are the source of many major drugs and novel genes for disease resistance in crop plants, for example), and some of intrinsic beauty or interest (who would want to tell their children that the rhinos, elephants, and tigers of their story books were extinct?). The catastrophic loss of species expected in the next several decades is predicted to rival that at the Cretaceous-Tertiary boundary (Ehrlich and Ehrlich 1981).

We are also losing the "forest", the complex of many species that provide us with essential ecosystem services, as we change the planet's climate, dumping vast amounts of pollutants into the air, as well as into its rivers, lakes, and oceans. The five billion of us are already utilizing about 40% of the annual production of all terrestrial plants for our food, the food of our domestic animals, and for fiber products (Vitousek et al. 1986). This 40% is the "annual interest" in our "biological bank account"; we cannot spend more than 100% without starting to deplete our "capital". Indeed, there is abundant evidence that we are already depleting our capital: soils erode, finite ground water supplies are mined to water the crops, fisheries collapse, and finite fossil fuels are mined to produce the pesticides that increase yields (Ehrlich and Ehrlich 1990).

The critics of doom

There are a number of criticisms of this scenario of impending doom. People have argued that crop yields are increasing, and indeed they are. The problem is a matter of scaling, for every improvement has to be viewed relative to the scale of the population increase. On such a relative scale, advances in food production are no more than a slight tweaking of yields, and are far below what is required to match the population growth.

But why should we not continue to destroy areas, like tropical forests, so that we can produce more food? Why should we not destroy the "biological capital"? One justification for the depletion of "capital" is that it may represent the optimal economic strategy. We may destroy the whale or fishery stocks, or deplete the soil or the ground water supplies in order to gain greater economic wealth than if we exploited them sustainably. Then, goes the argument, this wealth can be reinvested in another exploitive venture: harvest the whales to extinction, then invest the profits into large drift nets to exploit other marine species. Similarly, why not substitute new resources for old ones? So, fell the forests for fuel; when they are gone we can burn fossil fuels, and when they are expended we can turn to nuclear power. We have some trees; should we worry that many different species of trees are being lost in the tropics?

What society has done for centuries is to make two assumptions about our interactions with the natural environment. The first assumption is that resources are effectively infinite; we have assumed that the air, lakes, and oceans are so large that we cannot seriously pollute them. We have assumed ecosystem services to be free; we have assumed that plants will always absorb all the carbon dioxide that we dump into the atmosphere. When we destroy a natural resource, we expect that there will be always other resources that we can exploit in its place. Exponential population growth inevitably falsifies this assumption of infinity. At some population size, there is too much pollution for the system to handle, and the effects are felt, first locally, and then globally. Or there is no longer a fishery that has not been exploited beyond the point of sustainability.

The second assumption is that we can always substitute one resource for whatever resource we destroy. The difficulty with the "trees, then coal, then nuclear" argument is that it views resources as being very simple things – energy, in my example. The resources that our society values have become progressively more complex. One may indeed be able to substitute one kind of tree for another if all one wants to do is burn it. One cannot substitute the bark of an oak tree for the bark of a cichona tree, if the aim is to make a potion that will prevent malaria. Similarly, all populations of the same species are not the same: some may contain the genes absent in other populations which confer resistance to assorted enemies. Simply, the species matter.

While in the past the important question was how to breed faster and live longer in an often hostile world, the new ecological question is how we can mitigate the damage our success has caused. How do we reduce the current levels of ecological damage within our generation, and what can we do to help the next generation, which will probably be twice as numerous as ours and have even greater aspirations for its living standards?

INTEGRITY, STABILITY, FRAGILITY, AND THE BALANCE OF NATURE

The central questions I wish to address in this section are: what are we doing to the environment, what is being changed, and how will those changes affect us? The preceding questions might be rephrased by asking: in what ways do we threaten ecological *integrity?* In listening to both other ecologists and the general public I hear other terms: the idea that the environment is extremely *fragile,* or that ecosystems have some capacity to resist change or to recover from it, that is, there is some *balance of nature* or *stability.* So what do all these terms mean, and what scientific principles do they encompass?

Perhaps because the word "ecology" has two meanings in English (the science, and the political activities that promote environmental concern), there seem to be a number of hypotheses about the nature of the environment that lie outside the science's boundaries. I do not find it easy to discuss these hypotheses, and perhaps they are straw men; but I will start with them so that I can then move on to more substantial ideas.

"The world is infinite" or "the world is fragile"

I am often amazed at the surprise others experience in finding that the world is both finite and connected. Ecologists *expect* that there will be consequences of changes to the dynamics of our own or other species. Where those consequences will be, when they will occur, and how severe they will be, is the very stuff of ecology. For example, the availability of simple means to reduce infant mortality dramatically in developing countries can no longer be viewed simply as a major advance, when its consequences may be even more deaths from starvation or disease when those infants become adults. This topic is discussed at length in Sharp et al. (this volume), yet time-delayed mortality is familiar ground for the ecologist. Ecologists also recognise that, as our species becomes more numerous, it will likely acquire infectious diseases that would not have been able to spread when our numbers were low. (Our species has gained many infectious diseases historically [Dobson and May 1986]. HIV may be the latest in the series, but it is hardly likely to be the last, given our population growth.) As Sharp et al. (this volume) point out, we can only be incredulous when politicians withdraw support for population controls, for there will be consequences, and they are likely to be severe.

At the other extreme is the "flower-child" philosophy exemplified by the idea that picking a flower may lead to the collapse of the entire forest ecosystem. Certainly, one of the features of dynamical chaos is that small differences between systems grow exponentially over time. This is the so-called *butterfly effect:* theoretically, the flap of a butterfly's wing can alter the path of a hurricane on the far side of the world (Gleick 1988). What this tells us is that ecological impacts *can* be felt far from their initial causes; it does not mean that effects *will always* travel far, or amplify as they do so. (Examples of amplification might include the catalytic destruction of atmospheric ozone by CFCs, the concentration of substances like DDT in food chains, or the extinction cascades which I discuss later.) Ecosystems can be fragile, and sometimes they can put up with remarkable abuse; we need to know how to predict which. Arguing that ecosystems are always fragile is disabling. It obviates the need to make decisions that might mitigate severe damage.

From homeostasis to Gaia

Another popular idea is that ecosystems possess a natural homeostasis that will eventually lead to their restoration following damage. Again, there is some merit to this idea, but there is a danger of drawing inappropriate analogies. At one end of the spectrum of possible time scales for this restoration are the physiological processes that insure individual organisms remain relatively constant internally, despite considerable environmental fluctuations. This homeostasis is compelling, because it is so readily observable, but its implications for longer term processes are far from clear. Over longer scales, it is not the individuals that concern us, but the fate of entire populations, communities, and ecosystems. On such scales, the failure of this homeostasis – death – is a frequent occurrence. While I, as an individual, can happily walk outside and experience a temperature drop of 15°C, if the long-term average temperature outside were merely 5°C lower, the view from my window would not be of deciduous trees, but of the edge of the permanent ice-sheet behind the dwarf birches and spruce.

Relevant to these longer time scales is the Gaia hypothesis (Lovelock 1982). I shall not discuss this at any length, in part because I find it difficult to decide which scientifically credible hypotheses belong under its protective umbrella and which ideas come from the lunatic fringe of science and are taking a ride as stowaways. Gaia clearly deals with "the balance of nature", but on a spatial scale (the entire planet), and on temporal scales much larger than my concerns. While global processes exist that will offset the catastrophic loss of species we are now experiencing, they are extremely slow. For example, from 50% to 100% of the vertebrate species have been lost from some Pacific islands in the last 10–100 years (Pimm 1991). Speciation may restore the original diversity, but over thousands to millions of years. Similarly, while global warming may be somewhat limited by Gaia, it is not limited enough to prevent it from being a major concern for conservation biology.

In sum, if one thinks of the planet as a giant organism and so inevitably homeostatic, there is the danger of thinking that this homeostasis will somehow mitigate the environmental catastrophes we now face. The reality is that the important aspects of ecological stability lie on scales between individual homeostasis and global Gaia. What lies between these extremes is more complex and far more important practically.

A HISTORY OF ECOLOGICAL STABILITY

The environment responds in different ways to the damage we cause it, and environments differ in their susceptibility to the damage. There is no simple solution to the question of what will be the consequences of our increased impact on the environment. We cannot avoid finding out about the way an ecosystem works. Ecological stability has a history, and I want to review it briefly in order to set the context of what follows.

In his 1958 book, *The Ecology of Invasions by Animals and Plants,* Elton wrote:

> I will now try and set out some of the evidence that the balance of relatively simple communities of plants and animals is more easily upset than that of richer ones; that [the balance] is more subject to destructive oscillations in populations, especially of animals, and more vulnerable to invasions . . .
>
> For if this can be shown to be anywhere near the truth, it will have to be admitted that there is something very dangerous about handling cultivated land as we handle it now, and [it will be] even more dangerous if we continue down the present road of "simplification for efficiency".

Elton then discuses the evidence for his assertion that simple communities are less stable than complex ones. His first evidence comes from "mathematical speculation about population dynamics". Simple models of a predator and its prey, or a single host and its parasite, have population densities that "fluctuate in numbers considerably". Such fluctuations would mean that in simple communities populations would "never have constant population levels, but would be subject to periodic 'outbreaks' of each species". Elton's second line of evidence is from laboratory experiments that made the same point. The third piece of evidence is that the natural habitats on small islands seem to be much more vulnerable to invading species than those of the continents. His fourth argument is that "invasions and pest outbreaks most often occur on cultivated or planted land – that is, habitats and communities very much simplified by man." His fifth argument is similar. Insect outbreaks, Elton claims, are a feature of simple temperate forests but not of complex tropical forests. His sixth argument seems to recapitulate the earlier ones: insect pests are more likely to invade and become a nuisance in orchards, especially following spraying designed to eliminate the pests.

Elton's arguments have received extensive discussion. The models and the laboratory experiments available to Elton looked only at simple systems: models of, or experiments with, complex systems were not available as a control. The data on invasions were compelling, but comparative studies of population fluctuations were nonexistent. Moreover, Elton discusses a range of very different phenomena: population variability, the potential abundance of a species, the rate at which a pest species might increase in numbers after removal of its predators, how readily communities may be invaded, and the extent of damage the invading species cause. Yet this is perhaps the most remarkable part of Elton's argument. Ecologists should be thinking about different kinds of stability: population variability, population recovery, the ease of invasion, and the consequences of invasion. All are ways of characterizing population and community changes to facilitate comparisons. Moreover, Elton considers a factor called "complexity" that should determine "stability" and the importance of their interdependence. Notice something else: the effect of humans is to destroy stability, and with the loss of stability comes the loss of species, and so less complexity. Systems that are less complex are less stable, more likely to lose species and so on.

Controversy

The relationships between stability and complexity have generated a large literature (Pimm 1984, 1982). Controversy arose because of the many meanings of "stability" and "complexity" (Pimm 1984). Elton and more recent theoretical and empirical ecologists have used the word "stability" to mean at least five different things (Fig. 1). Theoreticians have considered stability in its strict mathematical sense, where systems are either stable or not. So, given May's (1974) argument that what we observe most often in nature will be stable systems, the theoretical results make predictions about *the kind of communities we observe and those that we do not.* The other four definitions of stability do permit comparisons of different existing systems, however, and these have been the concern of empirical ecologists.

To further complicate matters, different studies looked at "stabilities" at different levels of ecological organization. Theoretical studies of whether interacting sets of species will be stable consider whether the densities of all those species return to equilibrium or not. To ask how long a community persists before it is invaded is to ask how long community composition lasts. Such a discussion ignores fluctuations in the abundance of species, and looks only at the species list itself. In contrast, McNaughton (1977) placed his emphasis on how much total biomass changed, and not on which species were involved in the changes. Therefore we can discuss the stability of populations, community composition, or community biomass. The ecological literature also tried to relate the different definitions of stability to different features

Fig. 1. The meanings of stability:

A system is considered to be *stable* in the mathematical sense if, and only if, the variables all return to equilibrium conditions following displacement from them. By definition, either a system is stable or it is not.

Resilience is defined as how fast a variable returns to equilibrium once displaced from it. Resilience could be estimted by a return time, the amount of time taken for the displacement to decay to some specified fraction of its initial value. Long return times mean low resilience and vice versa. Resilience is measured as a rate of change.

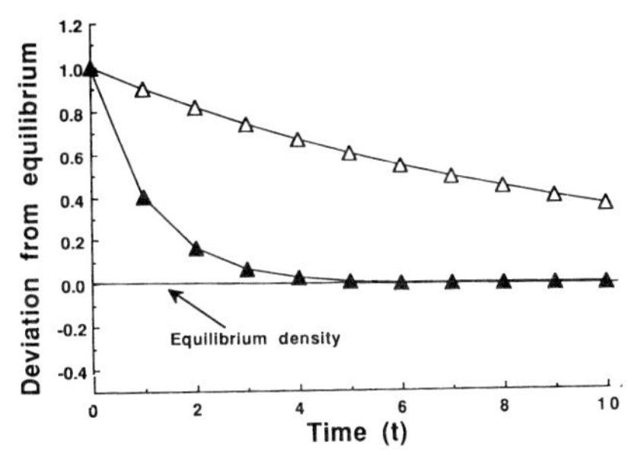

For example, here are two model populations returning to equilibrium density (set at a reference level of zero) from the same initial displacement at +0.5. One population (open triangles) is much less resilient than the other (closed triangles). The more resilient population returns to equilibrium much faster than the other.

Variability is the degree to which a variable varies over time. It will be measured by such statistics as the standard deviation or coefficient of variation of consecutive measurements of those variables that interest us.

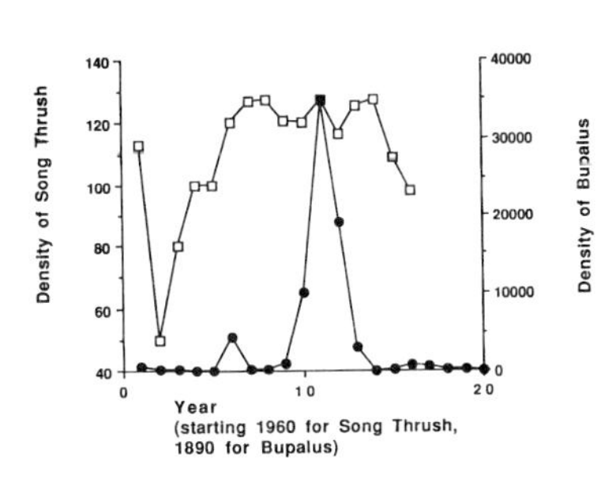

The figure shows how different year-to-year variabilities can be for different species: the song thrush in English farmlands (units of density are relative and scaled to one hundred in 1966) and the densities of the moth *Bupalus* in German forests (numbers are per hectare of forest floor). While the song thrush experiences a two-fold variation in density over about twenty years, *Bupalus* densities vary over nearly four orders of magnitude.

Persistence is how long a variable lasts before it is changed to a new value. Systems that change often could be described as having high turnover, so turnover would be the reciprocal of persistence. Persistence is measured as time.

Resistance measures the consequences when a variable is permanently changed: how much do other variables change as a consequence? If the consequent changes are small, the system is relatively resistant. Resistance is measured as a ratio of a variable before and after the change and so is dimensionless.

of community structure, "complexity" in particular. Complexity has been taken variously to mean the number of species, the degree of interspecific connectance, and the relative abundances of a species in the community.

Controversial though Elton's ideas may be, what he does so effectively is to concentrate our minds on ecosystem structure, the influence of that structure on the dynamics of the ecological community and its constituent species, and the overriding impact of our species, which is to simplify that structure. What Elton provides is a prescription for research into ecological stability; I will finish by considering some of the recent advances in this area.

THE STABILITY OF POPULATIONS

What happens to the dynamics of populations when we simplify the environment, either by causing species to become extinct, or by replacing complex, natural communities with simpler, artificial communities such as crops, orchards, or tree farms? Two ways of characterizing population stability are resilience and variability (Fig. 1), and both may be affected by such simplification.

Resilience

The example of pest outbreaks. A high resilience means a quick return to normal conditions following some disturbance. For species that we deem desirable, we equate high resilience with "stability". There is an interesting paradox, however. Consider an insect herbivore population increasing toward a high density that involves far more pests and far too few plants for our liking. A species with a resilient population will reach unacceptably high levels faster than a species that is not resilient. Resilience can thus be a measure of *instability* and also a measure of how often we may have to control the pest by chemical means (Fig. 2).

In a general way, it is their high resilience that makes insects such pests. Across animal species from aphids to elephants, small body size is coincident with high population growth rates, and insects are small. The resilience of a population, however, depends on more than just features of the species itself. It depends on the ecosystem in which that species is embedded. When we disturb the densities of a set of interacting species, no species' density can return to its normal value until all the others have done so. The resilience of each of the species in the community will depend on the least resilient species.

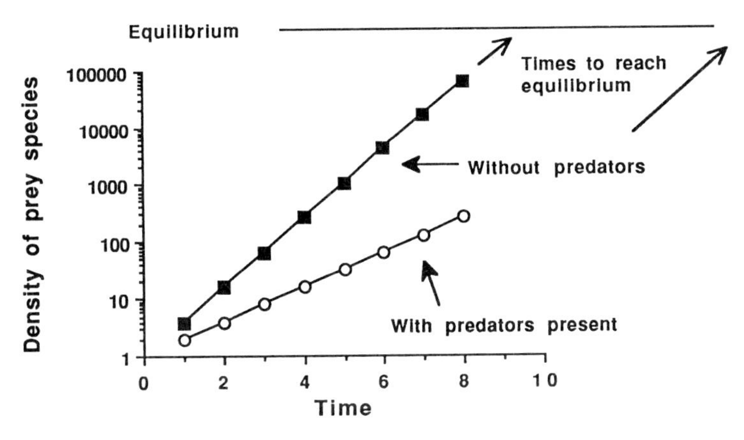

Fig. 2. Resilience and simplication. Systems may be more resilient when they are simplified. Consider, for example, a pest species. Pest outbreaks will occur faster when we simplify the systems (by removing the pest's predators) than they would in the presence of predators. (The final density of the pest will probably be higher, as well.) A way to evaluate these ideas may also give some insight into why insecticide applications often induce pest outbreaks. Insecticides clearly kill many pests, but many predators are often killed as well. The predators, moreover, cannot recover until the pests come back. This suggests that there will be a "time window" after insecticide application in which pests have very few predators and so, according to the theory, should increase more rapidly.

For example, suppose we reduce the density of phytoplankton in a lake. This will reduce the density of their zooplankton predators, and, eventually, the fish that feed on the zooplankton. The phytoplankton might have the capacity to return quickly to their former densities. But while the fish densities are recovering (slowly), zooplankton will be unusually abundant, and this will keep the density of the phytoplankton down. Simple systems, therefore, may be more resilient than complex systems. Systems may also be more resilient when they are simplified. Indeed, pest outbreaks occur faster after the pest populations had been treated with an insecticide, because the predators are removed as well, and take longer to recover (Fig. 2).

Year-to-year variability in densities

Highly variable species are more likely to occur at high enough densities for us to consider them pests, while for desirable species, high variability may lead to the species encountering such low densities that they become extinct. Like resilience, variability depends on a number of factors. Population variability will be greater in those ecosystems where critical environmental variables, such as rainfall in deserts, are themselves more variable. Variability depends on resilience, though the details of these relationships are very complicated (Pimm 1982, 1991). Variability also depends on the structure of the ecosystem in which the species occurs, and is altered when we simplify the system.

MacArthur (1955) was one of the first to argue that more complex communities were also more stable. He defined complexity as the number of pathways energy took to reach a given population and stability as the extent to which population density changed if one of those pathways failed. The argument is no more than the admonition "don't put all your eggs in one basket". If you drop your only basket, you will break all your eggs; if there is only one energy pathway, the energy supply is highly vulnerable. Watt (1964) produced a diametrically opposite argument. He argued that species with ready access to more resources would be *more* variable. With the economically important exceptions of species in monocultures (which have access to large areas of the same resources), populations of polyphagous species (those feeding on many species) will be more variable than monophagous species (those feeding on few species, scattered among many unpalatable species). Both MacArthur's and Watt's arguments are sketched in Figure 3, and there is field to support both of them (Redfearn and Pimm 1988; Pimm 1991).

What of predator diversity? Increased predator diversity ought to reduce the variability of their prey species, because a diversity of predators provide a more reliable control of their prey's density. A species preyed upon by many species of predator may be much less likely to escape the attention of those predators than a species exploited by only a very few species of predator. Increased predator diversity may reduce the likelihood of unusually high population densities. There is empirical support for this pattern in both insects and in mammals (Pimm 1991).

In sum, the structure of a system affects the population variability of its component species, but in complex ways. The more generalized species are in their diets, the more or less variable they may be. Even with Watt's argument, however, ecological simplification that leads to near monocultures will lead to more variability. The more predatory species a group of prey species suffers, the less variable it will be, so simplification leads to greater variability.

THE PERSISTENCE OF COMMUNITIES

How long community composition lasts depends on how fast the community is losing species and how fast it is gaining species. Species that gain and lose species the most slowly will be the most persistent communities (Fig. 1). We must understand both the processes of extinction and invasion.

Extinction

Rare species are particularly prone to extinction. One of the principal causes of extinction is that we either totally destroy species' habitats, or leave fragments behind which house too few individuals for each localized population to persist for long. There are many other factors that affect the rate of extinction, however. For example, species whose population sizes fluctuate greatly should be more likely to become extinct than those that vary little from year to year.

Invasions

Why do communities gain species at different rates? Physical features of the environment – such as how isolated the habitat is – play a part in determining the rate of invasion. Species differ greatly in their individual abilities to reach communities, and many species may fail because they land in places where the environment is unsuitable for them. Even when a few colonists find a potentially suitable environment, they may fail for one of the many reasons why very small populations become extinct.

Persistence depends on features of the community as well, and is likely to be altered significantly when the community is simplified. Models of how complex communities assemble from simple communities show a number of features (Fig. 4). Communities are easier to invade when they have few species, early in the assembly process, and when they have simple patterns of interspecific interactions. Moreover, model communities become progressively harder to invade as assembly proceeds, even when they are not changing their average numbers of species or connectance. Assembly itself increases community persistence – therefore, "old" communities are harder to invade than "new" ones.

When we destroy communities, we substitute communities that are easily invaded by other, often alien species, much as Elton (1958) suggested. Testing to see whether "old" communities are easier to invade than "new" ones is

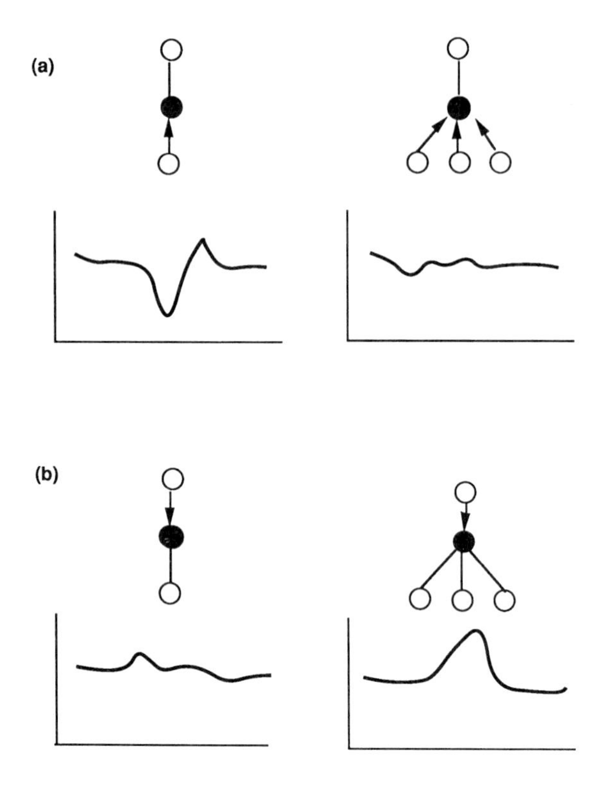

difficult. Ecologists readily accept that disturbed communities are more readily invaded than undisturbed ones. While "disturbed" means several things, one meaning uses "disturbed" to describe communities that are recent assemblages of introduced species. In Hawaii, relatively species-poor upland native (hence "old") bird communities appear harder to invade than the relatively species-rich lowland communities composed of largely introduced species (hence "new" communities) (Pimm 1991).

A ubiquitous feature of community assembly models is the existence of alternative persistent states – different combinations of species, each of which is persistent in the presence of the other combinations. In the laboratory, these alternative communities seem very easy to obtain, and there is abundant, if circumstantial, evidence of their existence from the field (Drake 1990). Another feature of the models is that it may not always be possible to create persistent communities from just their constituent species: other species (catalysts, if you will), that become extinct in the process, may be needed. There is

Fig. 3 (facing page). Two opposing arguments for the possible relationships between population variability and the degree of polyphagy. (a) MacArthur suggested that monophagous species (left) should be more variable than polyphagous species (right), because they will decrease most when one prey species declines in abundance. (b) Watt, in contrast, argued that polyphagous species may be more variable, because they may be able to reach higher numbers during periods when their predators fail to control their densities.

Circles represent species, and lines represent trophic interactions between species, such that species higher on the page feed on species beneath them. The species for which we measure the population variability is shaded. Arrows indicate which trophic interactions are most important in determining the density of the species of interest.

Compare two predatory species, one monophagous, the other polyphagous. These "predators" may be herbivores feeding on their plant "prey" without any change in the argument. In the kind of complex system envisioned by MacArthur, each predator would feed on many different species of prey, with much overlap between predators. In a simple system, the food chains would be more or less distinct. The specialized (monophagous) species is likely to vary more than the more generalized (polyphagous) species. If one prey species of the polyphagous predator becomes rare, then individuals can switch to alternative prey. In contrast, the monophagous predator would become rare if it were deprived of its sole species of prey.

In complete contrast, Watt (1964) argued that polyphageous species could be *more* variable than monophagous species. He suggested that species typically may be kept at levels below those set solely by the availability of their food supply, by the attentions of their predators, parasites, diseases, competitors, or generally inclement weather. Now suppose that occasionally these various enemies are themselves rare, or that the weather, for once, is favorable to the prey species we are considering. What will determine how abundant the species will become before the population declines to a more usual level? Two parameters seem preeminent: how fast the population can increase, i.e., its resilience, and how much of the habitat the species can exploit. What affects how much of the habitat a species can exploit? A monophagous species can only exploit a large proportion of its habitat, if the habitat is a monoculture of its food supply. A polyphagous species is likely to be able to exploit a large proportion of its habitat because of its diverse feeding habits. Therefore, populations that reach the highest relative densities will be *either* monophages in monoculture, or polyphages.

a *Humpty Dumpty* effect: even with all the pieces we cannot put the community back together again. If we are to restore damaged communities, we need to evaluate the importance of alternative persistent states that may prevent us from obtaining the desired persistent state. If there are "Humpty Dumpty" effects, then when we destroy communities, we may not be able to reassemble them, even if we have kept all the species alive in botanical gardens and zoos.

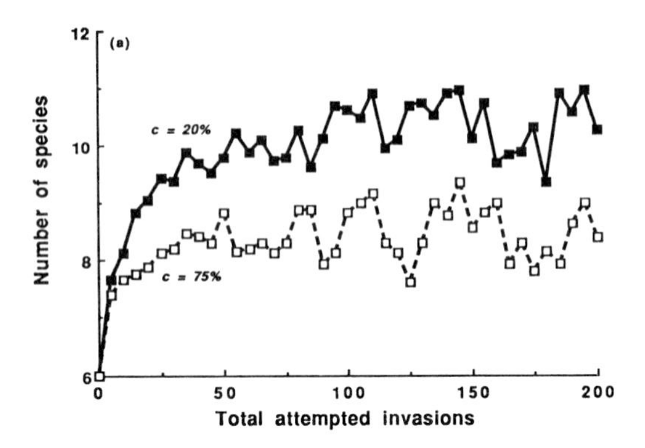

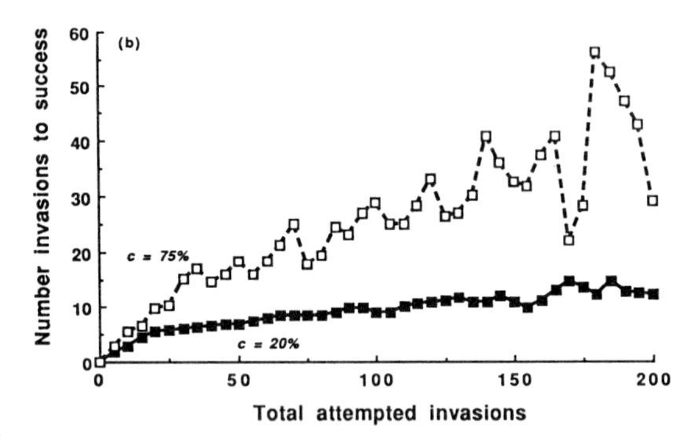

Fig. 4. Sample results from simulations of community assembly, as generated by a recipe that starts with a fixed number of species, then tries to add species to a community. Some species make it into the community, and they cause other species to become extinct when they do. Many species will fail to enter. At top, the number of species present increases quickly, and then tends to level off as the number of total attempted invasions accumulates. At bottom, the number of invasions attempted before one is successful continues to rise. The two sets of results differ in how connected the food web models are. Models that have a high proportion of their species interacting for the number of species present (dashed lines, open squares, connectance = 75%) tend to have fewer species, but are harder to invade, than those models which have a smaller proportion of the possible interspecies interactions (solid lines, solid squares, connectance = 20%). From Pimm (1991), based on an extensive set of simulations in Post and Pimm (1983).

In sum, simple communities are likely to be less persistent than complex ones, but simplified communities are very unlikely to persist – they are both simple and the product of a short assembly process. Complex communities, the end product of a long assembly are likely to be the most persistent. We may not be able to reassemble damaged communities. When we try to restore communities, we may end up with another type of community, either because the sequence of the assembly matters or because there may have been species, now extinct, that were essential catalysts in forming the original community.

RESISTANCE TO CHANGE

Species extinctions are caused by habitat losses and other factors. In addition, a major cause of extinctions is deliberately, or accidently, introducing alien species. Extinctions and introductions pose obvious questions involving resistance. Which invasions will cause most damage, and which species in a community are the "keystone" species, that is, those species which when lost, will cause the most *secondary* extinctions?

Figure 5 suggests that complex communities should be most sensitive to the loss of species from the top of the food web, because secondary extinctions propagate more widely in complex than in simple communities. In contrast, simple communities should be more sensitive to the loss of plant species than complex communities, because in simple communities the consumers are dependent on only a few species, and cannot survive their loss. Similarly, the introduction of trophically generalized species should have profound effects on community composition because such species can eliminate many of the species on which they feed, while introductions of specialized species should effect fewer changes.

There are examples that support these ideas. There were few extinctions, if any, following the loss of the chestnut tree from North America, yet many bird extinctions in Hawaii seem to follow the loss of a few "keystone" plant species. Presumably, the island communities had simpler food chains, with more trophically specialized species, and so were more vulnerable. Introductions of trophically generalized mammals to oceanic islands, where the communities are relatively species-poor, have had profound effects in a majority of cases; effects on introductions to continents are less noticeable, probably because there the introduced mammals were controlled by native enemies. In contrast to mammals, bird introductions are rarely reported as having effects. Birds are generally introduced for the purpose of brightening our lives, rather

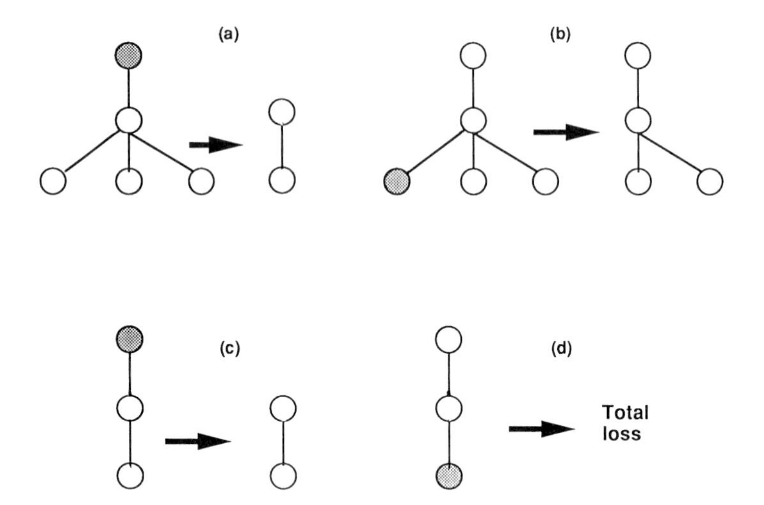

Fig. 5. The effects of species removals depend on the complexity of the food web and from where the species is removed: a) a predator lost from a complex web may release its prey, the herbivore, and so affect the loss of several species of plant; b) in contrast, a plant lost from a more complex web may have little effect, because the herbivore has alternative food sources; c) a predator removed from a simple food chain may have no effect (because the herbivore cannot eliminate its only food supply); while d) a plant removed from a simple food chain causes the entire food chain to be lost. (From Pimm 1991)

than as a source of meat. Consequently, birds are introduced into close proximity to us and thus placed into highly-modified, species-poor habitats where we should not expect them to have noticeable effects (Pimm 1991).

In sum, systems differ in how resistant they are to the introduction of alien species, or the loss of native species. Resistance depends on whether the system is simple or complex, and from where species are lost, or to where they are introduced. Often the consequences of the loss of one species are the loss of others.

CONCLUSIONS

Our effects on the environment are complicated but not incomprehensible. Certainly, when we destroy species, the simplified communities may more quickly gain new species, but these species may be undesirable, weedy species, the kind of tramp species that are human commensals worldwide. In creating communities that are "new" and not the product of a long process of community assembly, we may create communities that will persist for only short periods of time before new invasions will take place. Invading species may cause extinctions of native species. Communities are often not resistant to extinctions: one extinction may cause a cascade of extinctions, leading to a simple community. When we lose communities, there is a good chance that we will not be able to restore them, even if we have kept all the species in zoos and botanic gardens.

Similarly, species have the capacity to recover from losses, and that recovery appears to be faster in simpler systems. Sometimes we do not want fast recovery (e.g., pests in simple, species-poor agricultural systems). Simplification can also make populations more variable, leading to some species becoming extinct, others becoming pests. The properties of natural systems may sometimes mitigate the damage to which we subject them, but all too often that damage propagates widely.

Ackowledgements. I thank Mrs. Julia K. Pimm, Drs. D. Pauly and J.W. McManus, for comments on an earlier version of this paper, and Professor J.H. Lawton of the Centre for Population Biology, for his hospitality and use of facilities during my stay in England.

REFERENCES

Baxter, K. (1988). Future problems – Presidential address 1988. *Biologist* **35:** 173–178.

Dobson, A.P. and R.M. May (1986). Patterns of invasions by pathogens and parasites. In: *Ecology of Biological Invasions of North America and Hawaii,* eds. H. Mooney and J.A. Drake, pp. 58–76. Berlin: Springer-Verlag.

Drake, J.A. (1990). Communities as assembled structures: Do rules govern pattern? *Trends in Ecology and Evolution* **5:** 159–163.

Ehrlich, P.R. and A. Ehrlich. (1981). *Extinction: The causes and consequences of the disappearance of species.* New York: Random House.

Ehrlich, P.R. and A. Ehrlich. (1990). *The Population Explosion.* New York: Simon and Shuster.

Elton, C.S. (1958). *The Ecology of Invasions by Animals and Plants.* London: Chapman and Hall.

Lovelock, J. (1982). *Gaia: A New Look at Life on Earth.* Oxford: Oxford University Press.

MacArthur, R.H. (1955). Fluctuations of animal populations and a measure of community stability. *Ecology* **36:** 533-536.

May, R.M. (1974). *Stability and Complexity in Model Ecosystems. Second Edition.* Princeton, NJ: Princeton University Press.

McNaughton, S.J. (1977). Diversity and stability of ecological communities: a comment on the role of empiricism in ecology. *American Naturalist* **111:** 515–525.

Pimm, S.L. (1982). *Food Webs.* London: Chapman and Hall.

Pimm, S.L. (1984). The complexity and stability of ecosystems. *Nature* **307:** 321–326.

Pimm, S.L. (1991). *The Balance of Nature?* Chicago, IL: The University of Chicago Press.

Redfearn, A., and S.L. Pimm. Population variability and polyphagy in herbivorous insect communities. *Ecological Monographs* **58:** 39–55.

Post, W.M., and S.L. Pimm. (1983). Community assembly and food web stability. *Mathematical Biosciences* **64:** 169–192.

Vitousek, P.M, P.R. Ehrlich, A.H. Ehrlich, and P.A. Mateson (1986). Human appropriation of the products of photosynthesis. *Bioscience* **36:** 368–373.

Watt, K.E.F. (1964). Comments on long-term fluctuations of animal populations and measures of community stability. *Canadian Entomologist* **96:** 1434–1442.

Standing, left to right: Daniel Pauly, Michael Bernhard, Kenneth Hsü, Wolfgang Van den Daele, Peter Kafka, Horst Nöthel, William Jordan III
Seated, left to right: Robin Gordon, Stuart Pimm, Michael Crawley, Gary Sayler, Youssef Halim

Bioscience ⇌ Society
edited by D.J. Roy, B.E. Wynne, and R.W. Old, pp. 185–201
© 1991, John Wiley & Sons

Group Report
Does Bioscience Threaten Ecological Integrity?

Rapporteur: R.A. Gordon
M. Bernhard
M.J. Crawley
Y. Halim
K.J. Hsü
W.R. Jordan III
P. Kafka
H. Nöthel
D.M. Pauly
S.L. Pimm
G.S. Sayler
W. Van den Daele

INTRODUCTION

The condensed question in the title requires interpretation if it is to serve as the goal for a meaningful discussion. The first unclear term is "ecological integrity". Here the group addresses the question of how human action influences "nature" (1) and what we consider desirable (2). Before discussing possible threats posed by bioscience (4), the group attempted to sketch a perspective by describing the demands presently placed on our environment and how bioscience might reduce these (3).

ECOSYSTEMS AND HUMAN HABITAT: HOW CAN WE ASSESS HUMAN INFLUENCES?

Seen over a longer time scale, almost no part of the globe has escaped transformation by human influence. The only examples of comparatively undisturbed areas which readily come to mind are the deep sea and the Antarctic. Our question was meant to address relatively recent influence or relatively fast changes.

Assessment first requires tracing causal chains of effects back to human actions. For large parts of our immediate surroundings this is no problem, since they are so profoundly transformed by our purposeful actions as to be virtually artificial. Examples are our urban surroundings, highway systems, and most agricultural fields. For areas only partially transformed, like canalized river valleys or forests regrown after logging, influences are sometimes not so clear, but often identifiable (see box: ASWAN). In those areas not transformed by recent human actions (e.g., unlogged tropical forests and some tundras) subtle or distant changes may still have considerable, even irreversible effects.

ASWAN:

Since the High Aswan Dam became functional in 1965, the freshwater outflow to the Mediterranean with its usual load of nutrients has been reduced to some 6%. Egyptian marine fish landings dropped to about 25% of the predam level, and the Sardinella fisheries were even more drastically affected. The silt load is entirely trapped behind the dam. Shore erosion has become a problem, since it is no longer compensated for by the flood deposition of sand and silt. The loss in soil fertility in the Nile valley and delta is compensated for by intensified chemical fertilization.

However, any balance sheet of the positive and negative effects of the High Aswan Dam must include at least four major benefits: the magnitude of the hydropower generated; the doubling of the farmland area; the creation of a new fishery in Lake Nassar with catches similar to the last marine catches; and, no less crucial, the regulation of water supply to the country despite a decade of severe drought in the Sahel-Sudan-Ethiopian belt. Life in the Nile valley and delta is entirely dependent on the river as a water artery, rainfall being insignificant. The deficiency in the flood water supply necessitated a continuous emergency withdrawal of large volumes of water from the reservoir to protect the country from the catastrophic effects of drought. From 1978/79 to 1984/85, the total deficit compensated from the reservoir reached 73.5 km^3 (Halim 1990).

Three general approaches are often utilized to determine if a given ecosystem has been adversely influenced. One approach is aesthetic or visual and, while qualitative, is sufficient to discern large-scale changes or dramatic short-term effects. A second approach is quantitative determination of structural and functional changes in an ecosystem. This would be considered a professional ecological approach and is exemplified by measurement of species abundance, system robustness, quantified primary productivity, etc. The third approach could be called analytical and is best exemplified by chemical measurements of toxins, pollutants, or other environmental perturbants.

In ecosystems from which we harvest, a decrease in productivity (the ratio of harvest to input) is an indicator of stress. Even more sensitive may be an increase in the yearly fluctuation in harvest size. There are many cases (e.g., coral reefs and tropical forests) where systems will not return to their previous state when stress is relaxed (Goodland 1975).

In many cases, the influence is easy to assess but hard to convey to the public. Bioscientists understand the danger in destroying tropical forests well, yet Amazon voters have elected a public official who campaigned against conservation.

The problems with assessment which occur most often are that:

- we lack baseline information for recognizing a change (there are very few long-term data sets upon which to draw baseline conclusions, and these are often of obscure origin);
- the change is so subtle as to escape attention (e.g., despite very good records, there is still some debate as to whether we are observing a global increase in temperature);
- we do not know how to measure the phenomenon (e.g., "patchiness" in plankton density or in soil microbes);
- we do not know how to distinguish human from "natural" influences;
- the cause of change is multifactorial;
- we simply do not have enough data.

Although this last is a chronic complaint of scientists, our capacity to record experimental data in fact far exceeds our capacity to interpret it. Reinterpreting existing, published data can be scientifically quite rewarding. For example, the correlation between global warming (if it is occurring) and CO_2 concentration could not have been detected had the latter not been carefully recorded over years for entirely different reasons. This illustrates the potential value in monitoring systems and publishing the data, even when no apparent problem focuses our attention.

A suggested principle for assessing human influence is: if we observe in an ecosystem a change which we have never seen before and for which we have no alternative explanation, we should assume it is due to human influence (e.g., coral reef bleaching and red tides). This appears to be a conceivable strategy for dramatic changes, although it tacitly assumes that science has seen everything "old", so what is now discovered must be "new" phenomena. There was spontaneous disagreement as to whether we already follow this strategy, and whether we should.

Possibly, biotechnological tools could help in assessing ecological changes. DNA analysis as a tool to count microbes in soil is presently possible, and even species specifically engineered as indicators are feasible.

WHAT ARE THE ATTRIBUTES OF THE ENVIRONMENT WHICH WE WANT AND NEED?

We want and need environmental integrity, that is, a balanced environment. Although the notion of a "balance of nature" is useful (almost necessary) as an image around which we can organize our relationship with the environment, it is exceedingly elusive when we attempt to give it scientific meaning. Certainly, if we are to live within the "balance of nature", we must establish a sustainable interaction with our environment.

There are many attributes which we expect of our environment, ranging from the obviously essential to desiderata. There was general consensus that we expect our environment to provide the following:

a) ecosystem services;
b) production;
c) beauty;
d) preferred species;
e) biological diversity.

a) A balanced environment provides a host of ecosystem services: our sewage and innumerable other pollutants are degraded by organisms in our environment, the plant growth around us regulates the water cycle, breaks the wind, and filters dust from the air.

We tend to take many such ecosystem services for granted, and the majority of them cost us nothing. They can generally be substituted on a local scale, either by other system components or by artificial means. Artificial substitution, however, usually takes conscious effort and is sometimes quite expensive.

An example of an ecosystem service which probably cannot easily be substituted is the regulation of atmospheric CO_2. The concentration of CO_2 in the atmosphere is increasing at only one third the rate which would be expected from the amount of fossil fuel burned, and even less when the additional output from forest burning is considered. This probably reflects the absorption of CO_2 by the oceans and their phytoplankton and by terrestrial plants (Post 1990).

b) We depend on the material productivity of our environment for food, fiber, fuel and timber. Although habitual food preferences and material culture can offer strong resistance to change, there is still widespread substitutability of these products. For example, in the developed world, fuel and many other ecosystem products are at present derived largely from fossil sources. But production remains the most massive demand we place on our environment.

c) A sizable part of the tourist industry is based on the perceived beauty of landscapes. In most societies, people will pay quite a high premium to live in what is considered a beautiful environment. Substitutability depends here heavily upon cultural traditions, but is also limited by other factors, e.g., problems of transport.

d) We expect our environment to harbor certain preferred species. For example, the Great Lakes of North America produce roughly the same amount of fish today as in the early part of this century. However, the fisheries of commercially valuable species such as lake herring and lake trout have collapsed, and much less valuable species like alewife and rainbow smelt are fished today (Loftus 1987). Another example of a preferred species is the Madagascar periwinkle (*Vinca cataranthus*), a source of raw materials for the production of valuable drugs (Ehrlich 1987).

There is relatively little substitutability here, since specialized properties of the species are crucial. When preferred species are lost from an ecosystem and cannot be restored, the quality of the ecosystem is clearly degraded.

e) Biological diversity is meant here to include both diversity between species and within a given species. Its function in ecosystems is not well understood, but the loss of biological diversity may have many consequences for the functioning of natural systems, among which may be a loss of stability (Pimm, this volume). It seems plausible that biological diversity should increase the potential for a system to change in reaction to changing requirements.

Wild species provide not only genes for conventional breeding but also for transgenic work, making genetic diversity within a species desirable as a source of new varieties of crop plants and domesticated animals. A loss of biological diversity as the result of extinction must be considered irreversible, and there appears to be no substitutability here.

Some indications of a nonsustainabile interaction, and thus attributes we obviously do not want, are:

- soil erosion;
- loss of diversity (within species and extinction of species);
- harmful plankton blooms;
- epidemics/pest outbreaks;
- declining or more erratic harvests;
- chemical pollution.

Other possibly desirable attributes, such as continuity and self-management, were discussed but either defied clear definition or remained debatable.

In many situations, the attributes in the above list will overlap or will be in conflict with one another, as when gardening locally reduces biological diversity in favor of beauty. In particular, increased total demand for products is often met by trading more area away from the attributes a), c), and e) (although some will find endless cornfields beautiful). Economic priorities regularly conflict with the preservation (or achievement) of these attributes. The conflicts cannot be solved at a scientific level, but bioscience can make them visible, so that other social institutions can mitigate them.

Lacking a scientifically viable notion of the "balance of nature", a suggested guide is the "principle of precaution": do not change more than you must. But the term "must" is ambiguous at best, so that an alternative version appears both more rigorous and more practical. This is the "principle of environmental reciprocity": avoid any activity likely to bring about irreversible changes in ecosystems.

Natural processes can lead to recovery of a system; active effort can lead to its restoration. In either case, reversibility denotes not only the possibility of recovery or restoration, but also its feasibility or practicality. The time it will take, the likely cost in money, labor, and other resources, and the quality or authenticity of the resulting system are further important aspects (see box: FIRES).

HOW CAN BIOSCIENCE HELP TO ACHIEVE A SUSTAINABLE INTERACTION WITH OUR ENVIRONMENT?

In this section we will concentrate on sustainable interactions as a catch-all for desirable interactions. None of the group members questioned the fact that this (rather static) goal was desirable. Again, no rigorous definition of sustainability was presented, because this must include the question of population size and average consumption per capita.

FIRES:
The identification and assessment of human influences on ecosystems may seem deceptively simple in retrospect. A good example is the difficulty of determining how humans have influenced various ecosystems through the use or the suppression of fire. For example, the tallgrass prairies and savannas of central North America were probably burned more or less annually by fires set by Indians or by lightning, usually in the spring or fall. Various observers, including Thomas Jefferson, speculated that fire accounted for the scarcity of mature trees in these ecological communities, but the actual nature of the relationship between fire and vegetation was not clear. These speculations were supported by observation when the frequency of fires declined dramatically following settlement by Europeans in the 19th century. Fire was suppressed deliberately as a safety measure and inadvertently by plowed areas which served as fire breaks. Following this development, many Midwestern grasslands were rapidly invaded by trees and converted to the oak forest common in the landscape today.

Nevertheless, the relationship between fire and vegetation remained controversial among ecologists, and first attempts at prairie restoration, undertaken at the University of Wisconsin-Madison in 1935, did not employ fire. Only when these initial efforts were unsuccessful did ecologists begin experimenting with fire on prairies. This line of research has gradually worked out the relationship between humans, fire, and the vegetation of the area (Jordan, this volume).

Similar experiences have produced comparable insights in other North American ecosystems, including forest ecosystems of the West and the Southeast. This work teaches us the value of restoration attempts as a way of identifying causal factors influencing the condition of a landscape. Without understanding these causal relationships, we can often change a system easily enough – by burning it, for example, or by altering a historic pattern of fires –, but we are less likely to be able to restore it.

Maintenance of natural or historical ecosystems may also depend critically on this understanding. Thus, the prairies, and even more dramatically, the oak savannas of the Midwest, were almost entirely lost as a result of initial failure to appreciate their dependence on fires of human origin (Packard 1988).

Note that growth can only be sustainable if it is asymptotic, tending toward an absolute upper limit. Any given system of exploitation will eventually be outstripped if growth in consumption or population is not limited.

The image of the "noble savage" projects a sustainable interaction, but it is scientifically debatable whether this has ever been achieved. In particular, along with alleged examples of sustainability (Mayan society, Javanese society), there are many examples of societies which visibly did not achieve this goal (Greek and Roman societies together deforested the Mediterranean countries). The North American Indian cultures, sometimes cited as examples of sustainability, may have massively changed their habitat, driving to extinction many of the large game animals of North America (Martin 1984).

Alleged examples of sustainable societies tend to have been sedentary and relatively undisturbed by major wars, which suggests that cultural accumulation of knowledge about the environment may play an important role. Also, very little is known about the per capita consumption they had. Sustainability virtually assumes local production of food, fiber, and fuel. If at all, the "noble savage" probably existed at a much lower population-consumption level than we have today.

Humans directly or indirectly divert ca. 40% of terrestrial primary production to satisfy their demands (see box: NET PRIMARY PRODUCTION). Although other systems exhibit diversion rates between 5% (tundra) and 90%, those with a high diversion rate generally have a low-standing crop or cover only small areas.

There was consensus that the present human population-consumption level is not sustainable using presently known techniques. This was extensively debated, and there were many suggestions for increasing primary production, all relatively marginal. Some examples:

- Increasing the integration of agricultural systems may increase yields by a factor of 2–3, but this will apply to only a few areas.
- The potential of mariculture is generally overestimated (see box: FOOD FROM THE SEA). Only the integrated use by farmers of irrigation water for raising herbivorous fish promises a marked increase in local net productivity.
- Indirect consumption could be reduced if we chose to eat less meat. But most of the world is largely vegetarian already, and, far from exporting their present grain surplus, the meat-eating countries now import part of their animal feed.
- Despite mineral fertilizers and pesticides, primary production, e.g., in Great Britain, has been virtually constant during this century, although there has been a shift toward species which humans can better utilize.

These considerations all turned on the possibilities of providing consumption needs. During this discussion there was no reference to the effect this might have on the other attributes of a sustainable environment mentioned above in the section "What Are the Attributes of the Environment Which We Want and Need?" From the point of view of the ecologist, evolutionist, or geneticist, however, the preservation of biological diversity is by no means a luxury.

NET PRIMARY PRODUCTION:
Net primary production (NPP) is the amount of energy left after subtracting the respiration of plants from the total amount of solar energy that is fixed biologically. NPP provides the basis for maintenance and growth of all animal consumers, and so is a limit to the total food resource available to us.

The estimates of NPP differ widely across different kinds of ecosystems. Deserts and arctic and alpine regions represent a substantial part of the planet's terrestrial surface (37%), but contribute only a small fraction of the terrestrial production (<4%).

Similarly, the oceans constitute about 70% of the total surface, but are very unproductive with the exception of areas of coastal upwelling. In total, the oceans may produce only about 40% of the global NPP.

Vitousek and collegues provide three estimates of the degree to which NPP is used by humans:

- A low estimate as simply the amount of NPP used directly for food, fibre, fuel, or timber. Direct human consumption represents about 5% of terrestrial NPP, or 3% of global NPP.
- An intermediate estimate, adding to this the productivity of all lands devoted entirely to human activities. For example, although only a small amount of the biomass in a grain field is actually consumed, yet the entire area of the crop is devoted to our food production. With this in mind, the percentage of terrestrial NPP in one way or another co-opted by humans rises to about 30% (20% of global NPP).
- A high estimate, adding to this the production lost, for example, as a result of converting land to cities and highways, forests to pasture lands, or desertification and over-use leading to soil erosion. This estimate raises the figure to nearly 40% of terrestrial NPP, or about 25% of global NPP (Vitousek et al 1986).

Consensus is difficult to reach on what a sustainable population level is. An example is Europe, for which the group consensus was:

- in no way is the present European population-consumption combination sustainable;
- if consumption as measured by energy could be reduced to ca. 20% of its present level, then sustainability may be conceivable.

Some experiments in developed countries have shown only a marginal drop in agricultural productivity when use of artificial fertilizers and pesticides is discontinued. (However, some pesticides may have been used in adjoining crops, and the material export was compensated by organic fertilizers.) Consensus could not be reached on whether this implies the capability of the developed countries to feed themselves without fossil energy. (At present,

FOOD FROM THE SEA:
Confronted with potential food shortage, owing to growing human demand, many officialsd, managers and large segments of the lay public see the oceans as a major new source of food. This view is problematic for three reasons:

- The oceans are already exploited, and provided in the late 1980s about 80 million tons per year of high quality protein (FAO 1990).
- Sustaining this harvest under present fishing practices and lack of efficient management will be difficult, and increasing it to match a decade or more of predicted population increase appears downright unfeasible.
- Mariculture, like any farming enterprise, requires secure sites and inputs (feeds, fertilizers) which are in limited supply. It must compete with capture fisheries for sites and with agriculture for inputs. Also, most mariculture operations use soy-based pellets or cheap fish like anchovies to feed high-priced species such as salmon, seebass, or grouper, generally with losses of about 90% of the protein fed.

About 90% of the world's marine fish harvest stem from the shallow "shelves" with high primary production which gird the continents down to 200 m (Gulland 1970). All of the world's major shelves are now exploited, with serious overfishing problems reported from most of them, such as the North Sea (Gulland 1982) and the Sunda Shelf (Pauly 1988). Exceptions are possibly parts of the Patagonian, Antarctic, and Sahul shelves.

Attempts to identify "unconventional" marine resources have involved among others:

- lantern fish (fam. Myctophidae), of which billions of tons may occur in the bathypelagical zone of the world's oceans, but generally at very low concentrations (in the order of 1 gram per ton of water), precluding their commercial exploitation (Gjoesaeter 1980);
- oceanic squid, whose biomass and production appear to be very high (based on sperm whale stomach contents), but which are extremely difficult to catch and to market;
- Antarctic krill (*Euphausia superba*), the key trophic link in the Antarctic ecosystem. Krill feeds the antarctic fish, penguins, seals, and whales, now considered for protection. In any case, krill could not be exploited by (protein-)poor countries;
- fish or invertebrates in OTEC (Ocean Thermal Energy Conversion) plants, driven by artificially upwelled, nutrient-rich deep waters. This would require investments beyond the reach of countries now unable to secure sufficient protein (Linsky 1981).

Overall, no major marine resource appears to exist which would be capable of sustaining a new, economically viable fishery and whose landings would be cheap enough for consumption in developing countries, where most of the need will be.

no developed country does so.) Nor could consensus be reached on how this question could be resolved.

Advanced conventional biotechnology and new genetic engineering methods can play an important role in sustaining the interaction of human population with the environment at an acceptable level of ecological integrity. There is considerable literature on biotechnological approaches to agriculture which can be pursued to make developing countries relatively self-sufficient in the production of food, fiber, and fuel. While much of this biotechnology is conventional, recent advances in genetic engineering and plant cell culture technology will permit a more rapid pace of development. The goal is customizing crop species to increase productivity and reduce environmental impact. In this regard, engineered plants are likely to see widespread field demonstration within the next few years. If this technology proves safe and effective, applications for bioengineered animal products, feed additives, probiotics, and disease-resistant breeds may follow.

Turning away from plants, a host of biotechnological developments using both native and genetically engineered microbes is underway:

- for restoring damaged ecological and depleted agricultural systems;
- for reducing toxic and hazardous waste discharges and existing environmental contamination or for remote sensing of pollutant degradation;
- for cost effective, low impact resource recovery and recycling;
- for mixed organic waste treatment systems.

Already, synthetic plastics from bacterial biopolymers are a reality on a pilot scale. Realistic projections for the coming decade envision the combination of these with developments in composite material science and bioelectronic polymers lessening the demand for petrochemical processes. Plant and microbe-based systems are in development to improve soil fertility, composition, and texture, e.g., using nitrogen-fixing microbes or mycorrhizal plants.

A compilation of potential environmental biotechnology is beyond the scope of this report, but could also include bioconversion and biocatalysis systems, composting technology for wastes, solvent and biogas production, artificial photosynthesis, and CO_2 fixation systems.

Finally, strategies exist for development of gene storage banks as a last resort for conserving some of the biodiversity presently in jeopardy and for environmental health assessment using molecular detection and cataloging systems. The former cannot, however, realistically replace physical preservation of species, e.g., in reserves.

There is a major need to integrate these technologies with research on the dynamics and control of natural ecological systems for safe and effective applications. There is also a pressing need for evaluation of the indirect effects of these technologies, for example through the displacement of existing social and economic systems. Clearly the development costs are significant, but there is a likelihood of low cost application in an environmentally safe manner.

The ecology of developing countries is threatened most by the demands resulting from rising population, which can outstrip any increase in productivity we can create. There are two ways biotechnology can affect this problem: first, by contributing culturally more acceptable ways to control population growth; and second, by increasing primary productivity enough to bridge the time span until population has been stabilized. The techniques used must be robust and independent of sophisticated or expensive infrastructure.

The ecology of developed countries is threatened most by the demands resulting from rising consumption, which again can outstrip any increase in productivity we can create. Biotechnology can only affect this problem by giving us techniques to mitigate the environmental effects until consumption has been reduced to the sustainable.

One source of ecological problems is that external costs are often not included when actions are evaluated. In particular, simple economic models of exploitation tend to assume that any resource can be substituted and, by discounting future values, to overemphasize the present. However, shifts in exploitation mechanisms may prove less of an economic burden than first appearences would suggest, as illustrated by the ban on whale hunting (see box: WHALES). Bioscience can help make external costs visible which would otherwise go unnoticed.

HOW DO WE ASSESS THE ECOLOGICAL RISKS OF NEW BIOLOGICAL TECHNOLOGIES?

We mention first two indirect risks which are not peculiar to new biotechniques, though these may increase them:

- As new agricultural strains with better performance replace older strains, the latter may disappear, reducing future options.
- Breeding for resistance to adverse conditions can lead to a negative feedback cycle. For example, the breeding of salt-resistant plants may lead to irrigation with saltier water, thus, over time, creating demand for increasingly salt-resistant strains.

WHALES:

A number of natural living resources are presently under serious threat from an increasing world population; these include tropical forests, coral reefs, numerous species of large terrestrial mammals and birds, rare varieties of our major crop plants, etc. For most of these, prospects are generally bleak. However, the longterm prospect of one major group has radically improved in the last two decades: the great whales.

About these, Day wrote that the "rapid shift of attitude toward the killer whale from antipathy in the early 1960s to total sympathy by 1970 coincided with the same period of radical change in attitude toward all great whales. Opinion moved from indifference during what was the most wanton destruction of the whale in history (some 60.000 a year in the 1960s) to shock and alarm by the early 1970s, when some nations continued to hunt despite the obvious collapse of all whale populations" (Day 1987).

Changed public perception of the whales from mountains of lard (Gulland 1988) to sensible beings has gradually made their exploitation morally repugnant, thus changing the parameters of the very science investigating them, and rendering even "scientific whaling" inacceptable (Pauly 1987).

Quite aside from the tactics used by opponents of whaling, as detailed in Day's book, we feel there is a lesson to be learned about the conservation of other living resources. This applies particularly to the general public, which understands that the key issue was not one of conservationists fighting to "stop development" or to destroy jobs. The danger was one of irrevocably losing options, including that of watching whales, now a major industry.

More direct are the risks involved in the planned release of transgenic organisms which reproduce. Lacking direct observational data, we must depend on analogies, e.g., the introduction of alien species or genotypes. Many introductions have a negative effect on the ecosystem as defined by the criteria above in the section "What Are the Attributes of the Environment Which We Want and Need?" The risks which we might expect from this analogy are that the new organism will be a crop pest or invade the natural habitat.

For this analogy, we have many examples to draw from, such as the introduction of goats to islands lacking large predators, or the effects of introduced diseases like the chestnut blight. Note that Europe is a poor basis for intuition in this field, as almost no crop has become a pest there. This is in contrast to other areas (e.g., sorghum in the continental U.S.A., guava in Hawaii, carrot in several subtropical areas).

Theoretically, a simple invasion model would predict population growth from the intrinsic growth rate modified by such effects as competition, predation, parasitic diseases, and the like. Whether or not an invasion results depends on the values of the parameters, which are too numerous to be individually estimated. It is more realistic to determine the growth rate experimentally under controlled conditions. First results appear promising (Crawley, this volume).

It is debatable how good this analogy is for small transgenic changes in current crop plants. It may tend to overstate the risks, since transgenic introductions are derived from one individual and hence lack genetic variability. Self-destructing mechanisms may also be built into the genes of transgenic organisms, but these mechanisms might fail, e.g., the organism might mutate or find an unsuspected niche. It is thus safest to assume that release is irreversible.

A good way of freeing the analysis from emotion is to judge the product, not the process by which it has arisen. However, the ways in which techniques of genetic manipulation differ from "natural" evolutionary or breeding processes carry the potential for risks and for benefits:

- the rate at which viable mutants are created is higher than the natural mutation rate;
- the selection process is accelerated, at least with repect to the desired traits;
- recombined genetic material might later move to different sites on the DNA (newer technology might solve this problem);
- by combining genetic material which could otherwise not come together, adaptive peaks could be jumped.

It is also conceivable that natural processes might transmit genetic material from the introduction to other species.

Taken all together, the immediate risks to our environment arising from new biotechnology do not appear major when compared with those posed by present growth in human consumption and human population.

PRIORITIES AND CONTRIBUTIONS TOWARD THE COVENANT

Science is probably driven as much by funds as by curiosity. Funding policy can thus be used to direct effort toward projects which include analyses of potential environmental effects beyond those which are immediately obvious. The environmental impact assessments presently required by law in many countries are generally too narrow in scope, although they are a step in the right direction.

Societal fear of biotechnology is neither all unfounded nor all too justified. The discourse between bioscience and the lay public can be intensified to increase agreement on the significance of potential risks and benefits in research and in development projects. In particular, only sound, defensible arguments should be used when working in the political arena. A counterexample was the fight against the Tellico dam, centered around alleged impending extinction of the snaildarter (Disilvestro 1989).

Both at the level of teaching and of research there is need of a more integrated environmental science. Elementary textbooks generally contain only "consensus science", and debates on curriculum reform seem endless. Effort spent creating incentives for outstanding scientists to write interdisciplinary textbooks and participate more strongly in teaching would be well spent. Another way of improving the quality of ecology professionals and teaching could be through professional organizations. Present orientation in research is primarily toward pushing biotechnology to new frontiers. More emphasis should be placed on assessing what has been achieved and on developing an integrated, interdisciplinary approach. Funds are not the primary problem here, and rapid progress is improbable.

Aside from research directed toward increasing the options for controlling population, the most pressing research problems from the standpoint of ecological integrity are:

- To find ways of defining sustainable population-consumption levels in various settings.
- To make biotechnological results cheaply available, independent of sophisticated infrastructure. Particularly emphasize those which increase production of locally desirable food, fiber or fuel without drastic effects on the environment.
- To monitor apparently stable systems to gain baseline data. Top priority goes to systems with monitoring programs in progress which are in jeopardy. Also monitor variability within indicator species and the loss and gain of species.
- To systematically investigate theoretical and analogical ways of estimating the risk of transgenic and "conventional" introductions.
- To find ways of determining what levels of stress are reversible, depending on measurable system parameters.
- To find more ways of restoring systems after overstress or deliberate partial destruction (healing art).

Two points deserve mention which remained debated:

- What changes constitute damage to an ecosystem? Here opinions ranged from "potentially almost all changes" to "irreversible functional changes".

- Should the burden of proof be transferred to the party proposing an active change in an ecosystem? Here opinions ranged from "this would improve public acceptance without unduly hampering research" to "this would bring bioscience to a screeching halt".

REFERENCES

Berger, J.J. (1987) *Restoring the Earth: How Americans are working to renew our damaged environment.* New York: Anchor Press/Doubleday.

Day, D. (1987). *The Whale War.* San Francisco: Sierra Club Books.

Disilvestro, R.L. (1989). *The Endangered Species Kingdom.* Wiley Science, pp. 161–162.

Ehrlich, P.R. and A.H. Ehrlich (1981). *Extinction: The Causes and Consequnces of the Disappearnce of Species.* New York: Random House.

FAO (1990). *Fishery Statistics: Catch and Landings,* vol. 66. Rome: Food and Agricultural Organization of the United Nations.

Gjoesaeter, J. and K. Kawaguchi (1980). *A Review of the World Resources of Mesopelagic Fish.* FAO Fish. Tech. Pap. No. 193.

Goodland, R.J.A. and H.S. Irwin (1975). *Amazon Jungle: Green Hell to Red Desert?* Amsterdam: Elsevier.

Gulland, J. (1988). The end of whaling? *New Scientist* **120** (Oct. 29): 42–47.

Gulland, J.A., ed. (1970). The Fish Resources of the Ocean. FAO Fish. Techn. Pap. No. 97.

Gulland, J.A. (1982). Long-term potential effects from management of the fish resources of the North Atlantic. *J. Cons. Int. Explor. Mer* **40(1):** 8–16.

Halim, Y. (1991). The impact of man's alterations of the hydrological cycle on ocean margins. In: *Ocean Margin Processes in Global Change,* eds. R.F.C. Mantoura, J.-M. Martin, and R. Wollast, Dahlem Workshop Reports PC 9, pp. 301–327. Chichester: Wiley.

Linsky, R.B. (1981). Multiple use applications of Ocean Thermal Energy Conversion (OTEC). In: *Proceedings of the Workshop on Coastal Area Development and Management in Asia and the Pacific,* Manila, Philippines, Dec. 3–12, 1979, ed. M.J. Valencia, pp. 151–152. East-West Environment and Policy Institute, East-West Center, Honolulu, Hawaii.

Loftus, D.H. et al. (1987). Partitioning potential fish yields from the Great Lakes. *Canadian Journal of Fisheries and Aquatic Sciences* **44:** 417–424.

Martin, P.S. and R.G. Klein, eds. (1984). *Quarternary Extinctions: A Prehistoric Revolution.* University of Arizona Press.

Packard, S. (1988). R&MN **6(1).**

Pauly, D. (1987). On reason, mythologies of natural resource conservation. *NAGA, the ICLARM Quarterly* **10 (4):** 11–12.

Pauly, D. and Chua Thia-Eng (1988). The overfishing of marine resources: socioeconomic background in Southeast Asia. *Ambio* **17(3):** 200–206.
Pimm, S.L. and J.B. Hyman (1987). Ecological stability in the context of multispecies fisheries. *Canadian Journal of Fisheries and Aquatic Sciences* **44:** 84–94.
Post, W.M. et al. (1990). *American Scientist* **78:** 310–326.
Vitousek, P.M. et al. (1986). Human appropriation of the products of photosynthesis. *Bioscience* **36:** 368–373.

Bioscience ⇌ Society
edited by D.J. Roy, B.E. Wynne, and R.W. Old, pp. 203–210
© 1991, John Wiley & Sons

The Asilomar Conference: Was the Asilomar Conference a Justified Response to the Advent of Recombinant DNA Technology, and Should It Serve as a Model for Whistle-Blowing in the Future?

Roger Lewin

3802 Ingomar Street NW
Washington, D.C. 20015, U.S.A

Abstract. The 1975 Asilomar Conference focused on the potential health hazards of the newly emerging recombinant DNA technology. The constitution of the conference is assessed in the context of public policy formation. It is concluded that although Asilomar might be judged inadequate by today's standards on the need for public participation, the event was part of the process of raising scientists' awareness for such participation.

The Asilomar Conference of 1975 has been described variously as "an exemplar of social responsibility by the scientific community" and "a conspiracy by a small, elite, scientific group motivated by self-interest". Whatever history's final assessment of the event, there is no doubt about its significance in the process of development of biotechnology in the United States and, indeed, worldwide. It was the landmark event that brought recombinant DNA technology – genetic engineering – to wide public notice, and initiated local and national attempts to control the new technology.

Today, molecular biologists in university and commercial laboratories are able to carry out experiments using recombinant DNA techniques, relatively unencumbered by external restrictions. And a multibillion dollar industry has been built upon the technology, with established and potential applications in both medicine and agriculture. For instance, Genentech Inc., the first biotechnology company to be formed (in 1977), and the first to go public (in 1981), has recently been bought by Roche for more than $2 billion. Undoubtedly, biotechnology is an industry with much growth potential, making governments anxious to promote national commercial interests amid worldwide competition.

The rapid movement of recombinant DNA techniques from scientific novelty to practical exploitation has caused growing concern over public scrutiny – or lack of it – of the potential applications for these powerful techniques. This concern includes practical and ethical issues in medical genetics and in biological warfare. The question of ecological manipulation has also been raised as part of a currently vigorous awareness of environmental matters in general.

In addressing the questions of whether the Asilomar Conference was a "justified response" and whether it "should serve as a model for whistleblowing in the future", it is all too easy to allow the subsequent events, as we now know them, to influence our judgment. It is important to look at the circumstances of the conference itself, given what was known and could have been predicted at the time, and to examine the process underlying the events.

First, what was the Asilomar Conference? In one sense, it was the end-product of a rising concern among molecular biologists over potential health risks that might arise from the newly invented ability to manipulate genes and insert them into foreign "hosts". In another sense, it was the first step in public policy formation. Assessment of the success of the conference depends very much on which of these two views is emphasized.

The experiments that first provoked alarm involved using a known tumor virus, SV40, as a vehicle for introducing tailored pieces of DNA into a strain of *Escherichia coli,* the bacterium of the human gut. The first of these experiments was devised in Paul Berg's laboratory at Stanford University in 1971. Initially, apprehension was raised informally in private discussions among scientific colleagues. Very soon, however, the matter of potential health risks – specifically, the spectre of the spread of cancer genes – became discussed publicly, albeit within the scientific arena. As a result of one such discussion at the June 1973 Gordon Conference on Nucleic Acids, the co-chairs of the meeting, Maxine Singer and Dieter Soll, were instructed to write to the president of the U.S. National Academy of Sciences, explaining the potential problem:

> We are writing to you on behalf of a number of scientists to communicate a matter of deep concern ... Certain ... hybrid molecules may prove hazardous to laboratory workers and to the public. Although no hazard has yet been established,

> prudence suggests that the potential hazard be seriously considered. (Singer and Soll 1973)

The letter was also published in *Science*. The Academy responded by setting up the Committee on Recombinant DNA Molecules, headed by Berg, with ten other prominent molecular biologists. Within a year the Committee published the now famous "Berg letter", also known as the "moratorium letter":

> Our concern for the possible unfortunate consequences of indiscriminate application of these techniques motivates us to urge all scientists working in this area to join us in agreeing not to initiate [certain experiments] until attempts have been made to evaluate the hazards and some resolution of the outstanding questions has been achieved. (Berg et al. 1974)

The letter was widely reported in the newspapers, and for the first time the issue reached a general audience.

At first, most molecular biologists believed that a brief pause in certain potentially hazardous experiments was a reasonable and acceptable measure. But some became uneasy about the possibility of outside interference. An extract from one letter in the blizzard of correspondence that followed the moratorium suffices to exemplify the point:

> Let me conclude by stating that even *good intentions* could easily lead to the creation of a self-sustaining and even growing bureacratic monster which would discourage and delay very important research. (Szybalski 1981)

The moratorium was seen as an interim measure only, until the matters of health risk and appropriate laboratory practice could be resolved. This, primarily, was the function of the Asilomar Conference, held under the auspices of the National Academy of Science. "The purpose of the meeting is to review the progress, opportunities, potential dangers, and possible remedies associated with the construction, and introduction of new recombinant DNA molecules into living cells", said the letter of invitation to potential participants (Berg 1981). The conference agenda, spanning three-and-a-half days, focused overwhelmingly on technical problems of various kinds. Four papers touched on matters of public policy, legal liabilities, and ethics. A handful of lawyers was present, and a "pool" of journalists was invited.

"The significant events at the Conference took place during the last morning, which was a confused (some would say chaotic) affair", is how James Watson later described it (Watson 1981). During this session, guidelines were drafted which provided regimens for carrying out all but the most potentially hazardous experiments. "Having demonstrated their integrity, [the conference participants] naively believed that they would now be free of outside intervention, supervision, and bureaucracy", Watson added.

The two years following Asilomar produced a frenzy of local and national activity. Articles by prominent biologists (Cavalieri 1976; Chargaff 1976) pointed to possible dangers associated with various aspects of recombinant DNA research, and they triggered hearings in Ann Arbor, Michigan, and Cambridge, Massachusetts, in which the city councils sought severely to restrict the research. Hearings in Congress began, most notably by Massachusetts senator Edward Kennedy, who "saw his role as the defender of the people against powerful special interests" (Zimmerman, 1986). More than a dozen bills were introduced in both houses of Congress, designed to control by statute recombinant DNA work in industry and university laboratories. "The situation began to go downhill for us", says Norton Zinder. By "us" he meant "the organizers of the Berg letter, the organizers of Asilomar, and the vast bulk of the molecular geneticists" (Zinder 1986).

There is no doubt that most molecular biologists at first felt betrayed, like the person who is mugged by the old lady they have just helped cross the road. Very soon lobbying efforts began, which were designed to show that the hazards of the technique had been greatly exaggerated, and that U.S. research interests were in danger. Following the 1977 Gordon Conference on Nucleic Acids, its chairman, Walter Gilbert, wrote an open letter to Congress:

> We are concerned that the benefits of recombinant DNA research will be denied to society by unnessarily restrictive legislation . . . We feel that much of the stimulus for this legislative activity derives from exaggerations of the hypothetical hazards of recombinant DNA research that go far beyond any reasoned assessment. (Gilbert 1977)

There was an effort in the summer and fall of 1977 to produce a second "Berg letter", which would make three points: 1) that DNA recombination of the sort carried out by molecular biologists goes on naturally in nature; 2) that improvements in the technology allowed for safer work; and 3) that legislation would pose a threat to scientific progress in the United States. Many drafts of the second letter were produced, but a final version was never agreed upon or published, due in part to a greater perceived complexity of the issue compared with 1974, but also because the drive towards legislation in Congress suddenly lost momentum.

By this time the National Institutes of Health (NIH), the United States' largest source of funding for recombinant DNA research, had become involved, with the establishment of the Recombinant DNA Molecule Program Advisory Committee (RAC). The committee, which initially was made up only of molecular biologists, set itself three tasks: 1) to evaluate potential biological and ecological hazards of the new techniques; 2) to develop procedures that would minimize the spread of recombinant molecules in human and other populations; and 3) to develop guidelines for laboratory practice. After Asilomar the RAC effectively devoted itself to the last of these three, the development of guidelines, based on the outcome of the conference.

Later on, in response to widening public and congressional concern, the constitution of the RAC was changed to include one-third molecular biologists, one-third scientists who were experts in genetics, microbiology, and other fields directly applicable to recombinant work, and one-third experienced in related matters such as public health, law, consumer affairs, or public policy. The new RAC remained advisory to the NIH director. From the start, the NIH guidelines applied only to NIH-funded research, effectively leaving commercial research unhindered.

"There had been violations of the guidelines reported in university labs, but no one had even caught a cold or gotten a stomach ache", notes Burke Zimmerman, a legislative aide to Congressman Paul Rogers, sponsor of one of the major bills. With a lack of demonstrated hazard, passion for legislation was lost. "What really did it in was the attention span [in Congress] ... The period of interest certainly does not exceed two years and may well be somewhat shorter" (Zimmerman 1986).

As public policy analyst Susan Wright has observed: "One of the most remarkable aspects of the controversy about the hazards of recombinant DNA technology in the 1970s was the speed with which the whole issue faded away. Intensely debated in the period 1975–77, by 1979 the hazard question was almost a non-issue" (Wright 1986). With an absence of demonstrated danger to sustain it, public debate over the control of recombinant DNA technology withered away – no (short-term) health effects had been detected – and research in universities was permitted to proceed under weaker and weaker NIH guidelines. Commercial research had never been impeded.

How, then, is the Asilomar Conference to be judged? It had been convened because of a concern for potential health risks associated with a new biological technique. As a result, the conference focused almost exclusively on the issue of risk and on guidelines for laboratory practice. Viewed in this way, Asilomar was a resounding success. "Involved scientists recognized a problem, took an initiative directed to the public interest, devised an interim solution, and succeeded in defending it against significant opposition that would have imposed far greater restriction" (Grobstein 1986).

Given the absence of any demonstrated hazard to date, the molecular biology community might be viewed as prudent to the point of oversensitivity. Many of those involved in Asilomar, weary of the upheaval that followed it, in fact came to suggest that perhaps it should not have happened at all. "I see this view as understandable but misguided", says Clifford Grobstein, professor of biological sciences and public policy at the University of California, San Diego. "The public image of science was humanized and thereby enhanced. Asilomar is not interpreted as a brilliant conspiracy but as a conscientious effort by well-intentioned people who did not know what was best to do but did their best despite conflicting personal interests and inclinations" (Grobstein 1986).

On the other hand, what if Asilomar is viewed as the first step in public policy-making? In this case, the conference is open to many criticisms, the principal one of which is that it was much too narrowly focused, both because of the people who attended and the issues discussed. "Science is not ethically neutral", noted Donald Michael, a consultant in futures issues and technology policy, in reviewing the significance of Asilomar. "Thus decisions regarding science and technology cannot be made exclusively in terms of scientific and technical arguments, even though these are critical to the decisions" (Michael 1986). In the best of circumstances, the making of public policy should involve all those constituencies with an interest in that policy.

As it was, the Asilomar participants included a small number of lawyers, but it was dominated by a narrow section of the biological community, specifically, molecular biologists. And, even though the focus was on health risks, microbiologists, public health specialists, and epidemiologists were not present. A group of biologists from Harvard and the Massachusetts Institute of Technology, speaking for the Science for the People organization, sent an open letter to the Asilomar gathering:

> We do not believe that the molecular biology community, which is actively engaged in the development of these techniques, is capable of wisely regulating this development alone ... Scientific careers are not built solely on a concern for public health, for the well being of the underprivileged, or right action. (Beckwith et al. 1975)

In spite of these vicariously expressed sentiments, no consideration was given to nonscientific representation in scientific decision making, a position that was reflected in the initial composition of the RAC.

The issues of public policy associated with recombinant DNA technology include: the effects of scale-up in industrial contexts, ecological impacts of accidentally or deliberately released organisms, biological warfare, application to aspects of human genetics, and commercial control of a sensitive new technology. These issues were not addressed at Asilomar, but might have been, at least in a preliminary way, if it had been conceived and run as a public policy conference, not as a scientific conference. "The key defect [of Asilomar] was the failure to recognize that an issue large enough to halt, even temporarily and partially, the normal process of science requires consideration by more than the group of scientists involved" (Grobstein 1986). Had this been recognized by the organizers, Asilomar might then have been transformed into a true public policy-making event.

The questions of whether the Asilomar Conference was a "justified response" and whether it "should serve as a model for whistle-blowing in the future" are of course linked.

Given that those scientists most closely familiar with the new technique were sufficiently concerned about potential health hazards, and given the lack of information at the time about the magnitude of hazard that might be involved, Asilomar was surely a justified response in the narrow context of science. Equally, as the first step in public policy-making, Asilomar was inappropriate, because it did not adequately take into account broader public issues and did not invite broader participation. It therefore cannot serve as a model for whistle-blowing in the 1990s.

It is important to note, however, that the climate of science/society relations has changed in the past fifteen years, partly as the result of Asilomar. Although some molecular biologists felt scarred by the immediate post-Asilomar events, many took comfort from the fact that the public scrutiny that developed – chaotic as it was at times – in the end did not impose unreasonable restrictions. If "another Asilomar" were to be convened, public participation would be much more likely.

REFERENCES

Beckwith, J. et al. (1974). Open letter to the Asilomar Conference on hazards of recombinant DNA, reprinted in *The DNA Story,* eds. J.D. Watson and J. Tooze, p. 49. San Francisco: W.H. Freeman.

Berg, P. (1974). Letter reprinted in: *The DNA Story,* eds. J.D. Watson and J. Tooze, pp. 17–18. San Francisco: W.H. Freeman.

Berg, P. et al. (1974). Letter: Potential biohazards of recombinant DNA molecules. *Science* **185:** 303.

Cavalieri, L. (1976). New strains of life – or death. *The New York Times Magazine,* August 22, 1976.

Chargaff, E. (1976). On the dangers of genetic meddling. *Science* **192:** 938–940.

Grobstein, C. (1986). Asilomar and the formation of public policy. In: *The Gene-Splicing Wars,* eds. R.A. Zilinskas and B.K. Zimmerman, pp. 3–12. New York: MacMillan.

Michael, D.N. (1986). Who decides who decides? Some dilemmas and other hopes. In: *The Gene-Splicing Wars,* pp. 121 -138. New York: MacMillan.

Singer, M. and D. Soll (1973). Letter: Guidelines for DNA hybrid molecules. *Science* **181:** 1114.

Szybalski, W. (1974). Letter reprinted in: *The DNA Story,* eds. J.D. Watson and J. Tooze, pp. 15–16. San Francisco: W.H. Freeman.

Watson, J.D. (1974), The Asilomar conference. In: *The DNA Story,* eds. J.D. Watson and J. Tooze, p. 25. San Francisco: W.H. Freeman.

Wright, S. (1986). Molecular biology or molecular politics? *Social Studies of Science* **16:** 593–620.

Zimmerman, B. 1986. Science and politics: DNA comes to Washington. In: *The Gene-Splitting Wars,* pp. 33–54. New York: MacMillan.

Bioscience ⇄ Society
edited by D.J. Roy, B.E. Wynne, and R.W. Old, pp. 211–226
© 1991, John Wiley & Sons

Genome Project: HUGO, Big Brother is Watching You!

Hans Günter Gassen

Institut für Biochemie
der Technischen Hochschule Darmstadt
Petersenstr. 22
D–6100 Darmstadt, Germany

Abstract. Contrary to the popular belief that genetic research is something new, the elucidation of the structure and function of the human genome has been a continuous multidisciplinary activity for more than 100 years. It started with the discovery of nucleic acids by F. Miescher in 1868 and recently climaxed in the foundation of a worldwide organization for the deciphering of the human genome (HUGO).

Because of the breath-taking advances in molecular genetics and analytical instrumentation, especially molecular cloning techniques and automatic DNA sequencing, a goal until recently beyond imagination – sequencing the human genome – is becoming reality. To date the Human Gene Mapping Library (HGML) data base contains relevant information for 6,652 loci, 11,852 probes, 2,275 polymorphic loci, and GenBank records 5,066,049 bp of DNA sequence.

After an elementary introduction into the basics of genome mapping and DNA sequencing, the practical application of genome data is discussed in this paper.

INTRODUCTION

Before the dawn of modern human anatomy research during the Renaissance, the knowledge about the inner structure of the human body was close to nil (Meyer-Steineg 1965) (Fig. 1). It was the achievement of the famous Flemish anatomist Andreas Vesalius to provide the physicans of his time, and espe-

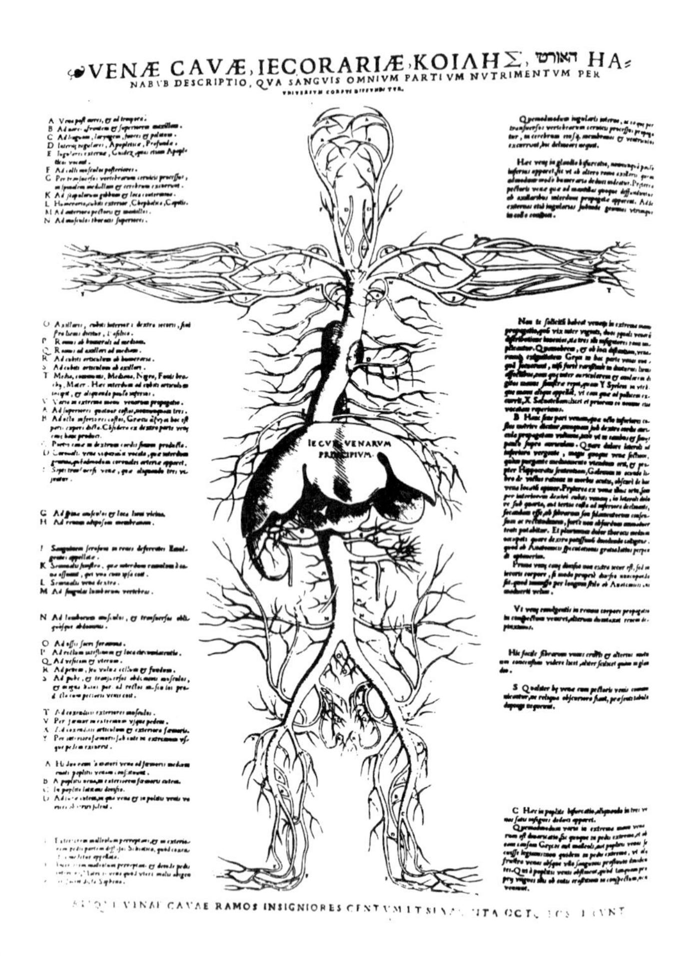

Fig. 1.

cially the surgeons, with a detailed atlas of the anatomy of the human body (*De humani corporis* 1543). It took a brave soul indeed to steal the cadavers from a churchyard or, even worse, from a place of execution. Michael Servet, a contemporary of Vesalius who described blood circulation, was condemned by Calvin and burned at the stake in Zurich in 1553 (*Christianismi restitutio* 1559).

Today, genes are being mapped on the 23 chromosomes at a rate of about a dozen a week – thanks to the tools of molecular genetics (Stephens 1990). McKusick, until recently head of the international Human Genome Organization (HUGO), predicts that by the year 2005 the location of every one of the 50,000–100,000 human genes on the chromosomes will have been determined, and the human gene map will be complete. As he puts it: "We have now arrived at a sort of 'last frontier' of anatomy – that of the human genome" (McKusick 1978).

To elucidate what follows, we define a *locus* as a part of the DNA on the chromosome which can be identified by genetic or biochemical techniques. The biological function of a locus may be known or unknown. A *gene* is defined as a DNA segment on a chromosome which codes for a protein or a nucleic acid of the RNA type (mRNA, tRNA, rRNA). The regulatory regions flanking the so-called structural gene are considered to be part of the unit "gene" (Gassen 1978).

Our understanding of the biochemical functioning of the genetic apparatus is still very limited at present. We know some details of how a DNA sequence directs the synthesis of a protein and how regulatory units determine the amount of protein to be made at a given time. On the other hand, we are ignorant about how the DNA thread folds into chromosomes, and do not understand the clock which determines which gene expression occurs at various stages of development (Gassen 1978).

Therefore, for the next decade the human genome sequencing project can only be aimed at two rather simple tasks: to assign each gene to a defined position on one of the 23 chromosomes (gene mapping), and to determine the base sequence of individual genes. When the base sequence has been determined, from that we can derive the amino acid sequence of the corresponding protein.

Defining the genetic constitution of an individual person can only be achieved in principle. The sequence of the genome differs from one person to another at roughly every 500th base. This phenomenon is called "genetic polymorphism", and explains why we are similar but not indentical in our phenotype (White 1988). This genetic relationship is comparable to the anatomical one. Basically we are all constructed similarly. However, the finger prints of of no two people on earth, for example, are indentical.

The DNA-sequences of a gene in different persons may deviate from one another in certain bases (genotypic differences). Nevertheless, the genes may code for either the same protein (silent changes), a protein with a different amino acid sequence but identical function, or even a protein with a modified (yet normal) function. Thus, it is not possible to conclude from the mere fact of alteration in the DNA sequence of a gene that it is "sick" i.e., that it is either functioning wrong or not functioning at all. The meaning of changes in the DNA sequences of a specific defined gene only emerges if these changes can be correlated to phenotypic evidence (Watson 1990).

A BRIEFING ON THE HISTORY OF HUMAN GENETICS

Recent publicity of the human genome project has tended to convey the impression that the deciphering of the human genome started very recently, and that methods and strategies are completely new. The important discoveries in that field, however, have been made continuously during the last hundred years. What is happening now is nothing less than the establishment of a worldwide organization to map and sequence the human genome. For this we need public support in order to finance the project.

No surgeon nowadays would operate without having a detailed knowledge of human anatomy. Using X-ray and other physical or biochemical methods, he or she will even examine to what extent his patient's anatomy follows the rules, and to what extent it is unique. Nobody argues about the right and obligation of surgeons to do this. In twenty years' time, physicians will examine particular genes of their patients to determine whether they are of the standard type or if they show irregularities which consequently require an individually tailored medical treatment. These genetic procedures and their diagnostic value may be compared to the present-day blood test. Before Vesalius, doctors treated their patients as mammals. Since Vesalius, they have treated them as humans. With the discovery of the human genome, patients can be treated as individual beings.

One of the first scientists involved in gene mapping was Johann Friedrich Horner, a professor of ophthalmology in Zurich. Horner mapped out pedigrees in 1876, showing that color blindness runs in families. Chromosomes in metaphase were visualized in tumor cells in 1877 by Walter Flemming of Kiel, and at the same time Gregor Mendel published his famous study of heredity in peas. Wilson mapped the first human gene for color blindness to the X-chromosome. Fluorescent banding techniques became available in 1970. T. Caspersson and L. Zech at the Karolinka Institute developed the staining of chromosomes with quinacrine mustard in about 1970, a procedure which enabled one to identify the chomosomes by their unique banding patterns. In 1966 McKusick published the first volume of *Mendelian Inheritance in Men – a Catalogue of 1487 Genetic Disorders*. The ninth edition lists 4,937 inherited characteristics.

The first workshop on gene mapping was organized by F. Ruddle in 1973 at Yale. With the assignment of small cell lung cancer to chromosome 3, it was proved that certain forms of cancer are somatic cell diseases (Culliton 1990). Today, approximately 1900 genes have been located on the 23 chromosomes (Stephens 1990). There are many scientists working on molecular details of the genetic constitution of the human organism. Part of that unique adventure is to determine the sequence of the nucleic acid constituting the human genome. In 20 to 50 years from now, every physician will have this sequence stored in his personal computer. Before commencing major medical treatment, he will be able to check his patients' genes for abnormalities and will adapt his treatment according to the individual genetic constitution of his patient. Furthermore, for the life sciences, especially for molecular genetics, the knowledge of the primary sequence of the genome of other organisms, such as *E. coli,* yeast, or plants, is a tremendous step towards the understanding of the correlation of structure with function (Table 1).

Table 1. DNA-Sequence data from different organisms

Organism	Genome [bp]	Completed [%]	Date
simian virus 40	5,000	100	1978
phage ΦX174	5,000	100	1978
phage λ	50,000	100	1982
phage T7	50,000	100	1983
Epst. Barr virus	100,000	100	1984
cytomegaly virus	250,000	100	1984
chloroplast and mitochondria	300,000	100	1989
E. coli	3 × 106	20	
S. cerevisiae	2.3×10^7	20	
Caenorhabditis elegants	8×10^7	10^8 cloned	
Arabidopsis	10^8	10^8 cloned	
Drosophila melanogaster	1.5×10^8	10^8 cloned	
Homo sapiens	3×10^9	1 to 2%	

Homo sapiens: 90% repetitive sequences, 50–100.000 gens, 5.066,049 bp known

THE HUMAN GENOME ORGANIZATION

The Human Genome Organization was established in 1988 as an international organization to promote collaboration on the mapping and sequencing of the human genome (Cantor 1990). One of the major objects of the program is to coordinate already existing activities at the CEPH (Center for the Study of Human Polymorphism) in France, the MRC (Medical Research Council) in the U.K., as well as at other institutions. Obviously the task ahead will require cooperation among many experts, as well as the sharing of equipment, biological materials, and data stored on computers. A particular problem is the computer storage and handling of the mapping and sequencing data. Although the potential for commercial gain raises difficult questions, it does not, however, preclude successful collaboration so long as prior agreement on the allocation of benefits is reached (OTA Report 1988).

Moreover, the fear that Big Science will invade molecular biology has often been voiced in criticism of a concerted international program of genome projects. This criticism relates to the problems of bureaucratic central management, the sacrifice of the individual scientist's freedom, and political and commercial interference.

Moreover, there have been very positive responses by well-known scientists in the field (OTA Report 1988). "The sequence of the human genome would be perhaps the most powerful tool ever developed to explore the mysteries of human development and disease" (Leroy Hood). "We will see a new dawn of understanding about evolution and human origins and totally new approaches to old scientific questions" (Allan Wilson).

Among the scientists participating in the program, there is practically no thought being given to the ethical consequences of the genome's being made available for commercial and institutional use.

MAPPING VERSUS SEQUENCING

In order to store the sequence of the human genome comprising 3×10^9 base pairs (bp) in book form, one would need 3,000 volumes of 1,000 pages each, with 1,000 letters per page. Mapping means knowing the volume, sequencing means knowing every single letter. The first goal of the human genome sequencing program is to locate every single gene on the chromosome (mapping).

Gene mapping, broadly defined, is the assignment of genes to chromosomes. A genetic linkage map permits investigators to ascertain one genetic locus relative to another on the basis of how often they are inherited together. Strictly speaking, a genetic locus is an identifiable region, or marker, on a chromosome. The marker can be an expressed region of DNA (a gene) or some segment of DNA that has no known coding function, but whose pattern of inheritance can be determined. Variation at genetic loci is essential to genetic linkage mapping. The markers that serve to identify chromosome locations must vary in order to be useful for linkage studies in families, because only when the parents have different forms at the marker locus can linkage to a gene be followed in their children. Alleles are the alternative forms of a particular genetic locus. For example, at the locus for eye color, there are blue and brown alleles. During meiosis, all of the genetic loci on a chromosome remain together unless they are separated by crossing over between chromosome pairs (Ott 1985).

Distance on genetic maps is measured by how often a particular genetic locus is inherited separately from some marker. This measure of genetic distance is called *recombination frequency*. The amount of recombination is expressed in units called centi-Morgan (cM). One centi-Morgan is equal to a 1% chance of a genetic locus being separated from a marker due to recombination events in a single generation. During the generation of sex cells in human beings, if a gene and a DNA marker are separated by recombination in 1% of the cases studied, they are on average separated by 1 million base pairs. Whereas distances between loci in kilobases of DNA are additive along a chromosome, this is not true for distances measured as recombination frequencies. Therefore, linkage maps are identical to physical maps when limited to small distances. The predicted total length for the sex-averaged linkage maps is 3,300 cM.

Recombination frequencies per megabase of DNA vary considerably by sex and by chromosomal region. Telomeric regions of the chromosome show proportionally more recombination in male meiosis, and centromeric regions have higher frequencies of recombination in female meiosis (Stephens 1990).

Physical maps can be based on cytogenetical or molecular data critera. Cytogenetically based physical maps order loci with respect to the visible banding pattern or relative position along the chromosomes, primarily by means of data from somatic cell hybrids and in situ hybridization. Molecularly based physical maps directly characterize large tracts of DNA by establishing molecular landmarks such as restriction endonuclease sites or sequence-tagged sites (STSs) (Olson 1989). Sequence-tagged sites have been proposed as the common reference point that could be used to coordinate information from different mapping strategies. There is no doubt that this where we are moving towards: locating loci on chromosomes. The short term goal for the linkage map is to enhance its resolution for each chromosome to a spacing of 2 to 5 cM. The more than 2,000 DNA polymorphisms already identified and catalogued would allow the construction of a linkage map with average size intervals between adjacent markers of about 2 cM. However, published maps have not yet incorporated enough of these markers to achieve a resolution of more than approximately 5–10 cM. Furthermore, since the distribution of known polymorphic markers is far from uniform, many gaps will remain until new markers are added. For physical mapping the short-term goals are to assemble STS maps of all human chromosomes with the aim of having markers at approximately 100,000 bp intervals (Kidd 1989).

The highest level of resolution for a molecularly based physical map is the DNA sequence which gives the linear arrangement of nucleotides for each of the 23 distinct human chromosomes. Leaving aside the question of sequence polymorphism, a complete reference sequence will contain roughly 3×10^9 bp of DNA.

The information of all molecular mechanisms within a cell is stored in the four-letter code of the DNA. The biological data bank uses the same principles as digital techniques with a binary code. The elucidation of a structural model for deoxyribonucleic acid (DNA) by Watson and Crick (1953) and the genetic code by Matthaei and Nirenberg (1961) were the fundamental discoveries in biology that paved the way for DNA sequencing (Gassen 1988).

Exactly twenty years ago (1971), Ray Wu determined a DNA sequence of the bacteriophage lambda 12 bases long. This took about two years with four persons working on it. With the development of recombinant DNA techniques, defined segments of DNA in practically unlimited amounts became available. The so-called polymerase chain reaction even allows the millionfold amplification of DNA segments in a test tube. These new methods stimulated DNA sequencing efforts tremendously. Nowadays, a well-equipped and experienced laboratory can sequence about 1,000 base pairs per day and per person (Zinke 1990).

DNA sequencing methods which revolutionized the procedures were published by Maxam and Gilbert, and also by Fred Sanger, in 1977. Details cannot be given here but those interested are referred to a biochemical text book (Stryer 1988). Chemically, the procedures are so simple that they are prone to automation. The latest method in the field is so-called *multiplex sequencing,* fully automated high-speed sequencing for the mid '90s. About 1% of the human genome has already been sequenced (Table 2). It is postulated that in ten years specialized laboratories will increase the sequencing speed to 1 million bp per day and per person.

Table 2. Mapping and sequencing data on the human genome: The status of mid-1990.

		Loci					
Chromo-some	Length	All loci	Genes	Sequenced genes	Polymorphic loci	Probes	Sequence (bp)
1	8.3	311	192	82	146	677	688,576
2	7.9	196	116	50	90	522	538,478
3	6.4	786	75	29	130	872	112,721
4	6.1	242	73	34	138	461	199,261
5	5.8	192	74	28	112	382	157,066
6	5.5	207	110	55	86	620	451,606
7	5.1	555	121	50	189	965	285,589
8	4.5	172	58	25	55	332	190,517
9	4.4	110	65	24	47	206	151,098
10	4.4	156	62	28	88	253	156,856
11	4.4	624	140	55	189	1,191	336,252
12	4.1	155	103	45	56	402	276,461
13	3.6	122	29	12	53	265	76,751
14	3.5	98	56	33	51	493	243,892
15	3.3	126	52	20	49	163	118,313
16	2.8	335	59	25	122	457	155,443
17	2.7	451	99	47	150	662	312,904
18	2.5	55	23	10	32	143	100,080
19	2.3	194	82	38	59	462	243,674
20	2.1	64	37	17	22	141	106,462
21	1.8	202	34	7	60	308	32,675
22	1.9	238	57	22	99	396	101,325
X	4.7	730	179	31	235	1,245	317,945
Y	2.0	231	13	5	17	234	15,105
Totals	100	6,552	1,909	772	2,275	11,852	5,066,049

(Taken from Stephens et al. (1990), pp. 237–243)

GENETIC KNOWLEDGE WILL INCREASE RAPIDLY

Neither the knowledge of a detailed map of our genome, nor even the availability of the complete sequence of the human genome, will explain per se the phenotypic characteristics of a human being (Grunbaum 1990). Qualities such as behavior, character, or intelligence are results of the hitherto unknown interplay of many hereditary traits as well as environment and education (Motulsky 1983). The "not yet" theory, however, cannot serve as a convenient excuse to postpone the discussion of the correlation between gene structure and human behavior until we reach the experimental stage. Fifty years ago, no one, not even the so-called experts, believed that the genetic information in any type of organism could be stored and expressed in a simple four-letter code. Even today most scientists do not realize how primitive in chemical terms the DNA molecule is. With an automated device which is commercially available, one can synthesize the DNA segment representing a gene in one week and use it as a genetic program in a living organism. With more than 100,000 biologists working worldwide on the elucidation of the link between chemical structure and biological phenomenon, the "not yet" theory may soon collapse. The breeding of animals to fit the needs of mankind will always serve – as far as techniques are concerned – as a forerunner to application on human beings. In vitro fertilization has shown this very clearly. John Steinbeck's novel *Of Mice and Men* could be retitled *From Transgenic Mice to Human Gene Therapy*. However, it must be clearly stated that DNA analysis can definitely not be used at present as a tool to predict human behavior or even moral qualities, nor is there any prospect of it being so used in the near future.

The discussion about the so-called XYY men, i.e., men that have an additonal Y-chromosome, may serve as a warning to those who prematurely try to link DNA sequence and behavior (Breuer 1976). Nearly all XYY men are rather tall and often of strange appearance. It has been suggested that the rate of criminality among them is above average. However, in careful studies of more than 4,000 men, no correlation between an extra Y-chromosome and unusual aggressiveness could be found. The discussion on this issue as outlined in reference may serve as an example of how wishful thinking and overinterpretation can cause much harm and create unnessary human tragedy (Witkin 1976).

However, knowledge about so-called "normal" genes will increase the efforts to detect and interpret the nonnormal genes. In this context it will be extremely difficult to separate polymorphism from abnormalities, especially if we have to rely on only the genotype without correlation to the phenotype.

MOLECULAR GENETICS AND COMPUTERIZED DATA HANDLING

When Johannes Gutenberg invented the art of printing with movable letters about 1450, this technique served to spread medical knowledge rapidly to a broad public. What the printing press was to the fifteenth century, the computer is to the twentieth. The advances in molecular genetics are intimately bounded up with the progress in computer science. Without computers, from PCs to worldwide data networks, experts would be unable to process genetic data. Without the aid of computerized methods, the interpretation of sequence information is extremely difficult, if not impossible. Only by the use of DNA and protein sequence databases, can one identify the meaning of a DNA sequence, e.g., regulatory parts, reading frames, or folding domains. Without data processing, any type of DNA sequences remain nothing but an array of four types of characters. We have said that storage of the human DNA in traditional form would require 3,000 volumes, each with 1,000 pages and 1,000 characters per page. A computer with only a capacity of several megabyte could fulfill the same task, and any scientist, physician, administrator, or businessman with a personal computer is able to extract information from DNA data banks (Table 2).

GENETIC SCREENING IN MEDICAL HEALTH CARE

Within its program *Predictive Medicine,* the European Community in Bruxelles decided to support research in molecular diagnostics with some 10 million DM. The justification for establishing the program came very close to positive eugenics: "Since it is very unlikely that we will be capable of abolishing all environmental safety risks, it becomes increasingly important to elucidate the factors of genetic disposition. It is intended to identify persons at risk, to protect them from diseases, and if advisable prevent the transmission of genetic disposition to the following generations" (paraphrased from memory by the author). This type of reasoning is strongly reminiscent of the arguments of Francis Galton (1883), the inventor of the term "eugenics". After an intensive discussion of this issue in the European Parliament, the European Community was asked to redefine the research program. Nonetheless, the public could well ask why so much attention and finance should be devoted to a scientific goal which is controversial even among biologists. Maybe it is the man-on-the-moon mentality, again.

Screening of newborn children for metabolic diseases has been practiced for a long time in western countries (e.g., phenylketonurea and hypothyreosis) (van den Daele 1985). Genetic tests on the basis of the DNA sequence will enlarge the scope and increase the predictability of illnesses that occur later in life (e.g., Chorea Huntington). However, the rapid progress in techniques, e.g., the isolation of fetal cells from blood samples of the mother, which reduces the risk of infecting the embryo, will allow further commercialization. The easy access to prenatal diagnostics may lead to a testing practice that could easily yield to the temptation of "quality control". According to pediatricians, any suggestion to a pregnant mother that their child might be born with a disability or illness increases the likelihood of the mother requesting an abortion (Table 3).

SCREENING OF EMPLOYEES

In the period from 1972 to 1981 the U.S. Air Force tested black soldiers for the presence of the sickle cell trait, since it was believed that persons who are heterozygous for this disease suffer from oxygen deficiency under low air pressure (Lavine 1982). Similar tests were performed by the chemical company Dupont. These tests have been stopped worldwide, partly because of public concern, mostly, however, because there is no satisfactory scientific basis for predicting the suitability of an employee to a given working situation (e.g., metal dust, organic chemicals). However, with increased genetic knowledge, this reason for not using public screening tests will soon be untenable. (Porter 1982).

ADMISSIBILITY OF DNA TYPING EVIDENCE IN COURT

On several occasions, and in many countries, DNA subtyping has been used in addition to blood group analysis for the identification of a suspect in court (PGM typing) (Grunbaum 1990). In 1988, the Attorney General of California went on record with a proposal for the establishment of a DNA data bank containing a DNA identification of every convicted sex offender and murderer in California. Comparable to other scientifically-based evidence in court, DNA typing or matching will soon increase the analytical capability of the police force. The rapid development of new techniques, such as the polymerase chain reaction, i.e., the amplification of DNA in a test tube, has increased the application of DNA typing for forensic purposes tremendously (Mullis 1990). However, until criteria for ensuring the quality of the analysis are developed, accepted, observed, and strictly enforced, forensic DNA analysis should be considered unreliable and therefore inadmissible for the present. Nevertheless, because of commercial interest in these techniques, both for parenthood identification and for forensic investigations, experimental standards comparable to those in clinical chemistry will soon be established.

Table 3. Examples for single-gene diseases.

disease	description	genetic marker identified	gene cloned	protein identified
Duchenne muscular dystrophy	progressive muscle deterioration	yes	yes	yes
cystic fibrosis	lung and gastrointestinal degeneration	yes	no	no
Huntington's disease	late-onset disorder with progressive physical and mental deterioration	yes	no	no
sickle cell anemia	deformed red blood cells block blood flow	yes	yes	yes
memophilia	defect in clotting factor VIII causes uncontrolled bleeding	yes	yes	yes
beta-thalassemia	failure to produce sufficient hemoglobin	yes	yes	yes
chronic granulomatous disease	frequent bacterial and fungal infextions involving lungs, liver, and other organs	yes	yes	yes - tentative
phenylketonuria	enzyme deficiency that causes brain damage	yes	yes	yes
polycystic kidney disease	pain, hypertension, kidney failure in half of victims	yes	no	no
retinoblastoma	cancer of the eye	yes	yes	yes

(Taken from the OTA Report 1988)

WHY IS GENE ETHICS DISCUSSED SO INTENSIVELY IN GERMANY?

Many of our foreign colleagues in the social or in the natural sciences wonder why the discussion of the future consequences of modern biology is so intensive and sometimes ferocious in Germany. There is a variety of reasons for this, the most harmless one being that so many of our countrymen believe in mysticism. Moreover, the terrible experience of National Socialism is still felt today. One of the laws the Nazis introduced as early as June 14th, 1933 was the *Gesetz zur Verhütung des erbkranken Nachwuchses* (Law for the Prevention of Progeny with Hereditary Diseases) (Siemens 1933). Enforced sterilization of all medically and socially handicapped persons was one of the milder consequences of this law. Fifty years later we are still very sensitive when it comes to the issue of positive eugenics in the context of the well-being of society.

CONCLUSION

In fifteen years' time, each gene will have been assigned to an individual chromosome, and possibly 10% of the human genome, including all genes coding for proteins, will have been sequenced. The data pooled by the Human Genome Organization and the numerous scientists working outside this organization will provide molecular biology and medicine with a new set of structural information which should be beneficial to biological research and medical practice. Without functional data on how the genes work, however, this knowledge would be of rather limited value. Most information will come from regulatory sequences and those areas of DNA research which today are considered either self-indulgent or even worthless. The breaking and rejoining, folding and unfolding of chromosomes cannot be explained without structural data. There has to be an intimate correlation of the efforts in the broad field of molecular genetics and with the pace applied in DNA sequencing (Knippers 1988). Neither the mapping nor the sequencing of genes and other parts of DNA represent a risk to human integrity. Comparable to the first modern studies of the human anatomy during the Renaissance, gene mapping and sequencing will bring benefits to science and medical care. (Lewantin 1984).

The public always becomes suspicious when the governments of industrialized countries such as the United States, Japan, or Germany call upon their scientists to "serve the nation". The consequences of the Manhattan Project show that the fears of the public are often more well-founded than the predictions of the experts. Since a person's genetic is intimately connected with his or her individuality, it is understandable that the public will not easily be convinced of the benefits of establishing a globally active, well-financed research organization such as HUGO. It has often been suggested that positive eugenics may provide an answer to the problems of overpopulation, minorities, the decrease in the quality of the human gene pool, etc. The knowledge of the map or even the sequence of the human genome could provide governmental agencies with the analytical background for genetic mass screening and consequent selective actions. The experts know that there are no such intentions within the aims of the genome sequencing project. However, the matter is so delicate that declarations of good-will are not acceptable guarantees against public or private misconduct. The human genome project is to be flanked by a legislative body which will guarantee the confidentiality of all individual genetic data and protect minorities against selective screening procedures.

REFERENCES

Breuer, G. (1976). *Sociobiology and the Human Dimensions.* Cambridge: Cambridge University Press.

Cantor, C. (1990). Archestrating the human genome project. *Science* **248:** 49.

Culliton, B.J., (1990). Mapping terra incognita (Humani Corporis). *Science* **250:** 210–212.

Gassen, H.G., A. Martin, and S. Bertram, eds. (1978). *Gentechnik.* 2nd edition. Stuttgart: Gustav Fischer Verlag.

Gassen, H.G., A. Martin, and G. Sachse (1988). *Der Stoff aus dem die Gene sind.* Frankfurt a.M. / München: Schweizer Verlag.

Grunbaum, B.W. (1990). Comments on the admissibility of DNA typing evidence in American courts of law. *Biotech Forum Europe* **7:** 305–312.

Human Gene Mapping 10. Tenth International Workshop on Human Gene Mapping. *Cytogenetics and Cell Genetics* **51:** 1.

McKusick, V. (1978). *Medelian Inheritance in Men,* 5th edition. Baltimore: John Hopkins.

Knippers, R.. (1988) Die "schöne neue Welt" und ihre Folgen. In: *Grenzen der Sozialwissenschaften,* ed. H. Mading. Konstanz: Universitätsverlag.

Lavine, M. (1982). Industrial screening programs for workers. *Environment* **24:** 26–38.

Lewantin, R., S. Rose, and L. Kamin (1984). *Not in Our Genes: Biology, Ideology and Human Nature.* New York: Pantheon.

Mullis, K.B. (1990). *Scientific American* **262:** 36–46.

Motulsky, A. (1983). The impact of genetic manipulation on society and medicine. *Science* **219:** 135–140.

Office of Technology Assessment (1988). *Mapping Our Genes. The Genome Projects: How Big, How Fast.* OTA-BA-373. Washington, D.C.: U.S. Government Printing Office.

Olson, M., L. Hood, C. Cantor, and D. Botstein (1989). A common language for physical mapping of the human genome. *Science* **245:** 1434.

Ott, J. (1985). *Analysis of Human Genetic Linkage.* Baltimore: John Hopkins University Press.

Porter, I. (1982). Control of hereditary disorders. *Annual Review of Public Health* **3:** 278–319.

Siemens, H.W. (1933). *Vererbungslehre, Rassenhygiene und Bevölkerungspolitik für Gebildete aller Berufe,* 5th edition. München.

Stephens, J.C., M.L. Mador, M.L. Cavanaugh, M.I. Gradie, and K.K. Smith (1990). Mapping the human genome: Current status. *Science* **250:** 237–244.

Stryer, L. (1988). *Biochemistry,* 3rd edition. New York: W.H. Treeman and Company.

Van den Daele, W. (1985). *Genomanalyse, genetische Tests und Screening. Diagnose, Prävention und Selektion durch Anwendung genetischer Techniken auf den Menschen.* Bielefeld: USP Wissenschaftsforschung Universität Bielefeld.

Watson, J. (1990). The human genome project: past, present,and future. *Science* **248:** 44.

While, R., and J.M. Labouel (1988). Chromosome mapping with DNA markers. *Scientific American* **258:** 40.

Witkin, H. et al. (1976). Criminality in XYY and XXY men. *Science* **193:** 547–555.

Zinke, H. und H.G. Gassen (1990). Sequenzanalyse von DNA. *LABO* **2:** 71–16.

Bioscience ⇌ Society
edited by D.J. Roy, B.E. Wynne, and R.W. Old, pp. 227–246
© 1991, John Wiley & Sons

AIDS: Targeted Research and Public Expectations

Deborah M. Barnes

The Journal of NIH Research
2101 L Street, NW, Suite 207
Washington, D.C. 20037, U.S.A

Abstract. The AIDS epidemic constitutes an international health crisis and a challenge to the research community. Scientists around the world have focused their efforts on AIDS and learned much about the AIDS virus and how it causes disease. Despite the progress, many questions remain. Drugs are available to inhibit the AIDS virus and treat associated infections, but none is a cure for AIDS. No vaccines are available to prevent AIDS. Targeted scientific research is necessary to meet the goals of finding a cure and a vaccine, but the system for funding targeted research is, perhaps unavoidably, encumbered by politics and poor communication.

Many people would like to believe that science will somehow resolve the AIDS crisis. This view is fueled by less-than-ideal interactions among scientists, elected officials, and the public. The current system of targeted research for AIDS encourages a cycle of unrealistic promises on the part of the scientific community, unrealistic demands by elected officials, and unrealistic expectations by the public. Science alone is not likely to resolve the AIDS crisis; reducing the threat of AIDS worldwide will require setting realistic goals and policies, improving communication and education about the disease, and making a financial commitment to support relevant research in the social sciences as well as the biomedical sciences.

"There is presently no cure for AIDS. There is presently no vaccine to prevent AIDS."
[The Surgeon General's Report on Acquired Immune Deficiency Syndrome. October 1986.]

"The AIDS pandemic, an issue that may rank with nuclear weaponry as the greatest danger of our era, provides ... striking proof that mind and technology are not omnipotent and that we have not canceled our bond to nature."
[S.J. Gould. April 19, 1987. The New York Times Magazine.]

Depending on who is asked, the response of the scientific community to the AIDS epidemic is a colossal failure, the best that could be expected, or an expensive overreaction. Whatever their opinions, however, most people would agree on one thing: AIDS research has been dominated by political pressure and public expectations.

Over the ten-year course of the AIDS pandemic, AIDS research has become a prototype for evaluating the success of targeted research* programs. Some question whether targeted research or research designed exclusively by individual investigators is the best means to accomplish particular goals, in this case a cure or a vaccine for AIDS. Others contend that the focus on scientific research as a means to resolve the AIDS crisis is inappropriate. Finding a cure and a vaccine for AIDS are complex objectives that may never be realized. Predictably, the public is frustrated by an apparent lack of progress.

Nevertheless, targeting funds for AIDS research has permitted rapid progress in understanding much of the basic biology of the virus and the disease process, even though clinical solutions to AIDS – a cure or a vaccine – still do not exist. Meanwhile, the emphasis on targeting money for biomedical research activities has probably diverted attention and funding for activities that might reduce the threat of AIDS more immediately. These include studies to determine the most effective means for educating the public about AIDS, implementing educational programs that are identified as effective,

---

* The term "targeted research" as used here refers to research directed toward a specific problem – namely, the AIDS epidemic. In the context of this chapter, targeted research on AIDS includes both basic and applied research. Basic research – on the molecular biology and pathogenic effects of the AIDS virus, the functioning of the human immune system, and the study of related animal viruses, for example – addresses fundamental questions in biology, often without regard to potential applications. Applied research is designed specifically to result in useful products, such as diagnostics, drugs, or vaccines for AIDS. Definitions of targeted or strategic research are the subject of much debate, however. For a discussion, see Johnston 1990.

establishing large-scale programs to test people for evidence of infection by the AIDS virus, and implementing legislation to protect the civil rights of infected people.

THE EXPANDING AIDS EPIDEMIC AND THE INCREASE IN FUNDING

AIDS is caused by human immunodeficiency virus-type 1 (HIV-1), and, in some African countries, by its less pathogenic relative, HIV-2. Both are retroviruses; their genomes are composed of ribonucleic acid (RNA) instead of the deoxyribonucleic acid (DNA) that makes up the genes of most organisms. Like all viruses, HIV must infect a living host cell before it can reproduce. It copies its RNA genome into DNA and usually integrates into the genome of the host cell – a stage called the provirus (Fig. 1). HIV can remain quiescent, or it can become active, producing infectious virus particles that bud from the host cell's surface. This often, but not always, kills the cell. HIV primarily infects human T4 lymphocytes and macrophages because these blood cells carry the CD4 antigen on their surfaces. CD4 molecules act as receptor sites for HIV particles, allowing them to dock on the cell surface, infect the cell, and – particularly for a T cell – ultimately kill it (Gallo and Montagnier 1988; Weber and Weiss 1988).

HIV is transmitted from person to person during homosexual or heterosexual intercourse, through infected blood or blood products, and from mother to child during pregnancy, childbirth, or, possibly, breast-feeding. People infected with HIV develop severe immune deficiency, because, over time, the virus destroys most of their T4 lymphocytes. These cells orchestrate many immune reactions that would normally protect a person from other infectious organisms – bacteria, parasites, and other viruses. Without a sufficient supply of T4 lymphocytes, people become susceptible to these opportunistic infections, and many die from them. HIV also infects the brain and other tissues (Redfield and Burke 1988).

Some AIDS drugs inhibit HIV at various stages of its life cycle and – at least partly – inhibit it from replicating or infecting cells. Other drugs treat opportunistic infections (Yarchoan et al. 1988). Researchers are currently trying to develop a mix of drugs that will keep the virus in check, because they are pessimistic that any drug will completely eliminate HIV from the body.

Researchers are also trying to design vaccines that will protect people from becoming infected with HIV. Vaccines work by stimulating the body's natural

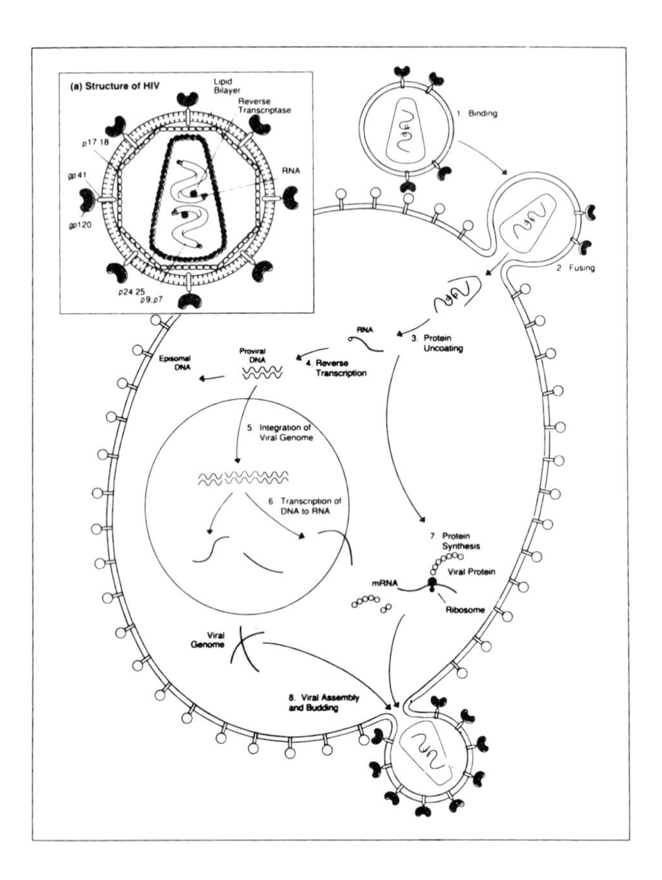

Fig. 1. Life cycle of human immunodeficiency virus (HIV). The outer coat protein of HIV binds to CD4 receptors on a cell's surface (top right). The virus injects its genome – made of ribonucleic acid (RNA) – into the host cell. Viral RNA is converted to deoxyribonucleic acid (DNA) by reverse transcriptase, a viral enzyme. The newly formed proviral DNA incorporates into the host cell DNA (middle). When the cell is activated, its DNA, including the HIV proviral DNA, is transcribed into RNA and traslated into proteins. Viral proteins and viral RNA self-assemble to produce infectious virions, which then bud from the surface of an infected cell (lower right). (from: AIDS/HIV Treatment Directory, vol. 4, June, 1990. Compiled by the American Foundation for AIDS Research (AmFAR))

immune reactions – antibody production and cell-mediated responses – that block the virus (Matthews and Bolognesi 1988). As yet, however, researchers are not certain which immune reactions will protect against HIV. To date, there are no drugs to cure AIDS and no vaccines to prevent it, although progress is being made on both fronts.

For example, recent studies on simian immunodeficiency virus (SIV), which cause an AIDS-like disease in monkeys, indicate that it may be possible to induce protective immunity by using killed whole virus particles as the basis for a vaccine (Desrosiers et al. 1990; Murphy-Corb et al. 1989; Murphy-Corb et al. 1990). Another recent effort is to test several candidate AIDS vaccines in a novel way – as therapies in people already infected with HIV (Barnes 1989). A complicating factor with all of these strategies, however, is that the virus – either HIV or SIV – mutates so rapidly that a vaccine developed against one strain may not protect against infection by another strain. Researchers are trying to address this problem by designing vaccines that contain immunogens from multiple strains and are therefore broadly protective (Matthews and Bolognesi 1988).

Public pressure to end the AIDS epidemic is enormous, and the increase in targeted research funds to meet that goal has been immense. Perhaps nowhere in the world are these elements so starkly represented as in the United States.

Although the AIDS virus probably originated in Africa sometime before 1950, it was not until the 1980s that researchers in the United States identified AIDS as a disease. In 1981, physicians in California reported an unusual surge of Kaposi's sarcoma (normally a rare cancer characterized by purplish skin lesions) and *Pneumocystis carinii* pneumonia (a parasitic infection of the lung that can be fatal) in homosexual men whose immune system function was severely depressed. The disease was named Acquired Immune Deficiency Syndrome (AIDS), and by December 1981 more than 200 cases had been reported in the United States. At the time, no one was studying AIDS, and no funds were targeted for AIDS research.

Since then, total spending by the U.S. Public Health Service for AIDS research, surveillance, regulatory activities, and health care has grown to nearly $1.7 billion (Fig. 2). The U.S. National Institutes of Health (NIH), which supports much of the country's biomedical research, spent more than $743 million on AIDS in fiscal year 1990 – about one-tenth of its entire budget. Federal funding for AIDS research is about one-half of that for cancer and nearly the same as that for heart disease. But the number of anticipated deaths from AIDS in 1991 is 37,000 to 42,000 (CDC 1990) – about one-tenth and one-twentieth of the expected numbers of deaths each year from cancer and heart disease, respectively. Why has the increase in support for AIDS research been so large and so rapid?

Paradoxically, the massive funding and the slow start of AIDS research may have a common cause. In the United States, AIDS initially affected, and continues to affect, an organized and politically powerful group – the homosexual community. In the early 1980s, at the start of the epidemic, members of Congress were reluctant to vote for targeted funds to support AIDS research, presumably because it was not politically popular to do so. But in the ensuing years, gay men, their friends, relatives, and political allies lobbied persistently and convincingly for federal support for AIDS research. Although people with other serious diseases – cancer, heart disease, and diabetes, for example – have private organizations to represent their interests, these groups have not achieved the political clout of AIDS activist groups.

Another factor that led to rapidly increasing federal support for targeted research on AIDS was the realization that the disease did not confine itself to gay men. In the mid-1980s, articles in the lay press highlighted a general fear of AIDS and a shift in public attitudes:

> The disease of *them* is suddenly the disease of *us*. The slow death presumed just a few years ago to be confined to homosexuals, Haitians and hemophiliacs is now a plague of the mainstream, finding fertile growth among heterosexuals. It is today a crisis for the U.S. more deadly than many wars of modern times. In just four years, the disease will have killed more Americans than the Vietnam and Korean wars combined. "It will probably be the most important public-health problem of the next decade and going into the next century", warns microbiologist and Nobel laureate David Baltimore. "It threatens to undermine countries."(McAuliffe 1990)

The horrific nature of AIDS further heightens public fear, resulting, perhaps, in psychological denials: "It can't happen to me", while at the same time making pleas for a cure even more desperate. People with AIDS suffer from multiple viral, bacterial, parasitic, and fungal infections. In addition, AIDS patients frequently have cancers, such as lymphomas or Kaposi's sarcoma; they may have movement disorders, dementia, or painful damage to sensory nerves. And, in the terminal stages of illness, most AIDS patients are wasted, weighing only a fraction of their normal body weight.

From infection to death, the course of disease in an adult varies, but probably takes an average of 8–11 years (Lifson 1988). Infants with HIV infections who acquire the disease from their mothers die far more quickly than adults. One-half of infants who develop clinical symptoms of AIDS before six months of age die within the next six months. One-half of infants diagnosed with AIDS after one year of age die within the next 20 months

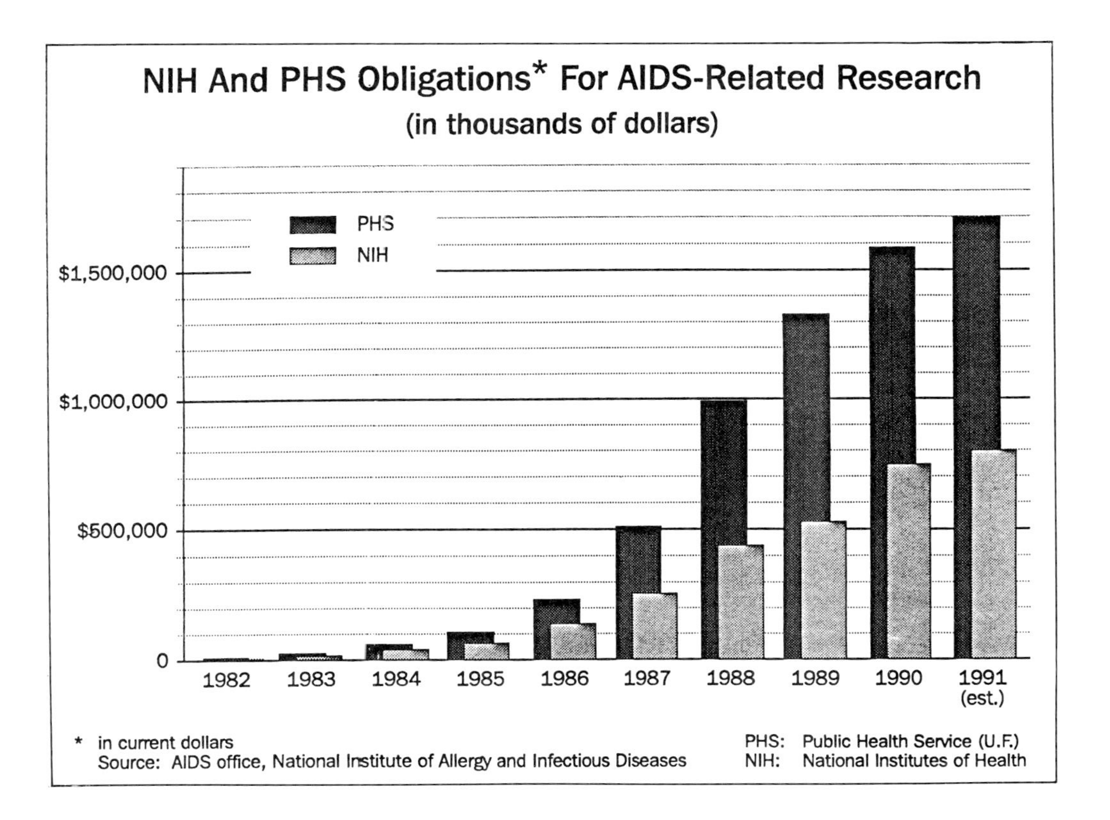

Fig. 2. Growth of funding for AIDS research in the United States from 1982 to 1991

(Rogers 1988). Many of these children, particularly those born to high-risk women in urban areas, spend much of their lives in the hospital.

Another reason for the rapid increase in AIDS funding is the state of scientific knowledge. The AIDS epidemic struck at a time when the techniques of molecular and cellular biology were well developed. As a result, scientists learned more about the AIDS virus, and learned it faster than they have for any other infectious disease (Barre-Sinoussi et al. 1983; Popovic et al. 1984; Gallo et al. 1984; Schupbach et al. 1984; Sarngadharan et al. 1984). The rapid advances in basic research on AIDS helped to justify the rapid increases in funding. But they also fueled public expectations, and, in turn, public expectations fueled more funding.

In the United States, the result today is a clash between public expectations and scientific reality. Many members of the public seem to expect that scientists will develop drugs and a vaccine for AIDS, and that these therapies will eliminate the disease. The scientific reality is that a cure for AIDS and a vaccine to prevent its further spread are goals that may never be fully realized.

Meanwhile, steps that could be taken immediately to stem the spread of the disease – assessing the number of people infected with HIV, educating them about ways to avoid transmitting the virus, passing legislation to protect the civil rights of those infected, and educating the public at large about ways to avoid becoming infected – are sporadic at best. The striking fact is that the United States, which has the world's largest number of reported cases of AIDS, still has no national policy on how to cope with the disease.

THE GLOBAL EPIDEMIOLOGY OF AIDS

According to a report by the World Health Organization (WHO) in Geneva, which tracks the AIDS epidemic internationally, the AIDS virus now infects an estimated 8–10 million people worldwide (WHO 1990a). WHO estimates that approximately 3 million of these HIV-infected people are women of childbearing age. Depending on a range of factors, between 30 and 50% of sexually active African women infected with the AIDS virus will pass it on to their infants during pregnancy, childbirth, or early postnatal life.

The key word in WHO's statistics – indeed in all AIDS statistics – is "estimate". No one really knows how many people are infected with the AIDS virus, because no country, including the United States, has a nationwide, mandatory program for testing people to determine how many are HIV-positive.

Most countries do have programs to report cases of AIDS, however. A tally of actual cases reported to WHO over the past four years illustrates the explosive growth of the international AIDS pandemic (Fig. 3). By November 1990, WHO had cumulative records of approximately 300,000 AIDS cases from more than 150 different countries. Again, however, the tallies are not hard numbers. WHO acknowledges that most countries underreport cases of AIDS, and it estimates the actual cumulative number of cases worldwide to be more than 1.3 million, roughly four times the number actually reported (WHO 1990a). WHO projects that some 400,000 of these cases occur in children, most of whom live in developing countries and acquired it from their mothers.

Africa now accounts for approximately 25 percent of the world's 300,000 reported cases of AIDS, and the Americas account for more than 60 percent. But WHO estimates that developing countries account for at least half of those infected with HIV, and that by the year 2000 these countries will be home to more than two-thirds of those infected (WHO 1990a). The obvious consequence will be a shift in the current global pattern of people with full-blown AIDS, and an increasingly serious world-wide public health care problem.

The patterns of HIV transmission vary among different countries. WHO estimates that, worldwide, more than "60% of all HIV infections have been acquired through heterosexual transmission". Heterosexual transmission is particularly prevalent in developing countries in Africa and the Caribbean, where the ratio of AIDS cases in men to women is approximately 1:1.

Already, in some urban areas of sub-Saharan Africa, 20–30% of adults aged 20–40 years are now thought to be infected with the AIDS virus (WHO 1990b) According to a recent article in *The New York Times*, the government of Uganda estimates that

> ... a million Ugandans of a total of 16 million are infected with the AIDS virus. That is one of every eight people 15 years old and over nationwide, men and women from all sectors of society. In the United States, a comparable rate would mean 25 million infected people, instead of the estimated one million. (Eckholm 1990)

The expanding number of AIDS cases worldwide contrasts with a more uneven epidemiological picture in the United States, where the Centers for Disease Control (CDC) in Atlanta, Georgia tracks the disease. As of January 1991, more than 160,000 cases of AIDS had been reported to the CDC, and more than 100,000 people had died from the disease (CDC 1991). The

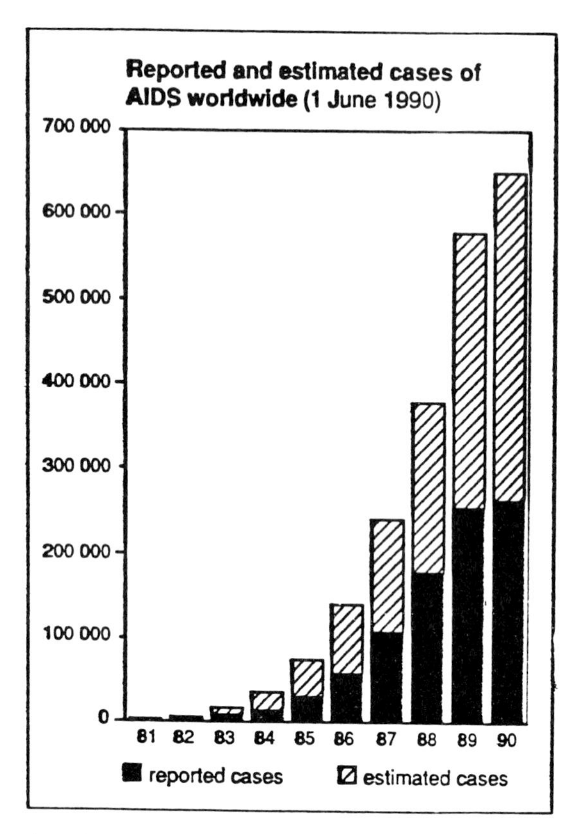

Fig. 3. Expansion of the number of cases of AIDS worldwide between 1981 and 1990. Figure shows the number of cases reported to the World Health Organization (WHO) (dark bars) and the number estimated by WHO. (WHO data as of June, 1990) (From: Global AIDS Factfile, June, 1990, WHO Global Programme on AIDS)

cumulative number of AIDS cases in the United States more than doubled from just over 29,000 in December 1986 (CDC 1986) to more than 80,000 by December 1988 (CDC 1988). The rate of increase slowed beginning in the mid-1980s, and in 1990 rose again (CDC 1991).

Unlike the dominant pattern of heterosexual disease transmission in developing countries, homosexual intercourse and intravenous drug use (among both males and females) account for more than 80% of AIDS cases in the United States. The ratio of men to women with AIDS is therefore unequal; approximately 90% of AIDS cases in the U.S. have occurred in males, more then three-quarters of whom are between 25 and 44 years of age. This male-to-female ratio may be changing, however, because recent surveys indicate that relatively more women are becoming infected with the AIDS virus than in the past. Heterosexual transmission in the United States is becoming an increasing risk factor, particularly among teenagers. Overall, heterosexual transmission alone now accounts for about 5% of all AIDS cases in the United States (CDC 1991), a two-and-a-half-fold increase since 1986.

The demographics of AIDS within individual countries also vary. WHO reports that in Africa the AIDS epidemic is no longer confined to urban areas and is spreading quickly into rural communities (WHO 1990a). In the United States, however, AIDS is still primarily an urban disease; more than 85% of cases occur in metropolitan areas with populations of 500,000 or more (CDC 1991). The incidence of AIDS is increasing among women and children in U.S. cities, particularly among Blacks and Hispanics. During the next ten years the demographics of AIDS cases in the United States may shift somewhat as the disease spreads into smaller cities and rural communities.

WHAT DOES THE U.S. PUBLIC EXPECT FROM AIDS RESEARCH?

It is probably fair to say that the American public wants AIDS to go away, and many would like the research community to make that happen. Public attitudes about AIDS and about the likelihood that scientists will develop a cure or a vaccine are difficult to assess, however, because of the lack of surveys on the issue. Nevertheless, public and private organizations have conducted surveys of public opinion, and anecdotal accounts of public attitudes are also available (Dawson 1990). Collectively, they give the impression that the public favors more AIDS research and believes that more of the available funds should be targeted toward finding a cure and a vaccine. Whatever specific attitudes may be, it is clear that the demands of AIDS activists, the Congress, and the lay public continue to shape the U.S. federal budget for AIDS research and to influence profoundly the testing and distribution of potential drugs and vaccines.

In the United States, AIDS research is funded largely by the federal government, and the public's demands on the research community have become synonymous with demands that the federal government should solve the AIDS crisis. Recent surveys show that the American public advocates increased resources for AIDS research.

For example, a nationwide survey conducted in June 1990 of 1000 adults over 18 years of age reported that 61% think the government is not doing enough to fight AIDS, 29% think the effort is about right, 3% believe it is too much, and 7% don't know. A large majority of blacks – 77% – say the federal effort is insufficient. 75% of those polled do not think AIDS research is taking money from research on other diseases such as cancer and heart disease. And 62% say they would be willing to pay higher taxes to support more AIDS research (Knox 1990).

As compared with survey results obtained five years ago, these recent statistics show increased concern that the government is not doing enough. In 1985 the Gallup Organization surveyed 759 adults, of whom only 50% said the government was not spending enough money on AIDS research. That survey also reported that 56% thought that developing a vaccine should be the government's top priority, 28% favored finding a cure, and 10% wanted better tests to determine who was exposed to the disease. (The survey was reported in the August 12, 1985 issue of Newsweek.)

Public opinions are sometimes reflected in anecdotal accounts. Occasional newspaper columnists, for example, have said that AIDS patients bring the disease upon themselves, and that research efforts are overfinanced (Kilpatrick 1988). More recent commentaries signal a shift in this attitude, saying that support for AIDS research should be sustained because it will inevitably benefit many areas of biology and medicine (Schrage 1990; Freundlich 1990).

The scientific community itself is divided in its opinions about AIDS research. A 1990 survey by the U.S. Office of Technology Assessment (OTA) indicates that 48% of scientists polled agree or strongly agree that "too much research funding has been diverted to AIDS/HIV research from other fields." Nearly the same number, 44%, disagreed or strongly disagreed with the same statement (U.S. OTA 1990).

Physicians who care for AIDS patients do not always plead for more research funds. Instead, they ask for more hospital beds, nurses and doctors, social workers, and support staff (Heagarty 1987). Health policy experts acknowledge that a cure or vaccine, if possible, is far in the future, and they recommend extensive public education programs that teach people how to avoid becoming infected. They also urge the country's leaders, particularly the President, to coordinate and direct the AIDS effort (National Academy of Sciences 1986). Others try to ensure that HIV-infected people do not suffer discrimination – in housing, jobs, schools, or insurance benefits.

Activist groups have had by far the biggest impact on policy decisions about AIDS. One of the most vocal and politically powerful citizen groups today is the New York City-based *AIDS Coalition To Unleash Power* (ACT

UP), which identifies itself as a "diverse, non-partisan group of individuals united in anger and committed to direct action against the AIDS crisis". ACT UP's expectations from the AIDS research community are clear:

> On May 21 [1990], AIDS activists from across the country stormed the National Institutes of Health because we were angry. We were angry – and still are – because in ten years of research, the N.I.H. has all but ignored the real needs of people living with the disease . . .
>
> The N.I.H.'s AIDS program, mandated to study new treatments against the disease has produced precious little. Instead, the program has spent the bulk of its resources re-testing AZT, a drug approved [by the Food and Drug Administration (FDA)] for general use three years ago. To find new therapies as quickly as possible, the N.I.H. program must research at least 30 new treatments a year in small, quick initial studies. (ACT UP 1990)

ACT UP's statement is provocative and it is misleading. It is true that AZT (3'-azido-3'-deoxythymidine) was approved by the FDA in 1987, but it was approved only for people in the terminal stages of disease – not for general use. In high doses, the drug is so toxic to bone marrow that 40% of AIDS patients cannot tolerate it. The NIH, therefore, organized clinical trials to test the effects of low and intermittent doses, and has also tested AZT in HIV-infected people who do not yet have symptoms of AIDS. These more recent trials showed a clear medical benefit to those taking lower doses of the drug earlier in the course of disease.

ACT UP's views may not be universal, but they are reflected in countless news articles and protest demonstrations, and they have helped to reshape the federal government's procedures for testing and distributing potential drugs and vaccines for AIDS. In particular, the NIH and FDA have streamlined processes for testing potential new therapies. Through their new "parallel track" and "expanded access" programs, the two agencies have permitted the distribution of several unproven drugs to thousands of AIDS patients, who willingly accept the risk of toxicity (Hooper 1989).

The popular move to distribute experimental drugs more widely may have adverse consequences, however. It arose as a direct consequence of political pressure from AIDS patient organizations, not because it is a scientifically sound method for determining the safety or efficacy of new drugs. And although "parallel track" and "expanded access" programs help to satisfy demands for access to new drugs, in the long run it may become impossible to test newly developed drugs for long-term safety or efficacy. Many AIDS patients will already be taking multiple medications, and it may be regarded as unethical to withhold one potential therapy in order to test another one. The net result could be that AIDS patients continue to take drugs with relatively low efficacy and relatively high toxicity, which are precisely the opposite outcomes that AIDS activist groups – and the scientific community – are seeking.

CAN TARGETED RESEARCH ON AIDS MEET THE PUBLIC'S EXPECTATIONS?

In the United States, the solution to a problem – any problem – is to throw money at it. The American public is conditioned to expect that money translates into instant results. Therefore, the reasoning goes, if AIDS research is well-funded, a cure and a vaccine should be just around the corner.

Perhaps the public needs high, sometimes unrealistic, expectations about the outcome of a federal program – including AIDS research – to support it enthusiastically. If scientists were to explain too forcefully that AIDS research might never yield a cure or a vaccine, but that the public's money would not be wasted because other valuable research findings would inevitably result, public support for AIDS research might be impossible to obtain. Elected officials and scientists recognize that stating research goals conservatively or stating research results narrowly could produce an intolerable funding gap in AIDS research, so they enthusiastically proclaim goals for the research that might never be realized. But, to some extent, the strategy has backfired. The public's desires for a cure and a vaccine for AIDS have not been met. The predictable result has been confusion, anger, and frustration – perhaps most clearly demonstrated by AIDS activist groups.

Meanwhile, scientists who do basic research on AIDS have made considerable progress in understanding the molecular biology of HIV and how the virus causes disease. But the translation of findings from basic research laboratories to the clinic is a slow process, not an immediate event.

Poor communication among different sectors of society compounds the problem. Many scientists are not good communicators, and they fail to explain the caveats of their research and the halting progress that typifies science to the public. Too often the media report on so-called clinical "breakthroughs", and their stories do not include the intricacies of scientific research and the questions that remain unanswered. And many in the public fail to appreciate the distinction between advances in basic research and clinical solutions to AIDS. These distinctions are further muddied because federal funds are "targeted" to support the entire scope of AIDS research.

Thus, the AIDS crisis has become dominated by a cycle of inflated public expectations, increased funding for targeted research, and frustration that perceived research goals have not been met. The cycle is difficult to break. Although the scientific community is not by itself responsible for generating unrealistic public expectations about AIDS research, scientists and administrators of research funds have clearly benefited from those expectations.

Is targeted research always subject to these difficulties, or could something different have been done in AIDS research to avoid them?

The U.S. government has performed a balancing act in AIDS research, and the actions of AIDS activist groups, the public, and Congress have threatened this balance. The government has supported what it hopes is enough basic research to answer fundamental questions about HIV and the disease process, while at the same time targeting blocks of money for specific, more clinically-oriented projects.

This range of options is encouraged by the NIH's various mechanisms for soliciting research proposals and awarding research funds. At one end of the spectrum are investigator-initiated research proposals. As a group, the basic research scientists who submit these proposals favor an individualized, creative approach to science rather than goal-oriented research.

> In my view the best way to foster interdisciplinary creativity is not to impose new structures, but to liberate individual scientists to reconstellate themselves as called for by the scientific opportunity. (Lederberg 1987)

Left to their own devices, basic scientists pursue research directions that they decide are intellectually stimulating. For many, intellectual freedom is a major reason for becoming a scientist. But a public health crisis such as AIDS requires good minds from many disciplines, and the U.S. government uses money to attract scientists into goal-oriented research. So at the other end of the spectrum, the NIH makes the funds available in various targeted research programs: some solicit grant proposals in loosely defined areas of research, whereas others offer contracts to meet specific needs.

This system of inviting and funding research proposals has existed for many years. But when large amounts of new money become available – as they did for AIDS research – it takes time for the system and the research community to adjust.

Initially in AIDS research, some of the grant proposals were not rated as highly as were proposals for other research areas. An overabundance of resources in the early years of AIDS was probably necessary to attract new investigators into the field. Now that AIDS research has become an established field, many researchers are drawn to it because of its intrinsic intellectual interest. The quality of AIDS research proposals is now high, and competition for AIDS funding is as strong as that in other fields.

Nevertheless, some scientists have contended that the usual funding system should have been changed to meet the AIDS crisis. They argue that NIH officials should have aggressively recruited scientists to do AIDS research, and that the government should have immediately funded the construction of more laboratory facilities that could safely contain the AIDS virus. Other researchers have insisted that the government should have encouraged collaboration among government, academic, and private industry researchers much sooner (Barnes 1987). In short, say the critics, the mechanisms necessary to meet the goals of AIDS research should have been more focused at the beginning of the epidemic.

But research that is targeted to specific goals too early risks error. In theory, the government could support many groups of scientists all working to develop one particular kind of drug – one that inhibits the activity of an enzyme necessary for HIV replication, for example. Given the available scientific knowledge, ideas about how such a drug might work could seem sound and the drug could be developed quickly. If it turned out to be the right drug, the research would be hailed as a tremendous success. But if it turned out not to work – or to work only in vitro and not in an infected person – it would mean that other potentially valuable research projects had been abandoned in favor of an approach that proved to be fruitless. The risk is too great, so the NIH spreads the money around, gambling that the right mix of basic and applied research will result in a therapy and a vaccine.

Meanwhile, the public seems to expect a more rapid return on its investment. Sometimes, expectations about AIDS drugs and vaccines are fueled by overly enthusiastic news reports of progress. The political system in the United States also feeds unrealistic expectations about AIDS research. Rather than leading a nationwide effort to address the AIDS crisis, or funding behavioral studies or educational programs that might help reduce HIV transmission, elected officials apparently prefer to let the scientific community take the heat.

Perhaps this approach is tolerated because the public accepts targeted scientific research as a means to a solution to the AIDS crisis. Perhaps it is tolerated because a scientific solution could absolve people of responsibility to alter their sexual or drug-using behavior that put them at risk for acquiring AIDS. Perhaps it is tolerated because parents and schools would not have to take responsibility for educating children about AIDS. Whatever the reasons for tolerating the status quo, they are undoubtedly complex and certainly not well understood.

And so, a cycle has developed. Scientists have allowed unrealistic expectations by the public to develop because, in the end, they need the strong support of the public to feed the financial appetite of the AIDS research effort. Politicians astute enough to understand the cycle have not attempted to correct it, perhaps because, in a broader sense, public ignorance feeds their own political ambitions.

A U.S. government health official who is frustrated that politics rather than scientific judgment prevails in AIDS research summed it up this way:

> We can't do AIDS research without federal support. That means it is political. And that means it can't be done the way it should.

CONCLUSIONS

Certainly, when society is faced with a public health crisis such as the AIDS epidemic, it needs to identify goals that may resolve the crisis. Public funds should be allocated to try to meet those goals, and the research supported by those funds is, by definition, targeted research.

By the standards of basic research scientists, targeted research for AIDS has been a tremendous success. Within nine years they identified the virus that causes AIDS, unraveled its complex genome, identified the functions of its proteins, gained information about immune system function, learned how some cellular genes are controlled by viral genes, and identified mechanisms that induce the expression of viral genes. No basic research scientist studying AIDS today would ever claim that knowledge about the virus is complete, but virtually all would say that research has been successful. From a clinical perspective or a public health standpoint, however, targeted AIDS research has been far less successful.

Part of the problem may be that goals for combatting the AIDS epidemic have been defined too narrowly. The public apparently expects a scientific solution to the AIDS epidemic – a cure, a vaccine, or both –, but such a solution may never be achieved, despite the astounding progress in basic research.

What can be achieved, however, is a reduction in the transmission of the virus. This goal requires behavioral research, educational programs, and the active participation of the public. Another attainable goal is better and more compassionate care for people already infected with the AIDS virus. This objective requires more funding for health care and legislation to protect the civil rights of HIV-infected people (National Research Council 1989).

The scientific community has not met its challenge perfectly. It is likely that the funds for AIDS research could have been spent more effectively in the early years of the epidemic, and that scientists in other areas of research could have been recruited earlier to study AIDS.

But the AIDS crisis cannot be solved by targeted scientific research alone. It is incumbent on the public, the research community, and elected officials to redefine the goals for the AIDS crisis and collectively accept the responsibility for attaining those goals.

REFERENCES

ACT UP (1990). 10 steps towards a cure (advertisement). In: *The Washington Post*, May 25.

Barnes, D.M. (1987). Meeting on AIDS drugs turns into open forum. *Science* **237:** 1287–1288.

Barnes, D.M. (1989). AIDS vaccines: Exploring the options for protection and therapy. *The Journal of NIH Research* **1:** 81–88.

Barré-Sinoussi, F., et al. (1983). Isolation of a T-lymphotropic retrovirus from a patient at risk for acquired immune deficiency syndrome (AIDS). *Science* **220:** 868–871.

CDC – Centers for Disease Control (1986). AIDS weekly surveillance report – United States. December 29.

— (1988). AIDS weekly surveillance report – United States. December 12.

— (1990). Estimates of HIV prevalence and projected AIDS cases: Summary of a workshop, October 31-November 1, 1989. *Morbidity and Mortality Weekly Report* **39:** 110–119.

— (1991). HIV/AIDS Surveillance. January, 1991.

Dawson, D.A. (1990). AIDS knowledge and attitudes for July-September 1989. Advance data from Vital and Health Statistics, No. 183, published March 14, 1990 by the National Center for Health Statistics, U.S. Department of Health and Human Services. (Data are based on The National Health Interview Survey, compiled from continuous, household interviews of 10,277 adults in the United States over the age of 18 by selected characteristics.)

Desrosiers, R.C. (1989). Vaccine protection against simian immunodeficiency virus infection. *Proceedings of the National Academy of Science (U.S.A)* **86:** 6353–6356.

Eckholm, E. (1990). Confronting the cruel reality of Africa's AIDS epidemic. *The New York Times*, September 19, 1990: A1.

Freundlich, N. (1990). No, spending more on AIDS isn't unfair. *Business Week*, September 17: 97.

Gallo, R.C., et al. (1984). Frequent detection and isolation of cytopathic retroviruses (HTLV-III) from patients with AIDS and at risk for AIDS. *Science* **224:** 500–503.

Gallo, R., and L. Montagnier (1988). AIDS in 1988. *Scientific American* **259:** 40–48.

Heagarty, M. (1987). AIDS: A view from the trenches. *Issues in Science and Technology* Winter 1987: 111.

Hooper, C. (1989). Getting the drugs out before the data come in. *The Journal of NIH Research* **1:** 34–35.

Johnston, R. (1990). Strategic policy for science. In: *The Research System in Transition*, eds. S. Cozzins, P. Healey, A. Rip, and J. Ziman, pp. 213–226. Dordrecht / Boston / London: Kluwer.

Kilpatrick, J.J. (1988). Aren't we overreacting to AIDS? *The Washington Post*, June 9, 1988: A19.

Knox, R. (1990). Most favor bigger role in U.S. AIDS fight. *The Boston Globe*, June 17, 1990. (The 1990 poll of 1000 Americans over the age of 18 was conducted by KRC/Communications Research in Newton, Mass.)

Lederberg, J. (1987). Does scientific progress come from projects or people? Address to the National Association of State Universities and Land Grant Colleges, November 9, 1987.

Lifson, A., G. Rutherford, and H. Jaffe, H. (1988). The natural history of HIV infection. *Journal of Infectious Diseases* **158:** 1360–1367.

Matthews, T., and D. Bolognesi (1988). AIDS Vaccines. *Scientific American* **259:** 120–127.

McAuliffe, K., J. Carey, S. Wells, B. Quick, and M Dobbin (1987). AIDS: At the dawn of fear. *U.S. News and World Report*, Jan. 12, 1987: 60–70.

Murphy-Corb, M., et al. (1989). A formalin-inactivated whole SIV vaccine confers protection in macaques. *Science* **246:** 1293–1297.

Murphy-Corb, M. et al. (1990). Induction of protective immune responses to viral infection by immunization of rhesus monkeys with formalin-inactivated whole SIV and glycoproteinlinked SIV subunit vaccines. In: *Vaccines 90: Modern Approaches to New Vaccines Including Prevention of AIDS,* eds. F. Brown, R. Chanock, H. Ginsberg, and R. Lerner, pp. 393–399. New York: Cold Spring Harbor Press.

National Academy of Science (1986). *Confronting AIDS*. Washington, D.C.: National Academy Press

National Research Council (1989). Social barriers to AIDS Prevention. In: *AIDS: Sexual Behavior and Intravenous Drug Use*, eds. C. Turner, H. Miller, and E. Moses, pp. 372–401. Washington, D.C.: National Academy Press.

Popovic, M., et al. (1984). Detection, isolation, and continuous production of cytopathic retroviruses (HTLV-III) from patients with AIDS and pre-AIDS. *Science* **224:** 497–500.

Redfield, R., and D. Burke (1988). HIV infection: The clinical picture. *Scientific American* **259:** 90–98.

Rogers, M. (1988). Pediatric HIV infection: Epidemiology, etiopathogenesis, and transmission. *Pediatric Annals* **17:** 324–332.

Sarngadharan, M.G., et al. (1984). Antibodies reactive with human T-lymphotropic retroviruses (HTLV-III) in the serum of patients with AIDS. *Science* **220:** 506–508.

Schrage, M. (1990). Debate on AIDS research spending should focus on broader benefits. *The Washington Post,* August 3, 1990: F3.

Schüpbach, J. (1984). Serological analysis of a subgroup of human T-lymphotropic retroviruses (HTLV-III) associated with AIDS. *Science* **220:** 503–505.

U.S. Office of Technology Assessment (1990). How has federal research on AIDS/HIV disease contributed to other fields? April 1990. (The survey, conducted in February 1990, compiled the responses of 147 professionals from the basic sciences, medicine, epidemiology, public health and health services research, and other related disciplines.)

WHO (1990a). The Global AIDS Situation. In Point of Fact. November, 1990. No. 72.

— (1990b). Global AIDS Factfile. June, 1990.

Weber, J., and R. Weiss (1988). HIV infection: The cellular picture. *Scientific American* **259:** 100–109.

Yarchoan, R., H. Mitsuya, and S. Broder (1988). AIDS therapies. *Scientific American* **259:** 110–119.

Bioscience ⇌ Society
edited by D.J. Roy, B.E. Wynne, and R.W. Old, pp. 247–259
© 1991, John Wiley & Sons

Why and in What Ways Has Assisted Reproduction Provoked a Strong Moral Response from Society?

Elliot E. Philipp

46, Harley Street
London W1N 1AD, U.K.

Abstract. A short description is given of each of the currently available techniques for assisting reproduction. For each method an attempt is made in the text to explain how and why society reacts. These methods are, in chronological order of their first descriptions: 1) AID; 2) AIH; 3) chilling (and freezing) gametes; 4) stimulation of refractory ovulation; 5) in vitro fertilization and embryo replacement or transfer; 6) freezing of embryos for use in later reimplantation or in research; 7) research into methods of examining embryos so as to avoid replacement of genetically (chromosomally) defective embryos; 8) surrogate motherhood – natural and scientifically assisted "host mothering"; 9) oocyte donation; 10) embryo flushing; 11) genetic engineering to remove defective genes, possibly replacing them with good ones; 12) embryo division; 13) cloning.

INTRODUCTION

When a couple are unsuccessful in fulfilling their wish to have a family, they are infertile. Other couples may be prone to produce children with inherited, congenital abnormalities which will shorten their lives or handicap their existence. These couples can be helped by modern scientific methods in the very early stages of reproduction.

The usual normal reproductive process may be assisted in several ways; but all artificial methods provoke reactions favorable, puzzled, or unfavorable from society.

Infertility research and treatment is a limited branch of bioscience, directly affecting at most that 15% who are infertile (Dawson 1988; Singer and Wells 1984), but provoking a disproportionate quantity of discussion and reactions because the most important problem of society at present is to limit the explosive growth of the world's population. Infertility treatment directly counters the aim of population limitation.

The first accusations against us working in the infertility field are that we "play at being God" and the we "interfere with nature". The main reasons for these accusations are: first, ignorance; second, because we are dealing with sexual matters, and they always provoke reactions; third, because we invade a field that is normally private between a male and female partner; and fourth, because quite a lot of infertility is secondary to sexually transmitted diseases and abortions, both of which are considered to be antisocial. How true are these accusations?

Society as a whole is suspicious of the work of scientists, and has been for centuries, partly because much of it has been shown to be faulty. Newton's work was incomprehensible for most nonscientists, and it still is. It was considered correct, however, until Einstein showed there were faults in it. Galileo's work was so suspect that the church made a quasi-martyr of him, although he was scientifically right. Recently doctors and scientists have scored a goal against themselves by inventing and prescribing Thalidomide. Einstein's brilliant discoveries not only completely upset much previously accepted knowledge about physics and mathematics, but ultimately led to atomic weapons and nuclear energy, both of which have provoked a strong moral response.

In 1990 lengthy and full discussions took place in the British Houses of Parliament prior to legislation being passed in accordance "with the will of the people expressed through their legislators". These legislators are elected members of parliament; and in the upper chamber they are lords, some of whose titles are inherited, and some of whose titles are recent (life peers). The deliberations of the two houses of parliament taken together reflect in the main those of British society as a whole (Hansard 1989). The French debated the matter "les procreations artificielles" (Alnot 1989).

Assisted reproduction consists of the whole series of processes listed; some combined together; new ones are still being rapidly invented. The assistance that can be given artificially by medical and paramedical people to couples denied the easy, pleasurable, and "natural" ways of achieving pregnancies and having babies can be traced historically.

AID (ARTIFICIAL INSEMINATION FROM A DONOR)

The first mention of artificial insemination is to be found in the commentary on the family history of one of the biblical Hebrew prophets Jeremiah (Ben Sirah, 11th century). His daughter went to the public baths and bathed in water into which he is reported to have discharged some of his semen. She became pregnant and bore his child. The story is almost certainly apocryphal, but it is, all the same, remarkable that consideration had been given centuries ago to artificial insemination by donors in a commentary by the rabbis of the middle ages!

AIH (ARTIFICIAL INSEMINATION BY THE HUSBAND)

John Hunter (Graham 1950, p. 639) invented the procedure to relieve the infertility of a man with hypospadias. He made the man ejaculate into a glass, and with a quill he sucked up the semen and injected it into the wife's vagina, and a pregnancy resulted. This was reported in 1799, although performed much earlier.

CHILLING (AND FREEZING) GAMETES

The first record of deliberately chilling semen in order to preserve it – although short of actual freezing, as is done today – was recorded by Lazzaro Spallanzani in 1786 (Spallanzani 1786). There is no record of contemporary public disapproval of these very early descriptions of AID, AIH, and Cryopreservation of gametes, possibly because these procedures did not receive wide publicity, but now the Catholic church and the more orthodox Jewish community proscribe AID.

STIMULATION OF REFRACTORY OVULATION

Stimulation of ovulation in women who cannot without assistance release oocytes (eggs) by the administration of hormones (Crooke 1970, p. 36–41;

Gemzell 1970, p. 6; Insler and Lunenfeld 1977, pp. 629–649) made society, quietly at first and then more vocally, start to complain about "interference with natural processes". For a woman previously "barren" suddenly to be able to give birth to several babies at one delivery as a result of hormonal assistance put the mother's health at risk if the pregnancy was of more than two babies. That possibly made it unethical. There had been cases of unprovoked multiple pregnancies, such as the Dione quintuplets in Canada. As they were the result of natural conception, there was no public disapproval of the doctors who delivered and helped to rear them. Members of society, however, discussed the morality of inducing multiple pregnancies that were risky for mothers, and doctors, too, voiced their concern. Parenthood, it was said, was "a God-given gift". The dilemma of the infertile patient and the reaction of society, mainly from those members who have not suffered infertility, raise questions of responsibility to society.

EXTRACORPOREAL FERTILIZATION – IN VITRO FERTILIZATION (IVF)

When in the 1960s it became known that Robert Edwards had used human donor sperms to fertilize a human egg taken from a woman at an operation, volatile disscussions started in earnest, even though at that time there was no question of replacing any resultant embryos into a woman. I happened to be present one evening in 1968 at the Royal Society of Medicine in London and witnessed the tall, dark scientist Robert Edwards approach and meet for the first time the suave gynecologist Patrick Steptoe. The latter had that evening demonstrated to an unprepared audience of endocrinologists and gynecologists laparoscopic (Edwards and Steptoe 1980, p. 77) pictures of human ovaries obtained with the use of an instrument manufactured in France to the design of Dr. Raoul Palmer. In a very short time the instruments were greatly improved by the discovery and perfection of fiber optics by Hopkins (Edwards and Steptoe 1980, p. 74), a physicist working in Reading, England.

Few moral objections had been raised when the original animal work on mice was adapted to improve the quality and quantity of cattle, although thoroughbred race horse breeders have refused to accept this method of producing offspring from prize stallions. Farmers use fractions of an ejaculate from a high quality bull to inseminate several cows. It is cheaper than getting each bull to service an individual cow. It was a short step from the use of fractions of donated semen from bulls to the use of fractions of donated semen from healthy young men to inseminate up to twelve women with the product of just one ejaculate, and that is what is still done. By the development of that practice, the possibility of offspring who could be half brothers and half sisters meeting and marrying became worryingly real. It is several times more likely than when a single ejaculate is used to inseminate a single woman. There is an unconfirmed case of a man who nearly married his own daughter, but realized in time that she had been born nine months after he had anonymously donated semen that was probably used to fertilize the girl's mother. It is because of cases like this that society is now demanding that registers of donors of semen and the dates when they donated should be kept and made available to bona fide enquirers.

FREEZING OF EMBRYOS

The floodgates of highly vocal protest were released with the birth of Louise Brown. There had previously been one ectopic pregnancy and one triploid pregnancy (with 69 instead of the normal 46 chromosomes) in a second patient of Steptoe and Edwards.

Louise has a younger sister who was born by in vitro fertilization carried out with the same techniques that were used by Steptoe and Edwards for the birth of Louise. Edwards (Edwards and Steptoe 1980, p. 113) has described the often vituperative comments he has received from the hands of self-styled moralists, who have even told him to his face that his work resembled that done by doctors in Nazi Germany. He has become tired of arguing how moral his stance is and how strongly his work is controlled by an ethical committee. I have the honor of being a member of this committee, and that is why he has asked me to present this paper in his stead. Most of the objections raised against IVF were repeated in the recent debates in our parliament, which were the first to be conducted publicly in a legislature in Europe (but not in Australia).

RESEARCH INTO THE PHYSIOLOGY AND PATHOLOGY OF HUMAN EMBRYOS

It is clear that research into the physiology and pathology of human embryos has now reached such a stage of sophistication that many serious moral issues arise. These are not susceptible to facile answers. We gynecologists and our scientific colleagues now urgently need society's advice and guidance. The arguments about the morality of some of the procedures are sometimes exceedingly complex, and the procedures themselves are beyond the comprehension of many doctors who refer patients, on whose gametes and zygotes manipulations will be practiced. How much more are the techniques beyond the comprehension of lay people, who object to work they do not understand at all! That is why it is so important that these discussions are published.

A team at the Hammersmith Hospital (Handyside et al. 1989) has managed to determine the sex of embryos which are produced as a result of in vitro fertilization. They remove one cell of an eight celled embryo two or three days after fertilization in a plastic dish. This removed cell is allowed to grow and multiply, while the remaining seven cells of the embryo are placed in a freezer at very low temperatures. The one cell kept in a culture medium is examined for its chromosome content. If it contains two X chromosomes (female), it is possible to replace the thawed-out seven remaining cells of the embryo into a mother with the assurance that the child will not be he bearer of a disease which affects only boys such as hemophilia. Still more recently (Sinclair et al. 1990; Gubbay et al. 1990) the position of the gene in the Y chromosome that produces males has been identified. The aim all along has been to eliminate certain inherited diseases by using in vitro fertlization techniques on embryos produced by carrier mothers. With their first achieved pregnancies Steptoe and Edwards had already insisted on amniocentesis being carried out on all fetuses at about the sixteenth week of pregnancy, to ensure as far as possible by chromosome analysis that normal babies would be born. The incidence of abnormality from in vitro fertilization babies is no higher than in naturally conceived babies; i.e., about 2% for serious abnormalities. Amniocentesis or a chorionic villus biopsy, when followed by a so-called therapeutic abortion, is destructive of a fetus and cannot be said to be "therapeutic" for the mother. That is why it is desirable to avoid as much trauma as possible by examining single cells taken from artificially produced embryos. Amniocentesis carries up to 1% risk of abortion, and chorionic villus biopsy carries a 3–5% risk, depending in part on the skill and experience of the operator.

cWe look forward to elimination of such diseases as thalassemia inherited from both parents, or Tay Sachs disease, by their being easily identifiable in the earliest possible stages. Even such inherited diseases as Huntington's Chorea, which may not reveal itself for forty years, can be avoided. To some it may seem obvious that such research is moral and religiously correct.

High clerics of different faiths such as the archbishop of York, who is the second most senior cleric in the Church of England, and the chief rabbi have spoken in favor of such kind of research. The archbishop is, as it happens, a science graduate of Cambridge University. The chief rabbi has expressed similar thoughts to the archbishop's lucidly argued statement on the theological, scientific, and moral issues, but he sought to bring an amendment to prohibit the production of embryos purely for research. He was voicing a very well-known reaction of society to pure science for the sake of science: embryos produced solely for research cannot be of immediate benefit to the women themselves from whom they are derived. Exaggerated pictures have been drawn. I have heard one gynecologist, who received a great deal of support, state falsely that she had seen rows and rows of early embryos, whom she called "small babies", in bottles waiting to be experimented on, with never a chance of survival or of seeing the light of day.

The issue of experimentation on embryos has been muddied by the introduction of arguments in the same debate about the maximum duration of pregnancy up to when an abortion can legally be carried out. Both issues are serious, but the moral status of voluntary termination of pregnancy must be considered separately from the totally different issues of the morality of human embryo research, especially since the end of such research is to help achieve life. The confusion has arisen because all embryos on whom research is carried out must inevitably die or be destroyed. The "pro-life school" contends that such destruction of embryos is equivalent to an abortion or to the termination of life.

Now, to come to the word "life": arguments have raged about the "moment of the beginning of life". There is no such moment. Human fertilization is a continuous process, involving a) penetration of the zona pellucida; b) entering of the sperm head into the oocyte; and c) the expulsion of a second polar body (the first polar body contains cortical granules which are extruded before fertilization). By definition, the fertilization process involves the fusion of the chromosomes of the nuclei of the gametes of both sexes within the egg, and finally the formation of a cell with the double complement (in the human 46 chromosomes) (Edwards 1980, pp. 573–667). Even then the process is possibly not complete. The single cell divides into two. If these stay together and divide and subdivide countless times, a human being results. If they separate after the first segmentation, identical twins (which are never totally identical) result. This may be the start of two lives.

The British parliament has now legislated that research can take place until 14 days after fertilization – up to the development of the primitive streak and perhaps the outline of the nervous system, long before any truly sensate cells are laid down.

SURROGATE MOTHERHOOD

It is now possible to take oocytes from a woman's ovaries, fertilize them with her husband's sperms, and, if she does not have a receptive womb, transfer their embryo to a convenient receptacle, namely, another woman's uterus. The second woman becomes a host mother for the couple's fetus and delivers a baby that is genetically theirs, but she has carried, nourished, and lived with the baby. The moral and legal issue here has been argued in the media and in the law courts in different ways, in different countries and states.

The host mother is certainly "assisting" the infertile couple in their efforts to reproduce. What she does may *seem* to be something private, but it is not; it is public. The birth has to be registered. The baby has to have a legal status. Who is the legal mother? Some cases have become very public in the United States, France, and Great Britain. Society has become involved. The arguments have nothing to do with "science". Matters have entered the field of the emotions as well as of the law. The host mother may want to keep the baby that results from her pregnancy, the nine months during which she has not only nourished the baby, but felt its kicks and other movements, excreted its waste products, breathed its oxygen through her lungs, eaten its food through her stomach, and delivered it. May she keep the baby? In October 1990 a judge in California ruled that she may not. It is not genetically hers. Society can begin to understand the arguments and to feel for both sides if there are conflicts. The origin of the baby by in vitro fertilization was purely a scientific one. A new dimension of "hosting" has been added. Have the scientists now logically to let go, and to let society make the decisions? Have the arguments helped the dialogue between bioscience and the public?

In host motherhood a woman is artificially inseminated with the sperm of another woman's husband, but herself provides the egg as well as the womb, and this has a different dimension, because here the host mother is the true mother (Editorial 1989). In this reference the moral, legal, and societal arguments are dealt with fairly fully.

In London I delivered the baby of an Italian woman who had been seduced into natural intercourse. The father was a rich tycoon whose wife was menopausally infertile. I watched the wife take the two-day-old baby from the natural mother. This was all in private, but the baby had to be registered as born to its natural mother, and then had to be legally adopted. However, even where both parents are genetically the parents of a product of in vitro fertilization and surrogate mothering, they may still have to adopt the baby, although there can be no doubt that the baby is "theirs".

OOCYTE (EGG) DONATION

To achieve this, a woman who is going to give her eggs must have ovulation stimulated and egg recovery performed. Her eggs may then be given to a woman who has a uterus, but is unable to ripen eggs because of defects in, or absence of, ovaries. The motives for giving eggs may be purely altruistic, as when a sister who has had babies volunteers to give some of her eggs to be fertilized by her sister's husband, for the resulting embryo to be implanted in the womb of the woman who cannot herself produce eggs. The donation may be anonymous, and often should be, and payment of expenses may be made. Whether this payment is ethical is being debated. The ethical problems are similar to those of sperm donation.

Surrogate motherhood has been severely criticized for the opportunities it opens up for exploitation of disadvantaged women. With a "convenient receptacle" there does not have to be any social or class difference between the surrogate mother and the woman for whom she is acting so benevolently, but inevitably there will be instances of women who are able to pay for "renting" the uterus of women who wish, or need, to be paid. Clearly it then becomes necessary to look critically at the question of whether social inequalities arise, not only in this field of using the new reproductive technologies, but in all the fields.

EMBRYO FLUSHING

It is now possible for a woman to become pregnant by natural processes, and when the embryo is about three to five days old, on about the 16th, 17th, or 18th day of her menstrual cycle, it can be flushed out of her uterus before it becomes implanted into the wall. It is then available to give or sell to another woman. What a trade that may be!

A woman can, in theory at any rate, have embryo flushing carried out six or more times a year. This then becomes a business involving the woman, scientists, and doctors. Some would argue that this is no worse than selling semen or oocytes, or blood, or a kidney for transplant. That is a subject for discussion.

GENETIC ENGINEERING

Scientists can now pinpoint the exact positions on chromosomes of many of the genes carrying, for instance, maleness, and hereditary factors, such as brown eyes and extra fingers (polydactylism); and they may eventually be able to manipulate them. In more than 98% of cases lethality of an allele is due to its nonfunctioning. Intelligence, however, is a multigenetic inheritance, and genes for intelligence are probably situated in several chromosomes. Intelligence is almost certainly not for manipulation, so one cannot visualize implanting any genes for higher (or lower) intelligence (Cusine 1988, pp. 129–166).

EMBRYO DIVISION (DAWSON 1988)

It is quite possible to separate cells in an embryo from one another, and in that way to produce identical twins.

CLONING

Cloning is the production of an exact copy of a living organism. A gamete has a haploid (half) complement of chromosomes contained in its nucleus. It will not segment (divide) until the number is made up to the double (complete) complement. Every cell in the adult human body has 46 chromosomes, except the sperms and oocytes, which each have 23 chromosomes. A method by which a clone has already been produced in lowly amphibian animals has been by removal of a nucleus containing the half number of chromosomes from an unfertilized egg, and replacing it with the nucleus of a poorly differentiated cell of the same species. This then is allowed to develop. A clone (an exact copy) of an individual can be produced by transplanting several nuclei from a single donor to several recipient eggs. A donor nucleus can be taken from a pro-embryo, or even from an adult cell. When the nucleus from an embryo cell is used, the egg will develop successfully right through to the adult stage. Many nuclei can be taken from the same donor pro-embryo, and thereby clone, so that several dozen genetically identical frogs can be produced. If the nucleus is taken from an adult animal, development is arrested at or before the tadpole stage.

CONCLUSION

A small fraction of society reacts very favorably to most forms of assisted reproduction, but a much larger fraction is frightened to allow scientists to go ahead without any curbs. There is a still larger fraction that voices its objections to interfering with nature in any way, and is really frightened. Many of the reactions are due, first, to ignorance of the processes involved in assisting childless couples to have babies; second, to a feeling that interfering with what should be a sexual function is "not quite nice"; and, third, to a feeling that it is against the will of God. Yet when information in a comprehensible form is delivered in unemotional terms by adquately briefed legislators, the vote, as demonstrated in the British parliament, is strongly in favor of most forms of assisted reproduction. Information suitably given to the right people, at the right time, and in the right place, counters ignorance. It needs careful public relations work to inform a larger and possibly less-educated section of the public. There is a very strong lobby of antiabortionists who confuse assisted reproduction with abortion because of the inevitable wastage of embryos that must occur during the research into assisted reproduction. It is difficult to get across that 2, 4, and 8 cell embryos are not little babies any more than spermatozoa are little men. There is life in every sperm, as there is in every egg, and both have the potential when they meet in suitable physical and chemical circumstances to produce a new life. Individual life starts with fertilization. This is not the same as saying that life starts with fertilization, merely that a new individual life starts at that time. The British parliament has clearly shown that it is as ethical to prevent unwanted infertility as it is ethical to prevent avoided congenital defects. The decision of the British parliament, and maybe of other parliaments soon, could be the most effective way of educating society and countering some of the visions and uninformed propaganda put about by those who oppose such life-enhancing measures as assisted reproduction. Ethics will clearly be decided greatly on the biological methods that become available. At present these consist of the methods outlined above: in particular, in preventing the birth of congenitally abnormal infants, either by aborting them as soon as possible after the abnormality has been diagnosed, or by avoiding transferring embryos in whom the diagnosis of a congenital abnormality has been made on examination of cells taken from the embryo. It is recognized that there may be dangers inherent in letting scientists experiment, particularly in the field of genetic manipulation, without the strict fetters of the criminal law. It is essential first, however, that effective communication be achieved between lawyers and scientists, who will have to make their decisions according to the ethical principals that are, or have been, considered.

REFERENCES

The Alphabet of Ben Sirah. Manuscript 11th Century, printed 16th Century. Tractate Niddah. In Babylonian Talmud. Page 13B.

Alnot, M.-O. (1989). *Les Procreations Artificielles.* Rapport au Premier Ministre. Collection Des Rapports Officiels. La Documentation Française.

Crooke (1970). *Developments in the Pharmacological and Clinical Uses of Human Gonadotrophins,* pp. 36–41. High Wycombe, England: G.D. Searle & Co.

Cusine, Douglas J. (1988). *New Reproductive Techniques – A Legal Perspective,* pp. 129-166. Aldershot: Garrard Publishers.

Dawson, Karen (1988). Segmentation and moral status in vivo and in vitro – A scientific perspective. *Bioethics* **2:** 1–14.

Editorial (1989). Surrogacy and autonomy. *Bioethics* **3:** 18–44.

Edwards, R.G. (1980). *Conception in the Human Female,* pp. 573–667. London: Academic Press.

Edwards, R., and Steptoe, P. (1980). *A Matter of Life.* London: Hutchinson.

Gemzell, C. (1970). *Clinical Applications of Human Gonadotrophins,* p. 6. Stuttgart: Georg Thieme.

Graham, Harvey (1950). *Eternal Eve.* William Heinemann Medical Books Ltd.

Gubbay, J. et al. (1990). A gene mapping to the sex-determining region of the mouse Y chromosome is a member of a novel family of embryonically expressed genes. *Nature* **346,** No. 6281 (July 19,1990): 247–250.

Handyside, A.H. et al. (1989). Biopsy of human preimplantation embryos and sexing by DNA amplification. *Lancet* **1:** 347–349.

Hansard (1989). Parliamentary Debates. Thursday December 7, 1989. In: *House of Lord's Official Report,* vol. 513, no. 11, pp. 1002–1114. (Human Fertilisation and Embryology Bill.)

Insler and Lunenfeld (1977). *Scientific Foundations of Obstetrics and Gynaecology,* 2nd edition. William Heinemann Medical Books Ltd.

Sinclair, A.H. et al. (1990). A gene from the human sex-determining region encodes a protein with homology to a conserved DNA-small binding motif. *Nature* **346** No. 6281 (July 19, 1990): 240–244.

Singer, Peter, and Ann Wells, (1984). *The Reproduction Revolution.* Oxford: Oxford University Press.

Spallanzani, Lazzaro (1786). In: *Encyclopedia Brittanica,* 15th Edition, vol. 2, p. 1022.

GLOSSARY

AID: Artificial Insemination (Donor). The fertilization of a woman or a woman's eggs by semen taken from a man other than her husband.

AIH: Artificial Insemination by the Husband's semen.

stimulation of ovulation: use of natural or synthetic hormones to release several eggs (oocytes) at a time.

in vitro fertilization: fertilization of an egg (oocyte) by a spermatozoon in a plastic plate (or test tube).

embryo transfer: transfer of an embryo that has reached an appropriate stage of development into the womb of a recipient woman.

segmentation: splitting of an embryo.

cloning: making of identical copies of a parent.

surrogate: a woman who agrees to carry a pregnancy for another couple. She may do this by being inseminated with the husband's sperm, but with her own egg or eggs being fertilized, or by receiving an embryo which is the product of the semen of a man and his regular partner's eggs that have been fertilized in a test tube.

oocyte: (Eggs) and spermatozoa are stages of meiosis, i.e., reduction from double ($2n = 46$ in humans) to single (23 in humans) chromosome sets. Males have XY chromosomes and females XX chromosomes in their double complements. Each embryo has an X or a Y chromosome inherited from the fertilizing sperm. There are, however, abnormalities in which an X or a Y is missing, or there are extra Xs or extra Ys.

oocyte donation: the giving of eggs by one woman to another in order to achieve pregnancy.

genetic engineering: the manipulation of the inheritance of an embryo.

embryo division: see segmentation.

Bioscience ⇌ Society
edited by D.J. Roy, B.E. Wynne, and R.W. Old, pp. 261–282
© 1991, John Wiley & Sons

Animal Experimentation and Society: A Case Study of an Uneasy Interaction

Andrew N. Rowan

Tufts Center for Animals and Public Policy
Tufts School of Veterinary Medicine
200 Westborough Road
N. Grafton, MA 01536, U.S.A.

Abstract. Although science is considered by the public as one of the most admired professions, scientists have been relatively unsuccessful in countering the allegations and protests in the controversy over the use of animals in research. The growth of the animal rights movement is examined from it's Victorian antivivisectionist roots, through the formation of humane organizations in the United States in the 1950s, to the almost exponential growth in membership in the animal liberation and welfare groups in the 1980s. The factors behind such growth are discussed, including the shift in the public's attitudes towards animals' cognitive abilities, the role of philosophers and philosophical arguments, and the women's movement. The concept of helplessness of the research animals as well as animal suffering vs. human benefit are also examined.

The public's perception of the personalities involved in research are shown to be stereotypical and often distorted. Scientists are seen as callous, unfeeling, and commonly mistreating the animals under their care. These stereotypes plus the perceived helplessness of the innocent animals make it difficult to persuade the public that the "coldly rational" researcher really does care about the animals he works on. Media images only help to reinforce this mistrust and misconception.

In conclusion, the paper examines recent strategies by biomedical societies to drum up support for animal research. It also offers reasons for why these strategies have essentially misfired, failing to slow the trends of public opinion on the issue. Finally, methods for resolving the conflict between animal protectionists and animal researchers are discussed. These include the need for intelligent and open debate, to allow a greater public role in research activities, and to upgrade the political sensitivity of scientists as a group.

INTRODUCTION

In the past two decades, bioscientists have been forced to confront an increasing variety of critics. One critical group that has grown tremendously in size and influence over this period is the animal protection movement. Protests over the use of animals in research, testing, and education have touched a responsive chord among the general public, as they did in the 19th century. The research community has been slow to respond to criticism of animal research and, for the most part, has not been particularly effective in countering allegations of inappropriate laboratory animal use. Despite this failure, the public still views scientists as belonging to one of the most admired professions. In the U.S.A., 88% of the public believe that the world is better off because of science, and scientists are second only to medical doctors in public prestige (NSB 1989). In the U.K., the three most respected public institutions are medicine, the military, and science, in that order (Kenward 1989).

Why then have scientists been relatively unsuccessful in countering the effects on public opinion and legislation of protests against animal research and testing, and allegations of laboratory animal abuse? Answers to this question can be derived from both the historical record and from a careful analysis of the reasons for the growth in the animal movement and the factors that influence the modern animal research controversy.

A HISTORICAL PRÉCIS

Despite the recent explosion in the number of books on the animal research debate, French's *Antivivisection and Medical Science in Victorian Society* (1975) is still the best book on the antivivisection-medical research controversy. Although he describes a conflict that occurred over one hundred years ago, his analysis is remarkably pertinent in helping us understand today's debate. According to French, some of the more important elements supporting the growth of the Victorian antivivisection movement were as follows.

First, the Darwinian revolution weakened claims about the uniqueness of human beings and blurred the absolute qualitative differences that had been considered to exist between humans and animals. The narrowing of this philosophical gap tended to support utilitarian arguments that animal suffering was morally important – as was human suffering.

Second, some of the new reform movements within protestantism also led to a weakening of claims regarding the uniqueness of human beings, by arguing that both animals and humans possessed souls. John Wesley preached specifically that animals had souls (a message ignored in modern methodism), and many of the early campaigners for animal welfare were clerics in the Church of England.

Third, the emerging public health movement (the sanitarians) promoted the development of better health and hygiene (e.g., cleaning up public water supplies) as a more effective way of improving public health than animal research. They did not oppose animal research, but neither did they support it very strongly.

Fourth, some in the medical establishment were threatened by the new "scientific medicine" based on experimentation, and some of these physicians became influential critics of animal experimentation. For example, Claude Bernard was criticized not only by the public (including his wife), but also by leading figures of the French medical establishment. Such criticism was, however, based more on professional jealousy than on concern for the animals. In addition, the physician-researcher was associated with two opposing public images. The physician was perceived to be a caring, humanitarian individual concerned with saving lives and alleviating suffering, often at some cost to himself. (There were very few female practitioners.) By contrast, the researcher was perceived to be an unfeeling individual who deliberately and coldly caused great suffering to animals. The Victorian novel *Dr. Jekyll and Mr. Hyde* illustrates the dichotomy.

Today, we see remarkably similar factors underpinning the modern animal rights movement and public worries about science and technology. The following discussion will draw on historical sources to explain some of the central features about the modern animal research debate and the reasons why scientists have so much trouble convincing the public of their sincerity and concern.

THE GROWTH OF THE ANIMAL PROTECTION MOVEMENT AND UNDERLYING FACTORS

In the past forty years, the animal protection movement has grown steadily in size and political influence. The present discussion is based mainly on the

American animal movement, but European groups have experienced similar growth in numbers and influence (Serpell 1990).

Before the 1950s, criticism of animal research was not taken seriously by most of the public and its legislators. The animal welfare groups tended to ignore research, and the traditional antivivisection societies were relegated to an ignored fringe of society. Then, during the 1950s, several new organizations were founded that criticized animal research and particularly the care of the animals. One prime factor leading to the formation of these new organizations (e.g., the Animal Welfare Institute, the Humane Society of the U.S., and Friends of Animals) was opposition to legally mandated acquisition of dogs and cats from community pounds and shelters for medical research. In 1966, criticism of research animal care and two public scandals involving research animal acquisition led to the first U.S. law regulating the acquisition of laboratory dogs and cats. At about the same time, growing public criticism in England led to a government committee of enquiry on animal research (Littlewood Commission 1965) – the first since the 1906 Commission.

Biomedical organizations tended to paint these new critics as neo-antivivisectionists and as sentimental Luddites, even though the campaigns focused largely on animal care and husbandry rather than animal use. Criticism from animal activists was thus dismissed as emanating from the fringe of society. However, the new animal protection groups continued to press for reform of laboratory animal care and use, and their memberships continued to grow, albeit relatively slowly. (The sixties and early seventies were a time of widespread activism in the U.S.A. with antiwar protests, civil rights activism, women's rights, Watergate, and environmental activism – e.g., Earthday, 1970.)

The next major milestone was Peter Singer's *Animal Liberation* (1975), a book that became a bible of the emerging animal liberation (in distinction to animal welfare) movement. Small radical activist groups began to emerge. For example, People for the Ethical Treatment of Animals (PETA) was started in 1980 in the U.S.A., while Animal Aid began to draw crowds to its demonstrations in 1978 in England.

In the early 1980s, membership of both animal liberation groups and animal welfare groups grew dramatically. The Humane Society of the United States increased from 35,000 members (people who paid the membership fee) in 1978, to 75,000 in 1982, and 550,000 in 1988. PETA had 8,000 contributing members in 1985, and now claims over 250,000. The membership totals now compare favorably with those of environmental organizations, while Greenpeace-U.S.A, a group combining activism, animals, and the environment, has expanded from 20,000 members in 1978 to over 2 million in 1990. Several factors appear to have played a role in the growth of the animal movement.

a) Animal cognition

In 1988, *Newsweek* published two cover stories on the subject of animals. One analyzed the animal rights movement, and the other discussed the public's shift in interest in the past ten to twenty years from the idea that animals are basically stupid – i.e., merely reacting, Cartesian, biological automata – to the notion that they possess a kind of intelligence. I would argue that it is not coincidence that these two themes should have been so closely linked. A shift in public attitudes towards animals – namely, from seeing them as dumb animals to seeing them as intelligent beings with emotions and drives similar to our own – is probably one of the major societal factors driving the growth of concern for animals. It is especially interesting to note how changes in our attitudes to animal intelligence over the past 150 years have preceded the waxing and waning of the animal protection movement.

In the 19th century, Charles Darwin shook up societal attitudes towards animals. His theories challenged the belief that animals were qualitatively different from humans, and they focused attention instead on quantitative differences, including mental capacity (Walker 1983). The interest in animal intelligence led to questions about the treatment of such "intelligent" creatures, and helped to spur the growth of the animal movement. In America, fascination with the humanlike emotions, desires, and reasoning abilities of animals was also widespread, as exemplified by the popularity of the somewhat anthropomorphic nature writings of Seton, Long, and London (Lutts 1989).

However, 19th century interest in the cognitive capacity of animals produced a backlash. Some scientists began to revolt against what they perceived to be uncritical "mentalism" that imputed reasoning and human motivation to animals on the flimsiest of evidence. By the 1920s, behaviorism – the study of pure behavior uncontaminated by any consideration of mental states – was taking over in many animal behavior and psychology laboratories. For example, Ivan Pavlov's studies in classical conditioning reinforced the view that animal and human behavior could be described as the response of a sophisticated but definitely nonconscious (and hence unfeeling) creature (Rollin 1989).

The behaviorist tradition dominated thought about animals from 1920 to the early 1960s when scientists again started to discuss and explore the cognitive and psychological abilities of animals. This period also marked the growth of ethology as a science and a re-awakening of public wonder over the natural behavior (and "intelligence") of animals. In 1976, Donald Griffin authored *A Question of Animal Awareness,* in which he argued that one could study the inner motivations and cognitive abilities of animals. Today, the book's arguments are unexceptional, but in 1976 it produced a very strong and often very critical response. Nonetheless, in an increasingly urban society, where attitudes to animals are shaped more by needs for companionship than was the case with the frontier and rural experience, animal cognition and intelligence again became a popular topic.

Television was also an important influence, starting in the 1960s. Millions of urban Americans delighted in shows depicting wild animals and their remarkable capabilities. *National Geographic* produced fascinating footage of the very humanlike behavior and reasoning of the chimpanzees in Gombe, and of Koko the gorilla and her pet kitten. Studies of "talking" chimpanzees raised uncomfortable questions about the uniqueness of human language. Oceanaria presented dolphins and killer whales as sweet, gentle, and very intelligent. The number of, and attachment to, companion dogs and cats grew. It is hardly surprising that an increase of public interest should have led the public to have become much more concerned about the way animals are treated.

Over the past two centuries, the rise and fall and rise of interest in animal intelligence and cognition followed a very similar pattern to the status of the animal protection movement in society. This does not prove causality, but the connection is suggestive.

b) The role of philosophy

Like Griffin, Peter Singer also produced the right book at the right time (Singer 1975). He articulated a relatively simple message underpinned by rigorous logic and the philosophical tradition of utilitarianism. For those who were already members of the animal protection community, Singer's academic standing and quasi-academic book (footnotes, etc.) provided a much appreciated sense of legitimacy. No matter how sure animal activists were that they were right to be concerned about animal well-being, as long as they could only justify their position using sentimental and emotional arguments, they felt insecure in the secular and rational modern world. Singer gave these people an easily understood argument, supported by logical reasoning, that proved very difficult for the research community to refute. Indeed, most research spokesmen still do not understand Singer's position, nor how they might attack it.

For example, David Jack, Chairman of the Research Defence Society (RDS) in the U.K., recently argued that Singer's real goal was the abolition of animal experiments, that his book provided the philosophical justification for the animal rights movement, and that the animals' rights case is based on the concept of granting animals rights that are essentially equivalent to those of human beings (Jack 1990). These points may be answered as follows.

First, it should be noted that most scientists would agree that they would prefer not to use animals if they did not have to (indeed one of the RDS's formal objectives makes this point). Therefore, logically it can be claimed that scientists would also like to see the abolition of animal experiments if, in their view, that were possible. However, they would differ with an animal activist on how soon such abolition might be achieved (if ever), and how much effort should go into pursuing such a goal. For the record, Singer's philosophical position is strictly utilitarian, and he does not support any such absolutist position. For example, if one could demonstrate that an animal research project would produce more good than harm, Singer would have to support it.

Second, while Singer's book has sparked the growth of the animal rights movement (more than 250,000 copies have been sold in the U.S.A since 1975, the bulk of which have been sold since 1984), Singer does not use "rights" terminology in his philosophy. He does not reject the title of "Father of the Animal Rights Movement" only because he sees the movement under that banner as a strictly political rather than philosophical force. In philosophical discourse, Singer rejects the concept of rights, whether it is applied to humans or animals.

Third, Singer does not consider animals to be equivalent to human beings, whatever some in the research and animal liberation communities might think. Singer argues that the morally important characteristic is that of sentience – the capacity to suffer and experience enjoyment. Thus, if an animal can suffer, its interests in not suffering must be taken into consideration. Singer does not argue that a rat's suffering is the same as an adult human's. He only says that if they are the same, then the rat's interest in not suffering must be considered equivalent to the human's interest in not suffering. Furthermore, it does not follow that, even under these circumstances, the rat and the human being should be treated the same.

It would not be difficult to mount a challenge to Singer's restrictive utilitarianism, and draw it closer to the more permissive utilitarianism that most scientists appear to use to justify their research (i.e., animal research will produce human benefit that outweighs the cost in animal suffering, which is, in any case, minimal).

The logical arguments developed by Singer helped to attract a wide range of street-smart and professionally trained individuals to the animal protection movement. Even one or two individuals with political and/or professional skills can be important in an activist movement. For example, the use of stockholder resolutions to bring pressure to bear on companies that use animals was developed and refined by one person – a lawyer with an interest in securities law.

It is, therefore, significant that in the past ten years seven national specialist groups advocating more attention to animal care or animal rights have been formed in the U.S.A. for the clergy, lawyers, physicians, psychologists, scientists, and veterinarians. The membership of these groups is still small, but their influence to date cannot be assessed by merely counting numbers. For example, the Physicians Committee for the Reform of Medicine claims the support of several thousand doctors, while the Association of Veterinarians for Animal Rights has a few hundred veterinarian members.

As an aside, it is interesting to note how little philosophical argument has been put forward to support the status quo of animal research. While there are countless books and articles supporting an increased moral status for animals, in the U.S.A. the research community has to draw its comfort and arguments from one short article in the *New England Journal of Medicine* (Cohen 1986) and a single book (Fox 1986), whose author rejected his own arguments three months after it was published. Fox is now an animal rights philosopher. A bioethical think tank, the Hastings Center, has recently issued a report on the topic. Their central message was that there is a "troubled middle" ground, where people recognize that animal research involves significant and important moral issues but do not think that an abolitionist argument can be sustained (Donnelly and Nolan 1990). In general, the lack of philosophical support for the scientific position has not helped spokespersons for animal research.

c) The Women's Movement

Carol Gilligan (an educational psychologist at Harvard University) has argued that nurturing and caregiving are important values for women (Gilligan et al. 1988). Indeed, it is true that concern for animals is higher among women than among men, and it has been argued that feminism and animal protection are closely linked (Donovan 1990). Certainly, many of the recently formed animal protection groups were started by women, and women continue to play a significant role in the movement.

A 1976 in-depth survey of a randomly selected national sample of over 3,000 persons reported that 2.0% of the female population has supported an animal protection group, while only 0.6% of the male population has. If women are more care-oriented, then it follows that "care" issues will receive more political attention in a society when the political status of women rises. It is noteworthy that the animal protection movement enjoyed relatively high social status in the 19th century when there was growing pressure to educate women and give them the vote (Elston 1987). In the 20th century, the increase in status of the animal movement similarly follows a push for greater equality for women. Singer's book was titled *Animal Liberation* at the height of the push for women's liberation, and this is probably more than simple coincidence. As with animal cognition, the link is interesting and suggestive, but does not necessarily prove causality.

ANIMAL RESEARCH: A SPECIAL CASE

While the modern animal protection movement addresses all human uses of animals, there is no question that animal research has always touched a particularly raw nerve. It is not entirely clear why this should be. However, in the 19th century, animal research was the only animal protection issue in which a social elite was accused of cruelty. By and large, all other significant animal protection issues involved the upper classes imposing their own mores on an "uncouth and uneducated lower class". It may well be that the deliberate infliction of suffering by a group perceived to be both educated and "sensitive" aroused particular horror. As Rowan and Rollin (1983) note, "The idea of medical scientists, who were looked up to as humane and cultured individuals, deliberately inflicting pain on animals for some tenuous, future benefit, touched a raw nerve ... At the most basic level ... the preoccupation with vivisection allegedly grew out of what the movement took to be a grievous outrage to the educated classes franchise upon moral leadership."

But there are also other underlying motivations that lead people to protest against animal research.

a) Helplessness and the research animal

One powerful factor appears to be an apparent empathy with the laboratory animal as a helpless innocent. The animal is viewed as an innocent victim (free of guilt) that does not deserve to be subjected to such treatment. As a corollary, one frequently hears an argument from individuals protesting against animal research that prisoners – i.e., those found guilty of some sin against society – should be used instead of animals in research. Experiments on humans without their knowledge or consent became part of public knowledge in the 1960s and 1970s (e.g., the Tuskegee syphilis research funded by the U.S. Public Health Service), leading to calls for "informed consent" guidelines. Animal protectionists have played on such inappropriate human experimentation in a variety of ways, but have also questioned how animals could possibly give consent to their use in research. The research animal is, thus, helpless in that it has no significant control over its environment, nor the way that it is used in research. Lansbury (1985), in an analysis of an animal research *cause célèbre* in London at the turn of the century, identified the image of helplessness as a key unifying characteristic that united antivivisectionists, suffragettes, and labor activists.

Today, empathy for laboratory animals as helpless victims in the hands of biomedical researchers may stem from the experience many people have in a modern hospital – at least, anecdotal cases indicate that this may be so. Modern medicine is often perceived as relatively impersonal and unconcerned. More often than not, hospitalization leaves the patient feeling frightened and helpless, at least for some of the time. It is only a short extension of one's own personal experience to an empathic concern for laboratory animals in a modern laboratory.

b) Suffering and benefit

Despite the fact that three-quarters of the public accept the need for animal research, it is relatively easy to persuade the public that a particular use of laboratory animals is abusive. In political parlance, the public's support for animal research is "soft". On the other hand, where there is a clear trade-off between a human and an animal life, the public will tend to support the use of the animal. This balance is clearly indicated when considering two paradigmatic cases, the use of a baboon heart as a replacement for Baby Fae, who was born with a fatal heart defect, and the use of baboons in head trauma research at the University of Pennsylvania.

Animal activists attempted to capitalize on the media attention paid to the Baby Fae case by demonstrating outside Loma Linda Hospital in California (where the xenograft was performed), and by attracting the attention of the news media with their protests. However, the overwhelming media response to such criticism was that the animal activists were misguided at best, and crazy at worst. One cartoon in the *Boston Herald* (November 6, 1984) captured this response. It showed Baby Fae in one half of the cartoon and a group of animal activists carrying placards in the other. The caption under

Baby Fae was "Born with half a heart", and that under the animal activists was "Born with half a brain".

By contrast, the exposure of the head trauma research on baboons at the University of Pennsylvania drew widespread media condemnation. The *Washington Post,* not known for its sympathy to animal activism, ran an editorial entitled "Animal Torture" (July 20, 1985), while Paul Harvey, a very popular and conservative Midwest radio personality condemned the research as an outrageous affront to decency and civilization. Admittedly, the head trauma research was publicized by a videotape compiled from video recordings made by the researchers themselves and stolen from the laboratory by members of the Animal Liberation Front. (The images, although very selective, were nonetheless indicative of flawed animal care and distorted attitudes to the animals. The scenes were graphic and shocking, and they drew a lot of attention.) Without the videotape, the media response would have been less widespread and dramatic, but they probably would still have raised questions about the care and use of the animals and the research.

The lessons to be learned from these two cases are that the public is much more concerned about the suffering of animals than about the killing of animals, and that the public is prepared to sanction animal use where the benefits to humans are readily apparent. Thus, in the Baby Fae case, there was no apparent suffering and, in theory at any rate, Baby Fae's life was being saved by the xenograft. In the head trauma research, the public could see graphic examples of apparent animal suffering and of apparent research callousness. In defense of the research, they were told that there are 50,000 head trauma cases in the U.S.A. annually, and that this research was providing important clues to possibly improved treatments and better understanding. This promise of potential benefits just did not carry the day against the perceived suffering of the animals, and the public rejected such research in no uncertain terms.

Basic research is therefore likely to run into significant obstacles among the general public if it is perceived to be causing significant animal suffering, especially to dogs, cats, rabbits, and nonhuman primates. Even the research community tends to defend animal research in terms of its practical benefits in producing better medicine, rather than in terms of the importance of advancing human understanding. John Orem, a sleep researcher in Texas, had his laboratory vandalized by animal activists. He later commented how much the break-in had disturbed him, but then also commented how he had felt almost equally disturbed by the medical community's defense of his work in terms of its importance to discovering a cause and potential cure for Sudden Infant Death Syndrome (SIDS) (Orem 1990). Orem argued that he was not seeking a cure for SIDS, that he was doing basic research on cats to try to understand the control of the mammalian sleep center. Orem is to be commended for his courage in demanding that his work be judged as basic rather than applied science, but this then raises numerous very difficult questions about how one might proceed to judge the value of the proposed knowledge (as opposed to the methodological design and rigor of the research protocol) (Rowan 1990).

PUBLIC ATTITUDES TOWARDS SCIENCE AND TO SCIENTISTS

The public's perception of science has been on a roller coaster ride since the 1950s, when it was felt that federally funded science could surmount any problem the country or world could throw at it. The development of the polio vaccine was a clear example. However, beginning in the late 1960s and lasting throughout the 1970s, science was perceived by more and more of the public as part of the problem rather than part of the solution. Problems arising from pollution, the destruction of the rain forests, and nuclear power have tended to undermine the public's confidence in science. More media attention that displayed both the human fallibility of scientists as well as their accomplishments left the public less confident in the pronouncements of science. Social researchers have also found that those who are well-informed tend to be more wary about science than those who are uninformed (Gerbner 1987). Initial public hope about the benefits to be derived from science rapidly gives way to fear of the risks involved and of losing control. Also, the swing towards more conservative values has tended to undermine support for science, because science is an agent of change and, therefore, antithetical to conservative values. Despite this, science is still considered a prestigious profession in most polls (Pion and Lipsey 1981).

The perception by the public of scientists' personalities has also changed over the past forty years, but it has always been stereotypical and somewhat distorted. In surveys from the late 1950s, scientists were seen as intellectual and dedicated, but difficult to comprehend and erratic in interpersonal relationships. A 1975 survey reported that they were seen as remote, withdrawn, secretive, unpopular, and singular-minded souls (Pion and Lipsey 1981). Other surveys identify qualities such as rationality, objectivity, and coldness with scientists (Gerbner 1987; Weart 1988). Gerbner reports that television images of scientists include positive roles, but there are many ambivalent and troublesome portrayals of scientists. He found that exposure to science and technology through television tends to cultivate a less favorable orientation towards science. However, television does not invent this ambivalent view of science. The caricature of the curious, if not mad scientist who ignores the dangers of his research in his relentless quest for knowledge is found throughout literature and other entertainment media. Several recent popular films (e.g., *Project X, Greystokes* and *Splash*) reinforce the image of the callous and unfeeling scientist mistreating the charges in his care.

When one combines the iconography of the laboratory animal as a helpless innocent with the stereotypical scientist, it is not surprising that it might be difficult to persuade the public that a "coldly rational" animal researcher would have concern for his or her research animals. Research scientists usually reinforce one or both of these images in the media by using dispassionate language and invoking rational argument. For example, one medical researcher commented during a public talk that she would use her own, much-loved pet cat in research if she thought it would advance her search for a therapeutic intervention for human disease. Her comment served to reinforce the view of scientists as cold, dispassionate, and unfeeling.

PUBLIC ATTITUDES TOWARD ANIMAL RESEARCH

Numerous polls of attitudes to animal research and testing have been conducted and the findings can be summarized as follows.

a) About two-thirds to three-quarters of the American public is prepared to accept the need for animal research.
b) The percentage that actually supports animal research is usually about 10 percentage points lower.
c) About 10–15% of the public actively opposes animal research.
d) The percentage opposing animal research changes, depending on the type of animal used and the type of research. Thus, most people support research that uses rats, but this figure may be halved if dogs are the research animal. Similarly, cancer research is considered very important by the public, but support drops off for alcohol and drug addiction research and product testing, especially of cosmetics and household goods.
e) So-called "basic" research does not receive as much public support as goal-oriented medical research.
f) About half the public are uncertain whether animal researchers treat their animals humanely.
g) It appears as though the public is becoming less tolerant of the use of animals in research. The biennial Science Indicators Survey commissioned by the National Science Board (NSB 1989) in the U.S.A. has begun to ask a question about support for the use of dogs and chimpanzees in medical

research to find cures for diseases. In 1985, 63% of the public responded that they supported such animal use, but this dropped to 53% in 1988 and to 50% in 1990. In the U.K., where a similar question was also asked in 1988, only 35% of the public supported the position.

While some in the biomedical community might feel that the above results are reasonably good, the actual support for animal-based research is relatively soft. As has been demonstrated, certain cases can easily mobilize widespread public outrage (from a majority of the public rather than just the 10–15% who are opposed to the use of animals). Also, the trends seem to be negative from the animal research perspective. It has also been demonstrated that younger people are more concerned about animals than older people (Kellert 1979). Therefore, one can only expect a continuing trend towards more criticism.

BIOMEDICINE, ANIMAL RESEARCH, THE MEDIA, AND THE PUBLIC

In the past year or two, biomedical professional societies, such as the American Medical Association and the Federation of American Societies of Experimental Biology, have begun to take a more militant stand towards their animal activist critics. Thus, the debate that was already sharply polarized has become even more so. The overall aim of these scientific organizations seems to be to persuade what they perceive as a relatively ignorant public that continued good health depends on animal research, and that there is a health-dependant choice to be made: animals or people, but not both. Thus, many news stories in print and in broadcast media about recent medical discoveries are now much more likely to mention the role that animal research plays.

Leaders in the biomedical community are also more likely to spend time and effort to counter the "threat" to biomedical progress posed by the animal protection movement. For example, Secretary of the U.S. Department of Health and Human Services Dr. Louis Sullivan reached out to the media before the "March for the Animals", a public demonstration by animal activists held on June 10, 1990 in Washington, D.C. As a result, most of the media stories on the march gave Dr. Sullivan's utterances about the need for animal research the major emphasis.

The strategy of aggressively taking the biomedical research message to the public is too new to judge its effectiveness, but some of the earlier campaigns and arguments in support of biomedical research have misfired and have failed to slow the trends of public opinion on the animal research issue.

An analysis of some of the arguments and strategies indicates why they may have misfired. The following examples are not based on a systematic reading of the literature. They are provided merely to focus and organize a discussion of the effectiveness of the research strategy.

a) Stressing the need for animals

Several years ago, the National Association for Biomedical Research (an organization similar to the Research Defence Society in the U.K.) released a film called *Will I Be All Right, Doctor?* The main theme was the importance of animal research in developing new therapies and treatments. A lesser theme was the good care that the laboratory animals received. Since three-quarters of the public already accepts that animals are needed in research and testing, the film is unlikely to address the real public concerns described above. On the question of animal care, I found the film to be accurate but unexciting and uninteresting.

b) Stressing the benefits of animal research

The biomedical research establishment commonly argues that animal research is done because of the benefits it produces for human and animal health. In so doing, the research community continues a long-standing tradition of science "education" (Birke 1990). Throughout this century, efforts to popularize science and to educate the public have tended to emphasize the benefits of science. As health care became more successful and more technical, and the public demanded more of those in authority, the public took purported benefits for granted. Furthermore, groups critical of science (e.g., environmentalists, animal activists, opponents of genetic engineering) started to question the benefit claims. In most instances, the scientific community did not address the criticisms carefully and fully, but tended to respond merely by emphasizing the benefits even more strongly.

Public accountability on what animal research is permitted and performed has become a major issue. At present the public has relatively little power to decide which research goals scientists will pursue, and how they will guard against animal suffering. Scientists have successfully created a separate system to support research creativity in which they are left alone to do their own thing. While this freedom has produced much success, and the public has been content to be served with a "collective production of knowledge", the times are changing.

During the past few decades, the public has demanded more accountability from all authority figures. The media has also knocked scientists from the pedestal they were on in the 1950s by reporting more on the details of science and its various controversies. Scientists are no longer larger-than-life heroes like Einstein, but ordinary human beings subject to all the usual faults and blemishes (Johnson 1990). This means that claims made by scientists are less likely to be seen as infallible, and are more likely to be questioned.

At the turn of the century, Walter Cannon of Harvard Medical School put much energy into fighting the antivivisection movement (Benison et al. 1987). He successfully challenged their claims and their credentials and continually pointed to the actual and potential benefits of medical research. The situation is different today. The public does not feel as threatened by disease and injury, it is more skeptical about blanket, one-sided claims of potential benefits from research, and it tends to consider the claims of the animal protection community more carefully.

From observations of the debate and the effectiveness of public relations pronouncements, the public tends to accept animal research and testing when it appears to be of obvious benefit and does not produce too much suffering. However, when the research is perceived to produce a great deal of animal suffering, then the benefits have to be significant, immediate, and self-evident if the public is to accept such research.

c) The media and the public are victims of a good public relations campaign by the animal protection community

One relatively common view among the research community appears to be that the animal movement has made very skillful use of the media to exploit a gullible public. It is certainly true that the animal research controversy makes for good media copy, but the animal protection groups have, for the most part, not been that skilled at disseminating their message. Henry Spira, the New York co-ordinator of the campaign against the rabbit Draize eye irritancy test had a field day with his advertisement in the *New York Times* which asked "How many rabbits does Revlon blind for beauty's sake?". People for the Ethical Treatment of Animals (PETA), the largest of the new animal rights groups, have done reasonably well in obtaining coverage of their stories (in part, probably, because they were based on "secret" information that was exposed by undercover or illegal activities). The animal movement has become more skilled in the past two decades at getting its message to the media, but this is probably due to the recruitment of more professionals and their skills into the movement, and is greatly helped by increased public concern.

An example of an earlier media campaign against animal research shows how difficult it is to mobilize an unconcerned public. In the 1940s, William Randolph Hearst put the muscle of his nationwide newspaper chain behind the antivivisection cause. But, apart from raising the blood pressure of some in the research community and stimulating some classic yellow press journalism, there is little evidence that the campaign by the influential Hearst newspapers had much impact on public opinion as regards animal research. In fact, public opinion in favor of animal research was extraordinarily high at the time, and remained so for another decade at least.

d) Animal activists and terrorism

In the past five years, biomedical spokespersons have frequently pointed to the terrorist activities used by members of the animal rights movement. Underground animal rights groups, e.g., the Animal Liberation Front, that break into and vandalize animal care facilities in research institutions have been identified as dangerous terrorists and equated with the IRA and other terrorist groups. At times, the linking of animal activists with terrorism is very broad, as though all activists are engaged in vandalism and life-threatening activities. This issue leads to heated debate within the movement, because the philosophy of animal rights respects all animal life, presumably even human being. Thus, any threat against the life of another human undermines the very philosophy that the activists are trying to promote. Nevertheless, there are still frequent anonymous threats made against researchers. In the past, these threats were treated as coming from disturbed and ineffective people, but they are taken much more seriously today, because of the use of bombs and other violent tactics by underground animal activists. Recently, Singer has argued that the moment the animal liberation movement loses the moral high ground (no abuse of animals or people), it will lose its legitimacy and its hold on the public (Singer 1990). However, the destruction of property is seen by many animal activists as falling outside the protection of the animal rights philosophy as long as no lives are threatened.

The argument over animal research has always been sharply polarized. Antivisectionists viewed their opponents as sadists and morally defective, while researchers viewed the antivivisectionists as irrational and sentimental at best, and insane at worst (Keen 1914). It is not easy to develop a productive dialogue under these circumstances. Today, we are in danger of falling into the same trap of polarizing the debate by categorizing the other side as either sadists or terrorists. Many in the animal protection and research community wish to avoid such counterproductive labelling but, with the media's attraction to diametric opposites, it is not easy. If the controversy becomes hopelessly polarized, the biomedical community will likely emerge as the victor, but at some unpredictable cost to its reputation and prestige.

e) Do not apologize for animal use

There are some in the research establishment who have decided that there is no need to be apologetic about the use of animals in research and testing. They even argue that any establishment support for the idea of "alternatives" to laboratory animals is inherently apologetic and should be resisted. However, opinion polls all indicate that the public strongly supports the search for and use of alternatives. People seem to believe that this is one way that they can have advances in health care without having to endure the psychic cost of animal research or the stigma of being labelled "antiscience".

There is one major problem with the 'no apologies' approach. Studies indicate that not only is the public uneasy about the cost in animal suffering and death involved in biomedical research, but so also are those laboratory personnel who are directly involved (Arluke 1988; Dodds 1989). While some research scientists might welcome the clarity that comes from dividing the animal research issue into opposing camps, many others will not be able to take this step, and will either stop doing certain types of research, or will refuse to accept the party line. This is already happening. The animal protection movement is certainly not short of sympathizers in the research community. Many of the laboratory break-ins, for example, appear to have been accomplished with inside help – that is, institutional employees, not all of whom have been "plants".

CONCLUSION

In the modern animal research controversy, "many citizens have begun to judge science according to their own moral standards rather than accepting the measures of professional achievement that scientists apply to themselves" (Ritvo 1984). Thus, experimentation on animals has become a focal point for differing world views held by animal protectionists and scientists. The result has been little more than shouting matches, accusations of immorality by both sides, and a steady progression of onedownmanship, with little constructive progress or careful analysis of the central issues.

Ultimately, any resolution of this controversy must be based on an intelligent and open debate. The public should be offered a more meaningful role in the scrutiny of animal research. Various models have been attempted with varying success. In Great Britain, public accountability is largely obtained through the Animal Procedures Committee, which contains representatives of the various interest groups. However, much of the debate occurs behind closed doors, so, for those who do not trust authority figures or the particular representatives on the committee, public accountability is still lacking. In Sweden, ethical review committees have been established to oversee and approve research protocols. A wider range of representatives from the critical public have participated in this scheme, but it has still had problems – including the degeneration of committee debate into a confrontational power struggle between animal activists and scientists.

In the U.S.A., animal research protocols must be reviewed by committees containing a "member of the community". In the vast majority of cases, this community member is not drawn from the critical public. Therefore, the critics are still excluded from any meaningful role in debating animal research policies, and they tend to see the institutional committees as mere rubber stamps. In Australia, a similar system of institutional committees is being established, but the outside member must have some ties with animal protection. Time will tell if this structure is any more effective in developing meaningful and constructive dialogue.

For a more effective response to criticism from the animal protection community, the scientific community needs to pay attention to the following issues. First, scientific spokespersons must develop a much more sophisticated understanding of the philosophies that underly the animal protection movement. At the moment, most scientists are easy game for even the average animal activist, who is better versed in the philosophical arguments.

Second, scientists must recognize the central role played by the spectre of animal suffering and the helplessness of the laboratory animal. It is probable that, if the public could be convinced that animals do not suffer at all in research, much of their concern would fall away. However, at the moment, many in the public "know" that animals are "tortured" in laboratories, so it will be difficult to convince people that animal suffering is rare. It should also be noted that relatively little attention had been paid to the alleviation of laboratory animal pain and distress until the past decade. For example, much had been written about housing for animals that is durable, readily sanitized, and free of disease, but the first helpful guidelines on the overall well-being of rodents used in cancer research did not appear until 1988 (UKCCCR 1988).

Third, scientific spokespersons need to display more obvious concern for the necessity to use laboratory animals. For example, they could take a lesson from one scientist who was asked by a journalist if his neuroscience work with monkeys did not bother him. He responded by asking if the reporter was crazy. The scientist said that, of course the use of monkeys bothered him, but after much soul-searching he had determined that the research had to be done. This particular neuroscientist had been targetted by a local animal rights group in their campaign to highlight the need for more regulation of research but, after the above response, the neuroscientist's name no longer appeared in the media stories.

Scientists need to show themselves as caring, personable individuals, not as cold, superrational robots removed from sentimental and emotional animal activists.

Fourth, patronizing the public and talking down to an audience that "cannot understand the complexities" is a recipe for failure. It has been shown that the public can understand rather sophisticated science when it believes this necessary, and it does not need detailed scientific knowledge in order to give thoughtful consideration to a complex issue (Doble and Johnson 1990).

Fifth, the political sensitivity of scientists should be upgraded, and academic societies and associations should join the ongoing debate. This is happening more, but there is still a tendency to avoid any real debate and to stay with the more comfortable option of speaking just to like-minded individuals. Sometimes this is justified with the rationalization that the opposition's argument should not be dignified by giving it a platform. However, there are dangers in allowing the argument to be conducted in the open media. In an academic forum, one can challenge the accuracy and validity of an argument in a way that is just not possible in a ten-second sound bite. Although scientists may consider a particular argument to be nonsense, there are many others who will not (in part, because they do not have the appropriate knowledge and skill to judge). The biology research community successfully challenged scientific creationism by exposing the errors in its argument. If scientists believe that the arguments of animal activists are false, they must take the time to explain how and why. The counterargument may not make the evening news, but it will level the playing field.

Ultimately, biomedical science and its spokespersons need to avoid the arrogance and cloistered smugness that lurk in wait for intelligent and creative, but unwary professionals. In the world of politics, one is only as good as one's ability to make an argument and present oneself as a credible spokesperson. Self-interest, or condescension, or inability to produce a believable rebuttal to your critics' arguments will undermine credibility. Dr. Franklin M. Loew often uses the following story to expose the dangers of expecting any special favors in the political arena.

George Brown, a Californian Democrat in the U.S. House of Representatives, supportive of science but also a canny politician, was presiding over a hearing on the animal research issue. The usual array of animal protection and biomedical representatives gave evidence before the subcommittee. At one point, after a particularly impassioned statement by an animal activist, one of the scientists, in an aggrieved tone, asked Congressman Brown why he listened to people like that. "Doctor," responded Brown, "we have a word for people like that up here. We call them voters!"

REFERENCES

Arluke, A. (1988). Sacrificial symbolism in animal experimentation: object or pet? *Anthrozoös* **2:** 98–117.

Benison, S., A.C. Barger, and E.L. Wolf (1987). *Walter B. Cannon: The Life and Times of a Young Scientist.* Cambridge: Harvard University Press.

Birke, L. (1990). Selling science to the public. *New Scientist* (August 18):40–44.

Cohen, C. (1986). The case for the use of animals in biomedical research. *New England Journal of Medicine* **315:** 865–869.

Doble, J. and J. Johnson (1990). *Science and the Public: A Report in Three Volumes. Volume 1: Searching for Common Ground on Issues Related to Science and Technology.* New York: Public Agenda Foundation.

Dodds, W.J. (1989). A scientist's perspective. *Anthrozoös* **3:** 74–75.

Donnelly, S. and K. Nolan, eds. (1990). *Animals, science and ethics. Hastings Report* Volume 20: 3, Special Supplement.

Donovan, J. (1990). Animal rights and feminist theory. *Signs: Journal of Women in Culture and Society* **15:** 350–375.

Elston, M.A. (1987). Women and antivivisection in Victorian England, 1870–1900. In: *Vivisection in Historical Perspective,* ed. N.A. Rupke, pp. 259–294. London: Croom Helm.

Fox, M.A. (1986). *The case for Animal Experimentation: an Evolutionary and Ethical Perspective.* Berkeley: University of California Press.

French, R.D. (1975). *Antivivisection and Medical Science in Victorian Society.* Princeton: Princeton University Press.

Gerbner, G. (1987). Science on television – how it affects public conceptions. *Issues in Science and Technology* **3(3):** 109–115.

Gilligan, C., J.V. Ward, J.M. Taylor, and B. Basdige, eds. (1988). *Mapping the Moral Domain.* Cambridge: Harvard University Press.

Griffin, D. (1976). *The Question of Animal Awareness.* New York: Rockefeller University Press.

Jack, D. (1990). Letter – animal rights. *New Scientist* (September 1): 62.

Johnson, R. (1990). The public image of science and its funding. *FASEB Journal* **4:** 2431–2432.

Keen, W.W. (1914). Animal Experimentation and Medical Progress. Boston: Houghton Mifflin Co.

Kenward, M. (1989). Science stays up the poll. *New Scientist* (September 16):57–61.

Lansbury, C. (1985). *The Old Brown Dog.* Madison: University of Wisconsin Press.

Littlewood Committee (1965). *Report of the Departmental Committee on Experiments on Animals, Cmnd. 2641.* London: Her Majesty's Stationery Office.

Lutts, R.H. (1989). *The Nature Fakers: The Romanticizing of Nature.* Golden, Colorado: Fulcrum.

NSB (1989). *Science and Engineering Indicators – 1989.* Washington, D.C.: National Science Board.

Orem, J. (1990). Demands that research be useful threaten to undermine basic science in this country. *The Chronicle of Higher Education* (March 14) B1–B3.

Pion, G.M and M.W. Lipsey (1981). Public attitudes toward science and technology: What have the surveys told us? *Public Opinion Quarterly* **45:** 303–316.

Ritvo, H. (1984). Plus ca change: Antivivisection then and now. *Science, Technology and Human Values* **9** (spring):57–66.

Rollin, B.E. (1989). *The Unheeded Cry: Animal Consciousness, Animal Pain, and Science.* Oxford: Oxford University Press.

Rowan, A.N. (1990). Letter to the editor. *The Chronicle of Higher Education* (April 11): B2.

Rowan, A.N., and B.E. Rollin (1983). Animal research – for and against: A philosophical, social, and historical perspective. *Perspectives in Biology and Medicine* **27:** 1–17.

Singer, P. (1975). *Animal Liberation.* New York: Random House / New York Review of Books.

Singer, P. (1990). *Animal Liberation,* Revised Edition. New York: Random House / New York Review of Books.

UKCCCR (1988). UKCCCR guidelines for the welfare of animals in experimental neoplasia. *Laboratory Animals* **22:** 195–201.

Walker, S. (1983). *Animal Thought.* Boston: Routledge Kegan Paul.

Weart, S. (1988). The physicist as mad scientist. *Physics Today* (June): 28.

Bioscience ⇌ Society
edited by D.J. Roy, B.E. Wynne, and R.W. Old, pp. 283–291
© 1991, John Wiley & Sons

Bioscience as a Social Practice: Some Preliminary Remarks

Robert C. Solomon

Philosophy Department
University of Texas
Waggener Hall 316
Austin, TX 78712, U.S.A.

Abstract. In these preliminary remarks, I would like to mollify the antagonism between bioscience and society, arguing that bioscience must be understood as a social practice, and that scientists have social reponsibilities despite their "pure" pursuit of the truth in a free market of ideas. So understood, the question of who "controls" scientific research is both unnecessary and inflammatory.

"Dr. Frankenstein, I presume." So began an earlier meeting between bioscience and society. Mary Shelly's creation (that is, the doctor, not the monster) captures the fear and fascination with which the ordinary citizen approaches such topics as genetic engineering and ecological intervention. Of course, monsters make great reading and good theater whereas the depiction of a bioscientific paradise would inspire little but envy. (Simone Weil rightly observed that evil and not goodness makes great fiction, but only good makes a good life.) From the other side, bioscientists rightly complain of public lack of appreciation for their work and political interference, poor working conditions, erratic and often irrational funding procedures.

And yet, we are, of course, a bioscience society. The fruits of bioscience have virtually doubled life expectancy and (insofar as it can be measured) the material quality of our lives in less than a century – merely a nanosecond in the long history of biological evolution. The ideas and promises of bioscience help define as well as challenge our social policies and our ethical principles.

The newest techniques of bioscience provide the topic for endless social and political debates. The average citizen is grateful for bioscience breakthroughs and expects always more – a cure for cancer, a vaccine against AIDS, a solution to global warming, an indefinite reprieve from the increasingly receding sentence of the Grim Reaper – and all of this as if by magic. But the bioscientist is no longer the leisurely naturalist, making some great discovery while walking through the woods or taking a tour of the South Pacific. The bioscientist depends upon society for very expensive research labs and equipment, for experimental subjects, and – not least but often most difficult to obtain – emotional support and appreciation. The relationship between bioscience and ethics, accordingly, is both symbiotic and mutually antagonistic. Society expects a great deal from bioscience, and yet every innovation is greeted with suspicion, a Jeremy Rifkin conference, an Arthur Caplan rebuttal, and the predictable made-for-television movie.

In these preliminary remarks, accordingly, I would like to mollify the antagonism between bioscience and society, taking their mutual dependency as something of a premise and begin a plea for mutual understanding and the social responsibility of scientists in a free market of ideas. But perhaps the first thing to say is that conflict can be healthy, and uninterrupted silence can be dangerous. The easily understandable attitude on the part of some scientists, "give us the money and let us do our research", is no more desirable than political control over science policy. (Brian Wynne has wisely pointed out that the absence of social concern, e.g., in the development of nuclear facilities in the 1950s, was hardly a sign of health.) What philosophers call "dialectic" (a sometimes heated conversation which moves from polarized disagreement to increased mutual understanding) is a sign of social health, even if annoying and cumbersome to the unwilling dialecticians. In the free market of ideas, it is competition and disagreement that gets rid of bad ideas and establishes good ones. In a democratic society, debate and doubt of authority – including the admittedly fallible authority of scientists – is essential. We distrust as well as depend on our experts.

In this sense, science is *not* in a privileged position in our society; it, too, is part of the chronic dialectical conflict that (within science) makes science work and (outside of science) is the essence of democracy. But there are healthy and unhealthy dialectics. What is unhealthy is polarization and extremism, those shouting matches in which neither side listens to or responds to the other, in which differences expand rather than contract and the voices that get heard are of the "more Mao than thou" variety, rather than the voices of reason (not to be confused with the voices of mere neutrality or disinterest). When irrational, uninformed, and sometimes insane positions come to dominate the debate, when the debaters are no longer critical of their own presuppositions, when the battle becomes political (literally, decided by power), when there is a total loss of compassion for either scientists or their subjects and the debate degenerates to moral name-calling ("Frankenstein!" and "Philistine!" respectively), we know that the dialectic has gotten derailed and a new "covenant" has to be forged.

THE PRACTICE OF BIOSCIENCE

Bioscience is a social practice. What this seemingly simple and hardly controversial statement implies, however, is that much if not most of the popular view of science – on the part of most scientists and science students as well as the "lay" public – is deeply flawed. That popular view prefers to take the scientist as a solitary cosmic wanderer (who may just happen to work in a laboratory with other like-minded scientists and just happened to study his or her subject at some more or less prominent university). The activity of science is that of an isolated thinker confronting the comprehensible wonder of the world. It is a rational relationship between a single mind and the Truth. The scientist, according to this popular view, is "dispassionate" and "objective" (though fascinated and enthusiastic, perhaps ambitious, too), working entirely on his or her own with only Nature as a constant companion. It is a heroic picture, but like most heroic pictures it is largely fantasy and inaccurate even as a description of the work of great geniuses such as Darwin, Mendel, and Einstein. True, they may have separated themselves from the world, insulated by the originality of their thoughts, but they were nevertheless very much a part of their times; they inherited their problems and their scientific vocabulary (however they may have subsequently changed it) from their teachers and colleagues. The structure of their discipline, the way they framed their questions, the way they considered their data: these were not personal inventions, but part of the practice of science at the time. (This seems obvious to us when we look back at the history of science, and it will be obvious again when future historians look back at us.)

To say that bioscience is a social practice is to say a great deal about its place in and its distance from the larger society in which it thrives. A social practice defines its own little world within the world. One might, without being demeaning, compare bioscience in this regard to American football. Each has its own rules, its definitive roles, its own language, aims and assumptions, its heroes and saints (Nobel laureates and Heisman Trophy winners respectively). Religious critics of science sometimes enjoy pointing out that science, too, is a "religion", on the grounds that it has its own internal assumptions and even saints. But the same could be said of any practice, even football (which is also taken to be a religion, for instance, in Texas). Every practice has its own "internal goods", the goals and satisfactions that make participating in the practice and being "good at it" worthwhile. It is participation that justifies the practice for its practicioners, not its products or results. The thrills of discovery and victory, for example, are internal goods, more aesthetic or even hedonic than "practical".

Practices also have external rewards, of course – the salary one receives for being a scientist or a professional football player, and possibly the fame (more likely for a football player). Practices can and must be justified "from the outside" as well as by the joys of participation. The practice must benefit, or at least not harm, the surrounding society. Football is entertaining for the public, and injuries are restricted to the players. Scientists work in their labs and sometimes produce socially valuable discoveries. But in this distinction between "internal goods" and "external rewards" and between the two sources of justification lies considerable room for misunderstanding. Scientists have difficulty explaining to the public that they are really interested in the questions more than the answers (and every answer leads to yet more questions), and football fans and players have a hard time explaining to the uninitiated why grown men would risk serious injury trying to place a bloated pigskin across an arbitrary line in the grass.

Bioscientists believe first of all in the intrinsic good of "knowledge for its own sake", and that belief, more than anything else, defines and justifies the practice. Those outside the practice, on the other hand, see bioscience and the knowledge it produces as a means, a way to increase crop and milk production, a way to cure or (better) prevent disease. This leads to some odd and uncomfortable consequences. Bioscientists need to seek funding to search for "knowledge for its own sake" by promising (only sometimes truthfully) that such knowledge will yield practical results. The public in turn expects miracle cures as if "by magic", ignoring the process that alone makes such discoveries possible and provides the meaning of the activity for the scientists who engage in it. The current clamor for an AIDS vaccine and cure is but the most visible example of such politically-directed "targeted" research. (One might also speculate how the misunderstanding between bioscientists and society originates in the process of bioscience education, where theories are presented as prepackaged products, and little or no attempt is made to convey the thrill of pursuit and discovery.)

One way of utilizing the distinction between internal goods and external rewards of a practice is to make some sense of the often abused distinction between bioscience on the one hand and biotechnology on the other. Bioscientists understandably see themselves as doing "pure" or "basic" research which may (or may not) result in some usable product for the public. The public, ignoring the process and practice, confuses or simply ignores the distinction, taking science as nothing (significant) other than the process of inventing and producing technology. On the other hand, some very sophisticated commentators on the bioscience scene have similarly objected – in agreement with the public but on much more substantial grounds – that the dichotomy between bioscience and biotechnology is artificial and ultimately insidious, that the very process of testing and experimentation is necessarily "applied" as well as "pure" (e.g., in the search for a vaccine). One very good reason for listening to such criticism is that it rightly rejects the idea that bioscience (unlike its technological cousin) is "value-free" and "noninstrumental", in other words, that bioscience is not a social practice. But one need not (and cannot) make the distinction between science and technology absolute, nor should one confuse the two. Biotechnology is essential *within* the practice of bioscience as well as a product of the applied significance of the practice. But here again we find considerable room for misunderstanding and antagonism. Insofar as bioscience can be defined in terms of its internal goods, there is always the political question of why society (which may not share the scientist's pursuit and joy of discovery) should fund it. And insofar as the new technology coming out of science serves policies and purposes which are themselves quite independent of science, why should they be given special priority just because they are the fruits of science? For example, even if the genome project results in genetic technologies that will alleviate suffering and save lives, a fraction of the billions of dollars that fund the project would allow public health officials to end much more suffering and save millions of lives through a worldwide public health, nutrition, and vaccination campaign requiring no new science or technology. Why do the more exciting scientific solutions take priority over old, established, and now-routine technologies that might, in fact, do much more good? In a world of limited funding, we find ourselves making deep ethical and political choices on the behalf of the new bioscience.

THE ETHICS OF BIOSCIENCE

In any discussion of bioscience and society, the concern for ethics is sure to take a prominent seat at the table. But there is considerable confusion here and, as in the practice of bioscience as such, much room for disagreement and mutual misunderstanding. In many of the discussions we read by bioscientists themselves, the defense of bioscience and biotechnology is couched exclusively in the language of *cost/benefit* analysis, or in more philosophical terms, a crude form of *utilitarianism.* Utilitarianism is an ethical view that evaluates actions and policies according to a single, seemingly simple criterion: "the greatest good for the greatest number". How many people are helped (for example, by a new vaccine or pesticide) and how many people are hurt, and how much are they helped or hurt? (It was stated as a much more sophisticated position by its founders Jeremy Bentham and John Stuart Mill, but that is not of importance here.) The argument on behalf of bioscience and technology, to be sure, is that its products over the past century or so have clearly led to a vast increase in human welfare, even in the face of nuclear holocaust, germ warfare, global warming, and other potential threats of apocalyptic dimensions. But the arguments against bioscience are typically not of this kind, and so the bioscientists and their critics speak past one another. The language of risk may be prominent (especially with the intrusion of product liability lawyers), but it is not the language in which ethics usually defines itself.

What one hears much more – in more or less primitive terms – are the accusations that bioscientists are "interfering with nature", "tinkering with evolution", and "threatening human integrity". These accusations may or may not be accompanied by religious doctrine. (Aristotle defended such a theory of "natural law" three centuries before Christianity; Thomas Aquinas, following Aristotle, wrote over a millenium later.) But what they point to is a dimension of bioethics (as opposed to bioscience) that is often uncompromising and, for many people, even more important than questions of physical well-being and potential harm. The wild-eyed and sometimes rabid reactions to the wide range of questions and techniques concerning human reproduction are the most obvious examples, and bioscientists often find themselves in a fluster facing seemingly irrational syndromes of attack by those who would outlaw not only abortions but all use of fetal tissue (however derived), embryo experimentation, and assisted fertilization. In the eyes of a cost/benefit analysis, the aim of which is the unquestioned concern for human well-being, those arguments may make little sense, but cost/benefit has very little to do with them. From the perspective of the admittedly obscure notion of human "integrity", however, they are all of a piece, however ill-considered. It is the process of "tinkering", not the results that are in question. On a more sophisticated and abstract plane, one may maintain the thesis that whatever humans do is part of nature and human integrity cannot be separated from the pursuit of human happiness. But the "gut reaction" of those very basic, even if uninformed, beliefs is not to be taken lightly, and in more thoughtful form they provide dimensions to the discussion that bioscientists will ignore only at their peril.

What is at stake here is not simply the familiar and often dogmatic confrontation between science and religion (knowledge and faith). It is the manifestation of what is probably the singularly most important feature of contemporary democratic society, and that is the *pluralism* of ethics. It is not just that people don't agree about various issues, such as the desirability of artificially inseminating infertile would-be-mothers, or intervening in the genetic structure of a newly formed zygote. There is a clash of ethical *frameworks,* and where sex, babies, and human bodies are involved – not to mention the suffering of animals who might otherwise be some family's pet, or the destruction of an environment that might well be a wall poster in the family dentist's office – one can be sure that much more will be at stake than either the excitement of science or the well-being of the lucky recipients of a new technology. To begin with, even utilitarianism has its obvious fractures, between the well-being of a single individual or a small group and the larger public good. What may well move us in an individual case can look more like a mathematical matrix when millions of people are involved, and the solutions to the larger problem may well offend us if considered individual by individual. (Thus epidemiology has with some justice been called "medicine without tears".)

The skewed debates and dimensions of pluralism are not confined to the molecular and the global; even where individuals are concerned (no matter how many of them), bioscience and scientists will find that they run up against the all-important conception of individual *autonomy* – a person's right to make his or her own decisions – and the charge of *paternalism* – the tendency of (scientific or political) authority to preempt such decisions. There are also questions of human dignity, intricately tied to such notions as autonomy and integrity, but also the much more specific and hard-headed language of *rights.* And there are questions of *ownership* – a concept tied in any number of complex and controversial ways to the notion of rights – including (especially) the odd idea of ownership of one's own body and questions about the ownership of one's children, animals, and of course the earth itself (or whatever more or less modest piece of it). There are *cultural* and *esthetic* values that may (and often do) conflict with these. (The practice of bioscience may be one example of such conflict.) There are complex questions of *justice,* which include some reference to rights, but much more complicated questions of comparative well-being and merit as well. There are sociopolitical questions about class interests and privileges, accusations of neocolonialism, and ethically-loaded distinctions between the "developing" and the "squandering" nations of the world. And then, of course, there are religious concepts and concerns, ranging from Hindu holiness of cows to Greenpeace holiness of whales, from the extremities of Christian Science and Earth-Firstness to the more mundane middle-class concern with cholesterol (by no means a merely medical concern).

All of these crisscross and clash with alarming frequency, giving the false impression to more hysterical society-watchers that we have lost our "values", that "relativism" runs rampant and "nihilism" awaits in the wings. But the truth is that democracy and science, too, thrive on such chaos. The greatest danger, much in evidence on all sides of the literature, is rather an undue fear of the treacherous "slippery slope", a form of argument that always suggests: "If you let *them* get away with *this* (fertilizing human eggs in vitro, cutting funds for this or that political reason), then it will not stop until it gets to *that* (Frankenstein monsters, no science at all)." The truth is rather that open dialogue and disagreement – what in American constitutional law is gracefully referred to as "checks and balances" – is healthy and productive for both bioscience and society. What is unhealthy and disastrous for both is that mutual suspicion which knows no outlet but avoidance and condemnation, that will not talk and mutually explore possible agreements or compromises, not only about risks but about values, too. The debate over the *control* of bioscience, whether by politicians and people who know nothing about it (how could they make informed and intelligent judgments?) or by the scientists themselves (how many professional groups have in fact shown the ability to "police" themselves?), is based on the wrong-headed asumption that scientists and nonscientists cannot and will not talk to one another about issues of mutual interest and importance. Undermining this assumption is what this conference is all about. It does no good if bioscientists see themselves as victims of the irrational whims of society, and society will not benefit if it maintains its "Frankenstein" fear of the bioscientist and his or her ability to convert life into monsters. We need to cultivate a society of bio-Einsteins, not in genius of course but in their ability to appreciate both the wonders and promises of science and its threats and ethical antagonisms. We need citizens who are not only better educated in science (it is by no means obvious that increased knowledge lessens one's fears of science) but properly torn and confused about an enormously complex and unsettled network of issues. Bioscience needs engaged, and not just appreciative, citizens.

REFERENCES

Callahan, Daniel (1973). *Bioethics As a Discipline.* Hastings Center Studies 1.

Engelhardt, H. Tristram (1986). Emergence of a secular bioethics. In: *Foundations of Bioethics.* New York: Oxford University Press.

Gustafson, James M. (1986). Basic issues in the bio-medical fields. In: *Foundations of Bioethics.* New York: Oxford University Press.

Howard, T. and J. Rifkin (1977). *Who Should Play God?* New York: Dell.

MacIntyre, Alasdair (1981). *After Virtue.* Notre Dame: University of Notre Dame Press.

The Monist issue on "Bioethics and Social responsibility". *The Monist* **60** (Jan. 1977).

Reich, Warren T., ed. (1978). *Encyclopedia of Bioethics.* New York: Macmillan.

Walters, LeRoy (1986). The ethics of human gene therapy. *Nature* **320** (March 20, 1986).

Bioscience ⇌ Society
edited by D.J. Roy, B.E. Wynne, and R.W. Old, pp. 293–303
© 1991, John Wiley & Sons

Conflicts Between Bioethics and Some Aspects of Traditional Ethics and Religion

Dietmar Mieth

Katholisches Theologisches Seminar
Abteilung Ethik II
Liebermeisterstr. 12
D – 7400 Tübingen, Germany

Abstract. Christian ethics is an ethics of abilities and potentialities, rather than one of duties. Because of man's ability to comprehend new potentialities in life, responsibility is placed on him. Despite our growing control over nature and ourselves, we experience our limits, finiteness, and social dependence, which is an experience of contingency. A "constructive contingency encounter" is envisioned with regard to some central ethical problems: the concept of personhood, of nature, and of freedom. Being aware of death is one of the most important presuppositions for a "constructive contingency encounter" to be successful. The theoretical conflict between this concept and bioethics is finally discussed with the example of termination of pregnancy.

The Christian ethic is not a duty-oriented ethic, but an ethic of human abilities and potentialities. One of the principles of the "bourgeois" ethics of Immanual Kant is: "Thou canst, for thou should." And one of the principles of the great tradition of moral theology is: "A 'thou shalt' presupposes a 'thou canst' ". In a basic orientation that relates closely to life[1], the question asked by many persons on the margins as well as in the center of society must be taken seriously: "I want to live, but I don't know how I can." This question challenges Christian bioethics to proceed from the "thou canst" and not

1 Viz.: Leben – philosophisch und theologisch (Life – philosophically and theologically). In: *Staatslexicon,* ed. by the Görresgesellschaft, 7th edition, vol. 3, pp. 852–857 (1987). Freiburg / Basel / Vienna: Verlag Herder.

from the "thou shalt". This also involves taking seriously the paradigmatic shift leading from the Ten Commandments, the second tablet of which is a summary of the concept of duty, to the Pauline ethic, in which man, who is embraced and permeated by spirit and by grace, also possesses new potentialities to live. Because he is able to comprehend these new potentialities, responsibility may be demanded of him. This shift of the paradigm of ethics from duty to potentiality – and only then to duty – delineates the way from Mount Sinai to the Sermon on the Mount, from the Law of Moses to the promise of Jesus of Nazareth.

There are broad spheres in which we are apparently inescapably trapped: the spheres of oppression by totalitarianism and fundamentalism, the poor distribution of goods, the consequences of overarmament. In all these spheres there has doubtless been progress in individual areas, but the fact remains that there is hardly a technology which we can employ to improve man's chances to live which cannot at the same time serve the end of destroying human life and refining weapons technologies, thus leading to a situation in which disarmament is often overtaken and surpassed by modernization. It is not necessary to hold forth at length on the poor distribution of goods in the world, but one can point out the fact that the basic principle of natural law is the right to property, that all human beings have a right to partake of the goods of this earth, to be nourished by them, and to be able to survive with their help.

Another sphere in which we appear to be inescapably trapped embraces the domain of medical high technology and the economic-ecological breakthrough technologies. The pace of development in both of these technological domains is quickening today. The pace is determined by the inherent necessities of economic life and success in research to such an extent that it is very difficult for us to establish with certainty whether the attendant measures we initiate, e.g., laws on security and safety, will make the resulting impacts weigh in favor of man. There is hardly an advanced medical technology, from organ transplants and reproductive medicine down to the new forms of genetic engineering, which is not on the whole ambiguous with regard to the question of a better life for humankind. Some people, for instance, ask themselves whether there is a meaning in prolonging individual life if this no longer contains the experience of freedom, and if the possibility of communicatively experiencing life is reduced to a genuine minimum.

Our economic-ecological breakthrough technologies – nuclear technology, information technology, karyotechnology – do not only solve problems, but they also cause new ones. I endeavor to summarize ethical responsibility in a "problem-solving rule": "Problems should not be solved in such a way that the problems created by the solution are greater than the problems solved." Scientists, engineers, and economists, who approach such questions under more pressure to act than does the theologian, generally respond to this rule by saying: "If we observed this rule, we would be unable to act at all." What could better express the hopelessness, the impossibility of perceiving opportunities to gain new ground, than such a fatalistic idea? Actions forced by necessities render the matter of responsibility to the status of pure theory. What has happened in general? Wherein does the basic question lie? What is our basic experience that is unveiled by such a situation of hopelessness?

These intractable issues reveal a basic experience that is familiar to all of us. Consider the biosphere surrounding human life, the realm referred to as "nature". This is the *first* nature human beings encountered. It held them captive in its bonds, and Goethe said that this nature dances with us and we remain subject to its metamorphoses. We have increasingly made this nature subservient to our own powers. We have learned to imitate its arts. We have gained power not only over the environment surrounding human life, but over human life as well. Biology, medicine, and the behavioral sciences have given us access to the secret structures of human life. This increase in our power over nature has extended our grasp of the causes of "natural" evil and is in the position to increasingly gain control over some of the "natural" causes of evil. All the changes mentioned above have on the other hand led to the consequence that we have increasingly become captives of our *own* nature. The more sovereign we become, the more we experience our own limits, because whenever we establish our own power, we are forced to recognize that humans are not perfect beings. They remain finite, limited, and socially dependent.

To be human, surely, means to be endowed with reason. This endowment with reason, however, calls for adequate communicative, social, and political structures. Aristotle saw that the perfection of human beings could only be advanced within fair political structures adequate to human need and human potential. He found this structure in the "polis". In a similar fashion, Immanuel Kant saw the possibility of human life as dependent upon the conditions of Just Law, the conditions within which human beings are able to develop their rights.

The sociologist Niklas Luhmann is right in designating the basic experience stemming from this finiteness and limitation as the "experience of contingency". The dilemma, "I want to live, but I don't know how to", is basically a question of "mastering contingency". The term "mastery" is, to be sure, presumptuous. It has something unruly, grasping, and manipulative about it. The better expression would be: "constructive contingency encounter". What is meant by this is the search for the balance that constantly eludes us; a balance between inside and outside; a balance between personal withdrawal into the realms of inwardness and external absorption in reshaping or maintaining the structures which form the basic conditions for human survival and the quality of life.

Within this great horizon of questions, a number of individual philosophical-theological questions emerge which are today particularly significant, and thus also particularly controversial. I wish to single out a few of these, without attempting to offer an in-depth analysis or entering a scholastic discourse on precise definitions.

THE QUESTION OF PERSONHOOD

One of the central issues in the search for balance in the "constructive contingency encounter" is the question of the human *person.* This question is controversial in philosophical ethics because utilitarians, like Peter Singer, and deontologists, like Tristram Engelhard, refer back to early stoic and boethian definitions and proceed from the assumption that "person" is characterized above all by self-determination and rational self-control. This means: the more a living being is able to determine itself, to control itself through its own reason, the more it becomes person. In this sense, one can indeed say that a developed primate is more of a person than a newborn infant. What remains unheeded in this concept of person is the fundamental philosophical and theological perception, going back at least to Thomas Aquinas, that person means "dependence on", that it includes a "constructive contingency encounter". The Christian concept of person is inconceivable without contingency and dependence in the sense outlined above. The concept of the *relatio subsistens,* as a relationship which supports itself because hope for self-maintenance is given through communication, represents such an essential component of the Christian concept of person *that nothing included within this communicative dependence can be excluded from human dignity.*

This denotes a parting of the ways today: the dignity of the person, in the sense of the constructive contingency encounter through communicative dependence, applies indivisibly to all human beings. On the one hand the personal dignity of *each and every* human being is embraced to the extent that it is part of personal communication by virtue of its communicatively acquired membership in the genus. Hence, every human being shares this personal dignity from the very outset. It is incorrect to separate human life, as life *sui generis,* from personal dignity, as some utilitarian and deontologist thinkers do.

But on the other hand, life, too, is indivisible. It is impossible to speak of the naked survival of man without, at the same time, also taking into consideration the personal conditions under which this survival is inspired by meaning and contains opportunities for the experience of meaning. Social questions and ecological questions are just as important to undivided life as the classical questions that, essentially, revolve around early human life. I mean precisely this when I speak of indivisibility. This indivisibility applies on both sides: on the side of dependence, and on the side of the communicative allocation of meaning.

THE CONCEPT OF NATURE

In the concept of nature we are confronted with a structually similar fundamental philosophical issue. It is no accident that the latest German congress on moral theology in Würzburg (September 1989) once again dealt with the topic of "nature in ethical argument". It did not focus on questions like birth control, "Humanae Vitae", and the ensuing inner-Church discussion, but rather on the phenomonon which has been termed by the sociologist Van den Daele the "moralization of nature" (Van den Daele 1987). He calls to our attention to a central question which confronts humans in the face of accelerated technological development as well as increasing differentiation of methods for the sustenance of life: Can our actions be stabilized by alignment with a "nature adequate to man? Can any concept of a constant nature provide an answer to the question of orientation, an element of foundation of what is morally right? After lengthy discussions we came to the conclusion that what is meant by "nature" can not be perceived and defined as something stabile, but rather the concept of nature has to be captured precisely as metaphysical entity and not as an ontological value orientation. Similarly to the concept of human persons the conceptualization of nature appears to be very formal. One instance of this would be the normative conceptualization of human nature as the superiority of personal freedom and dignity as well as the superiority of human mercy and empathy for others over other human drives. Such answers argue on a very general level and cannot remain without reference to freedom. But on the other hand such answers are necessary, and we have to fill them with content if we are to come to terms with problems in the specific spheres of life regarding the so-called human nature or the nature surrounding man., the jointly experienced as well as the lived-in nature.

THE QUESTION OF FREEDOM

A third fundamental question, the question of freedom, is closely linked to the two preceding considerations. In this question, too, I believe, we have reached a parting of the ways, and we shall see whether this parting is a path leading to a new start, or a path leading to new captivity and hopelessness. To begin with, I would like to reduce the question to the alternatives: freedom on the private level, or freedom in solidarity. This differentiation, too, emerges from the discussion of bioethics. Will the American Constitution prove right with its right to privacy? Or will the classical human rights structure, as it can be developed proceeding from Kant, prove right, in which freedom, as political freedom and as individual freedom, presupposes a *framework of conditions for freedom in solidarity* which will result not only in the common safeguarding of the greatest possible amount of private freedoms and individual decisions. From Australia to the U.S.A. the "right of privacy" and the individual "pursuit of happiness" predominate the discussion. The free and informed consent of the individual represents the fundamental paradigm of any ethic. This places the manipulation of one's own life, and the manipulation of the life of persons entrusted to our care, of disabled persons as well as of children, at the discretion of the individual person who happens to be responsible. Precisely the example of disabled persons should make it clear to us that freedom is *not private,* but that freedom must be comprehended in its most profound sense, theologically as well, as freedom in solidarity. The philosopher Krings has explored this problem (Krings 1973). Otherwise a constructive contingency encounter of the dependent creature, man, is impossible. Freedom in solidarity could also be another name for the Christian understanding of an experience of God: for that experience of the God who, for the sake of solidarity with creation, above all for the sake of solidarity with humanity, bound up his own freedom with specifications, with voluntary "limitations" down to the point of assuming the shape of human nature, thus at the same time elevating human nature.

If we desire a "constructive contingency encounter", we must then realize that we are increasingly ignoring the realities of suffering and death in our day-to-day experience of life. These suppressed realities are in part supplanted by *indirect* experience through the media. We term this experience *indirect,* because the TV set distances us from that which it communicates. To the extent that we accept this suppression, in everyday life we no longer encounter the consciousness of every moment potentially being the moment directly preceding death.

It is something of a spiritual exercise to become aware of the fact that life gains no intensity by being prolonged, because the last second will always remain unchanged. In view of the only too natural attitude that ties us to life, and is, in the biological sense, also right as an initial natural reaction, we will of necessity experience *every* last second as a loss of meaning, as a loss of identity. Only when we follow a hope that runs contrary to the prognosis and the spontaneity of our sentiment – a "second nature" of Christian hope – will we be able to recognize that this last second will remain unchanged in the structure of our experience of intensity, whether it occurs today, or in ten years. We can then endeavor to expand this experience into a readiness, such as that taught by the "ars moriendi" of mysticism. This readiness may help us to alleviate death's aggressiveness by means of an anticipation of the experience of contingency.

This, however, contradicts our intent to master our lives like lords of time – our own lives and the lives of others. However strong our passion for life, our commitment must be underpinned by a fundamental composure vis-a-vis life and death. We must rediscover the great traditions, like the tradition of finding one's way into life in a constructive contingency encounter, of being ready for life with, and including, its death, and in the end being able to speak the words of Meister Eckhart: "If one asked life why it lives, it would reply: I live for the reason that I live." This simple formulation of a constructive contingency encounter imparts more meaning to life than if I were to enlarge on it or extrapolate from it, thus transforming life from its mystery into some sort of thing that I must possess, that I must achieve, a *thing* without which life has no more meaning. In this will to have the "thing" of life, I will certainly fail in the short- or the long-term.

INVIOLABILITY OF LIFE?

Rigorism concerning the protection of life: Sole responsibility of women?

In society, a far-reaching consensus can be found on the proposition that the number of abortions performed should be lowered, without disregarding the responsibility of the woman in question. If, however, the objective is to help human beings to better accept unborn life, the personal responsibility of women is presented with the goal of protecting the life and incipient dignity of the person in any and every nascent life.

The statement that the life of man is inviolable need not imply an absolutization when it refers to unborn life. The reference to unborn life does not make inviolability absolute, but merely states what its subject is. The question: Who is meant? and the question: May the legal good of life be weighed against other legal goods? constitute two different questions. Hence, proceeding from the inclusion of unborn life into the inviolability of life, no conclusions are permissible in favor of rigorism vis-a-vis pregnant women. The self-determination of women on the question of abortion can only be asserted decisively if unborn life is not accorded the status of a subject in its own right. The legal formula of indications creating the legal basis for exemption from punishment is not a moral qualification; nor does it set aside the illegality of abortion *per se* (for instance in cases of social indication). An assessment of the rightness or wrongness of acts need not be extended to persons who perform such acts. Acts do, however, remain right or wrong, even if the need of the woman concerned significantly limits the moral responsibility of these acts.

The moral status of the human embryo

The recent and ongoing discussion of the biology of development attempts to come to terms with the assertion that the commencement of the life of a human being can be proven to be the point at which the germ cells fuse. This assertion proceeds from the assumption of a "leap" in the development of the cells, which is not significant in terms of developmental biology. It is, moreover, of no significance for the proponents of an effective and indivisible protection of embryos. Rather, this assertion proceeds from the assumption that proof cannot be given for the existence of a differentiable point of time between the commencement of human life, on the one hand, and the commencement of the life of an *individual* human being, on the other. The protection of life is thus not divisible.

Many argue in favor of the plausibility of this distinction and affirm, for instance, that individuation commences with nidation. They proceed from the assumption that the fertilized egg is not yet a human being, a person. Who could prove this without ending up in an endless circle of argumentation? If one rejects the empiricist fallacy (of drawing prescriptive conclusions from the descriptive level), biological observation can prove neither the assertion that the human embryo is an individual person, nor the opposite.

If, as many persons claim, the question as to when and under what preconditions human life must be termed the life of a human being is a "value-related decision" – I refer to this as a "decision in practical intent" – how can one make this value-related decision plausible? Those who place the burden of proof on persons in favor of indivisible protection, because they are in possession of no absolute certainty on the question of the point of time, argue on the basis of a degree of certainty which they, for their part, cannot prove. The thesis according to which the concepts borrowed from the terminology of experts (fetus, embryo) are employed to mark differences with regard to individual human life imputes to this terminology a value-related decision, and then finds confirmation in it. This is a dangerous course. It could, for instance, lead to a situation in which a "moribundus"[2] is no longer regarded as an individual person."

Reference to Thomas Aquinas' theory of successive animation sometimes seems to function as an argument with authority. However, if the theory propounded by Thomas should prove to be based on a value-related decision, this *observation* would offer no *substantiation* as to whether this value-related decision is right or wrong.

MORAL TOLERANCE AND THE POTENTIAL OF CONFLICT-HANDLING IN A DEMOCRACY

Moral tolerance in a democracy includes not only respect for persons who take different value-related decisions, but also the readiness to engage in conflict with regard to this value-related decision. Persons who hold ethical convictions different from one's own are entitled to one's respect, but in a democracy any person may assert his own conviction against another. Law will never be able to integrate *any and every* value-related decision. I do not see why it should be necessary to accord such high rank to a point of view such as "the maintenance of public peace" (in the context of heated debate on the pressing questions related to the protection of life). Democracy needs the conflict surrounding consensus, so that this consensus will be neither forced nor arbitrary.

In questions concerning the protection of life, the central issue is about what is, in human terms, generally right. If, as a Christian, I state from the very outset that hold a partial ethic that is not communicable, I resign in the face of the task of convincing others that I want, for all persons, what is right, and that I am of the opinion that this, as a matter of principle, is evident for all persons. In the political sphere, practical compromises affecting

2 Term used in medical language

legal codification may then result, which take their orientation in terms of the lesser of two existing evils and do not involve ethical compromises. Ethical conviction does not rule out practical compromise, but compromise does not suspend ethical conviction, either.

The social protection of women in difficult situations is the most important subjective element in the concern for life. But concern for life remains an objective goal. In practice, I must always view this concern concretely, and this view differs from that of ethical casuistry, even when this casuistry is struggling to come to terms with practical experience. The more individual cases are solved by rules, the more unsolved individual cases arise. We are not concerned here with a morality based on prniciples, even though principles like the inviolability of life do make sense and should not be abandoned, nor is it a matter of a lack of tolerance. The fundamental concern is a dialogue on the *right* courses for a safe and effective protection of life. My question for this dialogue is: how can one be in favor of the divisibility of human life and of a graded protection of life if the distinction between human life and the life of an individual human being cannot be proved? If the commencement of the individual is a continuous process, we should, in cases of doubt, opt for endangered life.

REFERENCES

Konsum und Tod. Die Herausforderung einer christlichen Lebensethik [Consumption and death. The challenge to Christian bioethics]. *Theologische Quartalschrift* **170:** 9–12 (1990).

Leben – philosophisch und theologisch (Life - philosophically and theologically), in: *Staatslexicon,* ed. by the Görresgesellschaft, 7th edition, vol. 3, pp. 852–857 (1987). Freiburg/Basel/Vienna: Verlag Herder.

Lebensförderlichkeit (Biophilie) [Biophilia]. In: D. Mieth (1984), *Die neuen Tugenden. Ein ethischer Entwurf,* pp. 94–106. Düsseldorf: Patmos Verlag.

Moral der Zukunft – Zukunft der Moral [Morality of the future – the future of morality]. In: *Kirche in der Zeit. Walter Kasper zur Bischofsweihe,* ed. H.J. Vogt in collaboration with C. Steiling, pp. 198–217. München: E. Wewel-Verlag 1990.

Schöpfung und Leben: Welchen Leitbildern folgen wir? [Creation and life: Which examples do we follow?]. In: *Ethische und rechtliche Fragen der Gentechnologie und der Reproduktionsmedizin (= Gentechnologie, Chancen und Risiken), eds. V. Braun, D. Mieth, K. Steigleder (eds.), vol. 13, pp. 327–332. München: J. Schweitzer Verlag (1987).*

Seelische Grundhaltungen unserer Gesellschaft in der Charakterlehre Erich Fromms und in theologisch-ethischer Reflexion [Basic psychic attitudes of our society in Erich Fromm's theory of character and as reflected in theological-ethical terms]. In: De Dignitate Hominis. Festschrift für C.J. Pinto de Oliveira, ed. A. Holderegger et al. 1987, (Studien zur Theologischen Ethik), pp. 317–335. Freiburg i.Br.: Universitätsverlag Freiburg/Schweiz und Verlag Herder.

Theologisch-ethische Ansätze im Hinblick auf die Bioethik [Theological-ethical approaches with regard to bioethics]. In: *Ethik in den Naturwissenschaften* (= Concilium) 25, vol. 3, 1989, ed. by the author, pp. 211–218.

Krings, H. (1973). Freiheit. In: *Handbuch philosophischer Grundbegriffe,* eds. H. Krings, N.M. Baumgartner, und Ch. Wild, pp. 507ff. München.

Van den Daele, Wolfgang (1987). Moralisierung der menschlichen Natur und Naturbezüge in gesellschaftlichen Institutionen. *Vierteljahresbericht für Gesetzgebung und Rechtswissenschaft* **2:** 351–366.

Bioscience ⇄ Society
edited by D.J. Roy, B.E. Wynne, and R.W. Old, pp. 305–313
© 1991, John Wiley & Sons

Scientific Responses to Moral, Cultural, and Religious Concerns: Three Case Studies

Baruch A. Brody

Center for Ethics
Baylor College of Medicine
One Baylor Plaza
Houston, TX 77030, U.S.A.

Abstract. This paper distinguishes three types of concerns about scientific and technological developments and offers an analysis of the appropriate response to each. The first type of concern, illustrated by concerns about in vitro fertilization, grows out of specific religious or cultural traditions, and is best handled by developing a respect for the divergent views in a single society. The second type of concern, illustrated by concerns about genetic testing, grows out of generally respected values, and is best handled by regulatory schemes which meet the concern while allowing society to reap the benefits of the new developments. The third type of concern, illustrated by some concerns about transgenic animals, rests upon idiosyncratic values and/or confusions, and is best handled by explicit challenges in the arena of public debate.

INTRODUCTION

Scientific research and its application in medical practice requires public support, in part to insure adequate financial support for research and for its applications, in part to avoid prohibitory or crippling regulations which make it impossible to carry out the research or its application, and in part to promote public use of the newly developed medical practices. For these three reasons, scientists and clinicians cannot ignore moral, cultural, or religious issues that are raised in the general community. Moreover, as moral agents, scientists and clinicians should want to insure that their work meets the highest moral standards. There is no doubt, however, that working researchers and clinicians nevertheless feel frustrated as they confront concerns which are not strictly scientific and which seem to threaten good scientific work. This paper attempts to provide a strategy for dealing with these frustrations by better understanding the appropriate responses to different types of challenges and concerns.

We need to distinguish three types of cases. In the first case, the concerns grow out of specific religious or cultural traditions and are seen as valid only by those who adhere to the viewpoint of those traditions. The proper response to those concerns is to recognize their validity for certain individuals, and to provide them with alternatives, while insisting that others be free to move forward in using the technologies in question. In the second case, the concerns are viewed by most members of the society as legitimate, and the proper response to these concerns is to fashion ways of modifying research practice and its applications to meet these concerns, while at the same time keeping the promise offered by the research and by its applications. In the third case, the concerns are based upon weak argumentation and/or idiosyncratic values and/or appeals to false claims, drawing support neither from specific traditions nor generally shared values, and the proper response to these concerns is to challenge them explicitly in the arena of public debate.

In this paper, I will briefly examine three case studies. The concerns about in vitro fertilization, I claim, are illustrative of the first of these types of cases. Concerns raised about it are rooted in specific traditions, and practitioners need to be respectful of those concerns in dealing with members of those traditions, while insisting upon the freedom to provide in vitro fertilization services to others who are not members of those traditions. Genetic screening in the work place, I claim, is illustrative of the second type of case. Concerns about this screening have general legitimacy, and the proper response is to recognize that legitimacy and find ways to meet those concerns, while still reaping the benefits of that newly emerging technology. The development of transgenic animals, I claim, is illustrative of the third type of case (although there is an element of the second type present as well), and the proper response is to defend this new technology publicly.

A paper of this type can obviously only present an overview of the issues. I believe, however, that this overview will, in any case, confirm the legitimacy of recognizing three different types of responses to moral, cultural, and religious concerns expressed about scientific and technological developments.

IN VITRO FERTILIZATION

In vitro fertilization is a very complex technique, because successful in vitro fertilization involves many components. These include the ability to obtain an ovum without damage, maintain the ovum extracorporeally until fertilization, successfully fertilize it, maintain the zygote extracorporeally until the blastocyst stage, and prepare the uterus for successful implantation at the right time. Much research, both of a theoretical and an applied nature, was required before in vitro fertilizations led to live births in the late 1970s and early 1980s, and even today the success rate for such attempts is far from satisfactory (Office of Technology Assessment 1988).

Although initial concerns about damaged children proved to be invalid, a number of concerns continue to be raised about this procedure (Office of Technology Assessment 1988). Among them are the following:

1) In vitro fertilization separates acts of procreation from acts of conjugal intimacy and is therefore morally illicit;
2) In vitro fertilization dehumanizes or medicalizes human sexuality;
3) Since some fertilized gametes are destroyed rather than implanted or frozen, in vitro fertilization involves illicit early abortions.

The first concern is, of course, a concern that is felt only by those who accept the Roman Catholic teaching (Sacred Congregation 1987) which insists that any morally illicit sexual act (as opposed to sexual relation) must involve both conjugal intimacy and reproductive potential. This is why Catholic teaching opposes in vitro fertilization just as it opposes artificial contraception. The second concern is found articulated in the Eastern Orthodox tradition (Harakas 1982), among others, although it has been articulated by some authors writing in a more secular mode as well (Kass 1985). The third concern is shared, of course, by all traditions which oppose abortion from the moment of conception.

Many working outside of these traditions will find these concerns less than persuasive. They will argue that in vitro fertilization augments the capacity for human autonomy, for it offers a chance to couples who wish to have biological progeny. They will argue that in vitro fertilization can greatly benefit these people and the child they will have, and will argue for the use of in vitro fertilization on the grounds of beneficence. They will, in fact, find it difficult to understand what the whole problem is about.

As I see the situation, it comes to this: standard bioethical analysis, appealing to such principles as autonomy and beneficence, will argue for the positive moral value of in vitro fertilization. It may, to be sure, worry about whether or not considerations of justice mandate that in vitro fertilization be paid for out of public funds, but it will be satisfied that there are no real moral problems in its use, so long as people are adequately informed about costs, success rates, etc. But members of certain traditions will oppose in vitro fertilization by appealing to nonstandard bioethical principles (to the unity of procreative potential and conjugal intimacy, to the eschewal of dehumanizing nonnatural reproduction, to opposition to abortion from the moment of conception) which are, however, accepted within their own traditions. The diagnosis of the situation is then clear.

How should researchers and practitioners of in vitro fertilization deal with these concerns? They need to begin by understanding why it is that the opponents of that practice oppose it. They need to accept the fact that there will be members of those traditions who will consider but reject the use of in vitro fertilization, because it violates the norms of their tradition. They need to encourage such potential patients to explore options (such as GIFT) which may be more compatible with the norms of their traditions, and if there are enough of these patients, they need to turn some research attention to developing these alternative options. In all of these ways, they need to be respectful of the opposition. At the same time, researchers and practitioners need to demand respect for the values of the many who do not share these special moral principles, and who therefore find in vitro fertilization to be morally acceptable. They need to resist unnecessary attempts to limit research, to limit insurance coverage for in vitro fertilization, or to limit its actual practice.

There is, of course, one special area of concern that does not lend itself that easily to this type of mutual respect. This is the question of the destruction of extra zygotes. Proponents of certain views, who see this as murder of innocent newborns, are not likely to be moved by the policy of mutual respect which I have advocated in the previous paragraph, while others will want to insist upon not reimplanting that many zygotes. The emergence of freezing as an alternative to destruction opens a window for avoiding this controversy.

In short, the controversy over in vitro fertilization gives us an example of concerns growing out of specific traditions, which are seen as valid only by adherents of the views of those traditions. Moreover, it provides us with a good example of how the proper response is one of mutual tolerance and support for divergent viewpoints. In this way, researchers and practitioners in this area can legitimately move forward while understanding and respecting the concerns raised by some.

GENETIC SCREENING IN THE WORKPLACE

One of the most exciting areas of current scientific and medical research is, of course, the genetic basis of disease. With the Human Genome Project moving forward, and with many other researchers working in this area, it seems nearly inevitable that we will soon understand, in ways that have never been understood before, the genetic basis of disease.

As early as 1973, H.E. Stokinger and his colleagues at the National Institute of Occupational Safety and Health argued for using genetic screening to identify workers whose "genetic abnormalities" rendered them unable to cope with levels of exposure to toxic substances which are safe for "normal workers" (Stokinger and Scheel 1973). More recently, Gena Kolata has persuasively argued that companies concerned about insurance costs may be more interested in testing for workers who are likely to develop expensive diseases which will impose heavy costs upon company-provided health insurance (Kolata 1986). Technological and scientific advances will surely make much of this possible in the next ten to twenty years.

Many concerns and questions have been raised about all of this testing capacity:

1) Can potential workers be forced to undergo such screening either to protect themselves or the companies that employ them, or would such compulsory testing be a violation of individual freedom?
2) To whom should the results of such testing be made available? Are the results governed by the ordinary principle of confidentiality? What sort of counseling should be provided for those receiving results of greater susceptibility?
3) Should companies be able to refuse to hire, or be able to fire, workers who show on testing that they are unusually susceptible to these illnesses, or would a policy allowing for that be an illegitimate stigmatization of such workers and an act of discrimination against them?
4) Can society require companies to lower the levels of exposure to toxic substance to make the workplace safer for very susceptible workers, or should companies be allowed to provide levels of safety adequate for average workers and refuse to hire workers who would not be safe in such an environment?

Many more questions of this sort can be raised, and several recent books have explored them (Nelkin and Tancredi 1989; Rothstein 1989).

One reaction to all of these issues would be to view genetic knowledge and its use in genetic screening as dangerous and as something to be avoided. I do not find such a response appropriate. Such knowledge and such screening may help us to counsel people who are at greater risk of disease to take measures which will limit their risk and the risk of their progeny. Another reaction would be to dismiss all of these issues on the grounds that worrying about them would stand in the way of scientific and medical progress. I find such a response equally inappropriate. The development of such knowledge and of such screening techniques could lead to terrible social problems if these problems are not faced and appropriately solved. Moreover, in time, a failure to address these issues will only lead to a social backlash against further research into, and clinical use of, genetic knowledge.

We have here, it seems to me, a very good example of the second type of case. The concerns which have been raised do not rest upon any special moral claims made by only some moral traditions. Concerns about privacy, about confidentiality, and about discrimination are common concerns shared by most citizens in developed democratic societies. The proper response in this case, as well as in others of the second type, is to accept these concerns and to develop regulatory frameworks which will address these concerns, while allowing us at the same time to reap the benefits of these scientific and technological advances.

What would such a framework be like? I have recently suggested the following for testing hypersusceptibility to occupational illnesses: a similar framework, modified by the social decision as to whether or not health insurance should be provided through the workplace, could be developed for general testing for hypersusceptibility to expensive illnesses which are not work-related.

1) The goal of any testing program should be to minimize occupational illnesses while protecting at the same time potentially conflicting values, such as maintaining equality of opportunity, and not excessively burdening employers with overly expensive modifications of the workplace.
2) Any program of genetic testing should be introduced only after appropriate social agencies are convinced that it can meet these goals. Once such a social decision is made, employers should be required to introduce such testing programs, and employees should be required to undergo such testing. Merely making such testing available may not lead to the desired minimization of workplace injury.
3) Part of the social decision is the decision about what to do with the results. If the results show that the health of many workers are threatened and if workplace modifications are available at a reasonable cost, workplace modification should be mandated. If only a few workers are threatened, and

if workplace modifications would be very expensive, worker reassignment is more appropriate.

4) Strict measures to protect confidentiality of results will be required, especially as individual results may need to be made available to some employers (if they need to reassign some workers) and all employees (so they can protect their own health), and aggregate results will need to be made available to relevant social agencies.

My concern in presenting this framework here is not to argue for its validity. It is rather to use it as an illustration of what I take to be the right type of response to a problem of the second type. Having accepted the legitimacy of the concern, one does not drop the scientific and medical advances, rather, one creates a framework which allows for their use in ways which are responsive to legitimate concerns.

THE DEVELOPMENT OF TRANSGENIC ANIMALS

One of the most exciting new developments in biotechnology is the ability to produce new forms of life through genetic engineering. These new life forms result from augmenting the DNA of an already existing organism by adding DNA from different animals or from human beings. These new life forms may be better models for studying diseases. Thus, Harvard University created a mouse that was engineered to contain a cancer-causing gene, making it very susceptible to developing cancer. This new mouse will be an excellent model in which to test carcinogens and cancer therapies. These new life forms may simply be better producers of food. Thus, many are working on new animals which produce more milk, or grow more quickly on less food. Still other possibilities have been suggested, and the development of transgenic animals clearly is an exciting new area of scientific research and technological development (Office of Technology Assessment 1989).

Many concerns have been expressed about this development. Some of them raise clearly legitimate points. These include the concerns that current federal regulations fail to protect the animals involved in this research adequately and do not attend to the environmental implications of introducing new animals into the wild. But these concerns are like the concerns we discussed in the previous section, and the appropriate response to them is improvements in regulatory schemes that enable us to get the benefits of this new technology, while avoiding unnecessary harm to animals and inappropriate threats to the environment (Office of Technology Assessment 1989).

I wish to direct attention to a series of concerns which challenge the very legitimacy of these new technologies, and I want to show that they illustrate our third type of concerns.

One such concern is that these technological developments show a lack of respect for the integrity of nature and are an abuse of our stewardship (rather than our ownership) of nature. Much as we have laws to protect endangered species and ban the destruction of certain animals, and environmental quality laws to protect the environment, so we must have laws that prohibit this disregard for the integrity of nature (Office of Technology Assessment 1989).

This concern seems to me to be one that we ought to reject because it is based upon idiosyncratic values and/or confusions. There is a widespread acceptance of the idea that mankind only has stewardship of nature. But the point behind that notion of stewardship (which explains our endangered species and environmental protection laws) is that we have to consider the impact of our use of nature on future human beings. We must treat nature, at least in part, as something we hold in trust for future generations. None of this, however, involves any belief in a "respect for the integrity of nature" which calls upon us not to transform nature by creating new species. Those who advocate this concern have either introduced without argumentation a new value (respect for nature's integrity) or have mistakenly derived it from the traditional notion of stewardship. The appropriate response to their concern is to show its roots and challenge its legitimacy. It provides no sound reason to ban the production of transgenic animals.

A second set of concerns is expressed by those who believe that the development of transgenic animals challenges the inherent sanctity of every unique living being (Office of Technology Assessment 1989). The thought seems to be that this type of manipulation has us using animals as means to human ends rather than as ends with some inherent sanctity. The conclusion that is drawn is that we should not be in the business of developing transgenic animals.

These concerns also seem to me to be based upon idiosyncratic values and/or confusion. A society like ours which eats animals, wears clothing derived from animals, and tests drugs on animals (to mention only three ways in which we use animals) clearly accepts the morality of treating animals as means to human ends, and rejects the idea that there is an inherent sanctity in every unique living being. None of this means that we are required to be Cartesians who are indifferent to animal suffering. There are many appropriate positions between Cartesian indifference to animal suffering and the acceptance of a belief in the inherent sanctity of each living being. Those who advocate this argument have introduced without argumentation an idiosyncratic value, and the appropriate response to their concern is to expose its roots and challenge its legitimacy. It provides us with no sound reason to ban the production of transgenic animals.

These sorts of concerns illustrate then the third type of case which I distinguished at the beginning of this paper. They have no deep roots in any particular religious or cultural tradition in modern developed societies, and they certainly should not be seen as having a clear-cut legitimacy. Instead, they grow out of new and idiosyncratic ways of looking at things, ways which are not even clearly recognized as new and which are certainly not rationally defended in any significant fashion. The appropriate response to them is to expose their shortcomings and to move forward to reap the benefits of the technology in question.

Acknowledgment. Much of the research on these three case studies was done under contract to the Office of Technology Assessment of the U.S. Congress. I would like to express my appreciation for that support, while making it clear that the support in no way extends to an endorsement of the views expressed in this paper.

REFERENCES

Harakas, S. (1982). *Contemporary Moral Issues Facing the Orthodox Christian.* Minneapolis: Light and Life Publishing.

Kass, L. (1985). *Toward a More Natural Science.* New York: Free Press.

Kolata, G. (1986). Genetic screening raises question for employers and insurers. *Science* **232:** 317–319.

Nelkin, D. and L. Tancredi (1989). *Dangerous Diagnostics.* New York: Basic Books.

Office of Technology Assessment (1988). *Infertility.* Washington: Government Printing Office

Office of Technology Assessment (1989). *Patenting Life.* Washington: Government Printing Office.

Rothstein, M.A. (1989). *Medical Screening and the Employee Health Cost Crisis.* Washington: Bureau of National Affairs.

Sacred Congregation for the Doctrine of the Faith (1987). *Instruction on Respect for Human Life and its Origins and on the Dignity of Procreation.* Vatican City: Congregation.

Stokinger, H.D. and L.D. Scheel (1973). Hypersusceptibility and genetic problems in occupational medicine. *Journal of Occupational Medicine* **15:** 564–73.

Standing, left to right: Jonathan Moreno, Roger Lewin, Walter Bechinger, Dietmar Mieth, Hans Günter Gassen
Seated, left to right: Elliot Philipp, Andrew Rowan, Deborah Barnes, Tristram Engelhardt, Jr.
Not shown: Robert Solomon, Brian Wynne

Bioscience ⇌ Society
edited by D.J. Roy, B.E. Wynne, and R.W. Old, pp. 315–324
© 1991, John Wiley & Sons

Group Report
What's Wrong with the Interaction Between Bioscience and Society?

Rapporteur: J.D. Moreno
D.M. Barnes
W. Bechinger
H.T. Engelhardt, Jr.
H.G. Gassen
R. Lewin
D. Mieth
E.E. Philipp
A.N. Rowan
R.C. Solomon
B.E. Wynne

INTRODUCTION

No satisfactory account of the relationship between the biological sciences (including their technological applications) and society can begin without acknowledging the many levels of the topic's complexity. First, "bioscience" does not denote a single science, or set of problems, or techniques, but rather many fields and approaches, some of which are only now emerging. Second, "society" is a notoriously ambiguous concept, one that attempts to capture interpersonal arrangements that are themselves undergoing rapid change. The very distinction between science and society is, it should be remembered, a conceptual convenience that can cause confusion. Finally, many of the problems most pressing in the biosphere, such as those associated with overconsumption and population growth, are taken for granted as part of the background of this report.

DISTINGUISHING SORTS OF PROBLEMS

A source of even greater difficulty is the logical problem of determining what is "wrong" with the relationship, without being quite sure what is, or could be, "right". That is, to assume for the moment an adversarial relationship, would a world in which bioscientists enjoyed perfect freedom to pursue their investigations exemplify the "right" relationship between bioscience and society? Or would it be "right" for bioscientists to be obliged to undergo constant external scrutiny? Whether we find one of these practically impossible extremes attractive or not, from some self-interested point of view we may still not be sure that either is the state of affairs that ought to be seen as the best one in the long run. Thus we seem to be driven to the preliminary conclusion that a relationship between bioscience and society that is without conflict would not necessarily be desirable.

When one introduces the further assumption that, on the whole, democratic social arrangements recommend themselves to all members of the polity, including bioscientists, then some degree of conflict is positively desirable. Without conflict, after all, alternative programs for action in the real world (including the introduction of one form of biotechnology or another), would never have the opportunity to overcome the status quo. Thus conflicts between bioscience and society (insofar as there are any) cannot themselves be "wrong" in all cases; they are "wrong" only when democratic processes that enforce the peaceable resolution of controversies have broken down. What is sought is the progressive resolution of conflicts, one that involves the transformation of the parties concerned, rather than a destructive outcome.

In this sense, the proposition that there is something wrong with the relationship is distinct from the proposition that there are tensions in the relationship. The latter refers to phenomena that are often normal and even desirable, while the former refers to deep structural inadequacies that seem more difficult to ameliorate. It is not always easy to provide examples of one or the other because their status may change in retrospect. For instance, charges by AIDS activists that traditional, controlled clinical trials unfairly and unnecessarily restrict access to new drugs will in the future be seen as an indication of something deeply wrong only if the tension between those demands and the views of most researchers cannot somehow be accommodated. On the other hand, when one considers the radically divergent views concerning the use of RU-486, the so-called "abortion pill", one suspects that there is more at issue than a matter of poor communication – something truly wrong. On the other hand, the familiar complaints among scientists about their inadequate or misdirected funding seems to be a normal feature of the relationship and not indicative of a deeper failing.

Not all conflicts between scientific and popular views, however deep and intractable, are of such gravity that they are evidence that something is seriously wrong, or even problematic. If the stakes are not high, there can be peaceful coexistence. For example, many people who believe in the predictive powers of astrology cannot be scientifically dissuaded from this belief. This would seem to exemplify a deep epistemological conflict between science and at least many members of society (probably including some scientists who share these or other nonscientific beliefs). Yet the institution of science is usually able to proceed without serious and immediate disruptions that can be blamed on this sort of nonscientific belief. In other cases, the tensions and conflicts can lead to positive changes in science – for example, what problems it addresses, or how it is directed or managed.

The remainder of this report will presuppose an appreciation of the distinction between what is wrong in the relationship between bioscience and society, and that which is (merely) a sign of tensions. It will also presuppose an appreciation of the difficulty of determining whether a particular controversy falls into one or another of these categories. Yet this very lack of clarity is instructive, for it explains in part why controversies can be so frustrating while they are taking place: often one does not know in advance whether they are more or less normal or indicate deeper and intractable differences. The situation is further complicated by the fact that assessments of the nature of a controversy are themselves subject to different social visions and values.

SYMPTOMS AND SYNDROMES

Conflicts in the relationships between bioscience and society of either sort manifest themselves in various ways. In considering these manifestations of conflict, it is useful to distinguish between those issues that necessarily engage the biosciences, such as reservations concerning the manipulation of the human germ line, and those that are only contingently so, such as the incidence of malpractice suits against physicians. An increase in malpractice litigation is not unique to the medical sphere, but germ line engineering is unique to applied bioscience. Nevertheless, a thorough inventory of "symptoms" of difficulties in the relationship between bioscience and society may well include examples that are not unique to the biosciences. The reluctance by American authorities early in the HIV epidemic to close bathhouses where gay men engaged in sexual relations with multiple partners could be seen as a civil rights matter. But it surely also represents a way in which conflict can emerge between applied bioscience, which proceeds partly from a public health ethos of limiting infection, and societal groups that have a different estimate of the significance of certain liberties.

Attempting to analyse the symptoms of more basic problems in the relationships between bioscience and society, three clusters or "syndromes" can be identified. These clusters can be said to define the territory of difficulties in the relationship, though again they do not map out a single space or even three discrete spaces. Such is the muddy nature of the region being described. Hence, it will immediately be perceived that a single symptom or conflict can be encompassed by more than one syndrome or cluster. Nevertheless, each cluster can illuminate some aspects of the conflicts. Furthermore, as should become evident, it is not possible to isolate in a rigorous manner the definition of the problems from an account of their sources.

Problems in communication

The problems in communication between bioscience and society stem from many sources. Four of the most important are: 1) the communication barrier produced by technical language; 2) the arrogant attitude among too many scientists that the public is unable to understand complex issues; 3) the lack of scientific literacy among many members of the public; and 4) the confusion that can result when there are conflicts between official statements and institutional behavior. All of these problems can be exacerbated by self-interest and a focus on short term goals, rather than on long-term consequences of certain actions. Adding further to the problem is the fact that, like the lay public, scientists also have problems in communicating with one another.

These problems in communication between bioscientists and members of society at large are also subtly bound up with the public image of science and scientists. For instance, consistent with the concern to avoid bias, scientists often (but in fact, of course, not always) employ a dispassionate tone, even when approaching matters of great human importance. This perceived detachment can be disconcerting to laypersons who tend to appreciate emotional engagement. The public image of bioscientists can also suffer in the long run when certain aggressively promoted and expensive projects (the "war on cancer") are later perceived to have been oversold. Finally, the news media in reporting bioscientific developments often emphasize dichotomies of opinion that fail to capture the crucial middle ground in which accommodation may be possible, or initial independence maintained. Journalists who specialize in reporting on scientific matters bear a special resonsibility to effect adequate translations for the lay public. Scientific institutions should be more aware of, and prepared to discuss, the tacit values which dictate their work.

Different ways of knowing

Like scientists generally, bioscientists believe that the world can be better understood by employing a scientific way of knowing, including tests of hypotheses under controlled experimental conditions. The result of these experiments is more information about a particular phenomenon or groups of phenomena. But the results are conditional and limited, for they depend on (or even stem from) the experimental conditions, which are necessarily always incomplete and provisional, given the qualifications and caveats that must apply. This is especially true in the many cases where laboratory knowledge has to be extended into the real world. Even the notion of "adequate" scientific knowledge is ambiguous; for example, experimental results that too closely confirm the predictions of a theoretical model are subject to suspicion. Scientists are not themselves aware enough of the negotiated basis of scientific knowledge.

The following example illustrates the informal character of knowledge acquisition in science. In the United Kingdom in 1986, after the fallout of radiocesium from Chernobyl on upland farms, scientists made erroneous public statements about the necessary duration of restrictions on sheep sales. They had falsely assumed that existing knowledge of rapid cesium immobilization (drawn from observation with clay soils), could be applied to the peaty, acid soils of the hills now affected. They therefore confidently predicted a short period of restrictions, approximately three weeks. Five years later, the same area of farms is still restricted, much to the detriment of public credibility of science.

The scientists did not appreciate that knowledge generated under one set of conditions – clay soils – was limited to those conditions. In acid, peaty soils, as they only discovered after making their mistake, cesium remains chemically mobile and actually recycles into edible vegetation via the roots, thence into the sheep, and back into the soil via eventual excretion and leaching, at which point it is available again for vegetation uptake.

This process of learning via mistaken extrapolation is not unusual in science – it suggests a less deliberate and more accidental process than conventional models of scientific knowledge generation. It also exposes the crucial point that scientists routinely assume their knowledge to be more widely reliable and less conditional than in fact it is. This is a syndrome that naturally generates deepseated public anxieties when science is seen as being granted power and indiscriminate authority in public affairs.

For nonscientists, as for scientists in their nonscientific lives, "knowing truly" is a far less systematic and conditional matter, and the motivations for seeking knowledge are usually more directly associated with prospective action. In addition to developing suspicions about the unconditional ways in which scientific knowledge is often expressed and used, lay people also may have an orientation to their environment, natural and social, which is simply incompatible with that of science. Thus, whereas scientific epistemology assumes control and active manipulation of the environment (including bodies and society) as the defining basis of knowledge, people in many life situations assume that adaptation and a more passive (defensive) relationship with the environment is appropriate: what counts as knowledge is accordingly fundamentally different. Thus, public reception of, and response to, science may be affected by basic dislocations of this kind before the question of the contents or effects of that scientific knowledge is even reached. There may, as it were, be a "credibility gap" even before acknowledged agendas or concerns are addressed. There may be tacit "agendas" buried in the very culture, or epistemology, of science and other bodies of knowledge. Furthermore, the presumption that the scientific epistemology is naturally superior to that of the lay public may be simply an authoritarian imposition that provokes the very response it is hoping to assuage.

Different visions of the world

This cluster of difficulties in the relation between bioscience and society must be carefully distinguished from different ways of knowing. At the extreme, different visions of the world can be incommensurable. An example of worldviews that seem hopelessly at odds would be the more radical animal rights view, which sees at least all vertebrates as members of the moral community and demands respect for them equal to that normally accorded to human beings, as against the view of many scientists and nonscientists that nonhuman animals exist wholly for the sake of human welfare.

It is often thought that there are differences between worldviews of lay people and scientists. However great these differences are in reality, the typical scientific biography does contribute to a tendency to see the world in certain ways. Early in life, scientists segregate themselves from nonscientists because of their personal interests and abilities. The formal education of scientists and their professional experiences probably further enhance these characteristics. Scientists tend to view the world narrowly and empirically, because they need this focus for professional success. Hence, many scientists may lack the ability to see issues broadly in terms of the humanities and of culture generally. Or they may reject the notion that they have special insight into public issues, or that the uses and power of science mean that they as scientists have a special social responsibility.

The socialization of bioscientists can contribute to public frustration and even fear of bioscience. In particular, biological scientists have in the recent past crossed a threshold from taxonomic description of a molar realm of living things (such as the classificatory scheme of genus and species), to nomological explanation (and manipulation) of a molecular realm of essential components of living things (providing lawlike generalizations embedded in theories about chromosomal material). It is difficult to estimate the profound changes in values that may eventuate from the achievements that follow from this passage, but the public is aware that traditional values are indeed at hazard in the process, despite presentation of the science as "value-free". Because new bioscientific information affects such fundamental conditions of human life – genetic individuality, the ability to reproduce, the safety of the environment – bioscience is regarded as powerful and threatening. At the risk of oversimplification, one may say that conflicts in this area arise from the tension between recognized social values and the unstated values embedded in scientific developments and technical possibilities.

This cluster of concerns is arguably that which sets bioscience's relationship to society apart from that of other sciences, and is worth special emphasis. Problems in communication and even different ways of knowing are surely familiar elements in the social history of atomic physics. But through genetic engineering, bioscience can in principle subject to direct, material manipulation the only feature of the universe that might be said to give it some intimacy and familiarity: self-identity or self-consciousness. Other scientific and technological programs impose largely nonnegotiable changes in social relationships and identities – think of the automobile, the television, and "the bomb". But they do so less directly and less intensely than biological forms of manipulation of our genetic constitution. When people oppose in vitro fertilization, they do so from a well-founded concern (which is not to say that their view should prevail) about how it will be socially regulated. The apparently emerging ability of the biosciences to create new forms of subjectivity by chromosomal manipulation not only changes our traditional view of the world; it also presents the possiblity that entirely new world views, new forms of subjectivity in genetically designed beings, can be deliberately created.

A CASE STUDY: BOVINE SOMATOTROPHIN

It may be helpful to examine in more detail a single controversy that contains symptoms that can be brought under the rubric of each of the three clusters that have been described: the bovine somatotrophin (BST) debate in Britain. BST, also known as bovine growth hormome or BGH, can be produced in great quantities by genetically engineered bacteria. When injected into animals, BST both greatly increases milk production and makes milk production more efficient, an apparently unalloyed public good. The developers therefore did not anticipate that BST would engender significant controversy. But some groups found BST very disturbing.

Activists asserted the dangers of genetic engineering, while animal rights groups objected to the continuing industrialization of animal agriculture, the mechanization of animal husbandry, and the potential suffering caused by forcing cows to produce even more milk. Scientists and industrialists claimed that the technical benefits of BST could be divorced from questions of superfluous milk production, of profit motives being given precedence over animal health, and of large-scale social and economic changes in farming if BST were used widely. The critics could not understand why the scientific dimensions should be separated from their real-world context.

Industry seemed incapable of responding to its critics' drastically different worldview and engaged in a dispute with the animal rights groups over the reliability of evidence concerning the impact of BST on the animals. Dairy farmers joined in the protest, fearing that a 15–30% increase in milk production would favor only large producers. However, the industry argued that well-managed farms would do well regardless of size, but this assumed that small farms were as technically sophisticated as larger farms. The general public was alarmed by the prospect of BST in milk, an icon of goodness and purity, in even small amounts, while the industry argued that the levels of BST would be tiny and would in any case be digested like other proteins. Critics also argued that the fat content and mineral balance of milk are changed by BST.

The companies involved in developing BST did not spend much time negotiating with the public over its introduction, and the opposition took them by surprise. Initially, they tried to brush aside criticisms as misunderstandings (problems in communication), as wrong (different ways of knowing), or as drastically wrong-headed (different visions of the world). Gradually, the industry increased its efforts to reach the public but continued to talk past its critics. To date the companies have been unable to break through these barriers. Currently, BST is not licensed in Europe, and it is under consideration in the United States. Some American states have placed a moratorium on its use, and large chains of food companies have said they will not use milk and milk products from BST cows.

Although submerged in the particulars of the debate about BST, the three clusters that define the territory of conflict between bioscience and society surfaced at various points in the controversy. Like all sciences, the biosciences inherently threaten common sense and familiar aspects of the world. Yet they constitute social practices that, although situated in principle in pluralistic value systems and multiple institutions, are often in practice owned and controlled by large private interests. Under these conditions, questions of risk and benefit, and the methods for resolving these controversies, are difficult to articulate, let alone definitively answer.

This controversy also highlights the role of industry in the development and exploitation of biotechnologies. One should note that invariably those who support research (whether industry, governments, or philanthropic individuals), tend to determine the field and the focus of its use. This is not necessarily wrong, as long as limitations and constraints are openly recognized. When certain research programs and approaches are selected, alternatives are, for the moment at least, postponed. All research bears the mark of the interests of those who support it and those who undertake it. Scientists, as the recipients of this largesse, bear a special responsibility to be aware of these forces.

A LOOK AHEAD

What new developments in the biosciences and in society will change the problems we have characterized? One can reliably predict that biological scientists over the next few decades will increase their ability to alter nature and to objectify intimate human events. These developments, impressive though they are likely to be, do not even include advances in other sciences, such as artificial intelligence, that are sure to affect advances in the biological sciences. At the same time, we may find that the terrestrial ecosystem is less able to sustain the lifestyle of the developed world. Even if this last problem can be ameliorated, the growing wealth and technological options available to the developed world will stand in stark opposition to the less developed world as population increases beyond control. The political and ecological consequences of these disparities in technological options will present new challenges for the biosciences and affect the course of their development.

Other changes in the relationship between bioscience and society may follow from social arrangements that provide new mediations between the two. An example might be communication centers in research institutes that are more skilled in explicating new results and applications and in understanding and addressing public concerns. New institutional forms of bioscience could emerge which would better translate public concerns into research agendas, overcoming the frequently negative spirals of polarization which exist nowadays. Another development could be the "normalization" of biotechnology in everyday life through the routine availability and use of genetically engineered products, and the familiarity of genetic engineering as a career path on a par with computer programming.

Even if these or other, as yet imponderable, developments mute the conflicts between bioscience and society, it will still be reasonable and important to wonder whether that is a desirable outcome.

Bioscience ⇌ Society
edited by D.J. Roy, B.E. Wynne, and R.W. Old, pp. 325–335
© 1991, John Wiley & Sons

How Revolutionary is the Biological Revolution? Notes Toward a History of the Future

Horace Freeland Judson

Program in the History of Science
Building 200-031
Stanford University
Stanford, CA 94305, U.S.A.

Abstract. Two different movements that can be called revolutionary are proceeding now: the one in the science of biology that is largely identified with the set of methods called recombinant DNA, or genetic engineering, and the one in the biological world around us that is driven by the massive growths in human populations and their concentration in cities. Both are altogether unprecedented: history offers no clues to their consequences or guidance in their management. Yet the dangers of the first, the new science, seem overrated, often under the influence of a false analogy with nuclear bombs and power; the dangers of the second, the growth of human populations, are terrible and probably unavoidable.

SOME PRELIMINARY DISTINCTIONS

How revolutionary is the biological revolution? The question is extravagantly broad. It invites two broad answers: hardly at all; and, on the other hand, apocalyptically. The organizers have urged us to include provocative statements and controversial issues, and above all to stimulate discussion. Yes; but first we need to clarify the discussion. We need distinctions and definitions. I'll begin with three. They will seem extremely obvious – at first.

Distinction 1: The biological revolution is not the same thing, not the same process, as the revolution in biology.

The revolution in biology I take to denote the changes in the last 150 years in the science itself, or to speak more precisely, in the various life sciences – in their content, splittings, comings-together, aims, methods, techniques, practitioners. At present we tend to identify these changes with the development of genetics from Mendelian to molecular and cellular, and with the development of evolutionary theory. We tend to think of these as revolutionary in themselves: but I question what that means. And we think of the social effects of these as great: but, I suggest, until recent years the effects have been more important conceptually, ideologically even, than practically and technologically.

The biological revolution, in contrast, I understand to be what (some people assert) is going on out there in the living world – the changes in the ways mankind lives and in the balance of our relations to the rest of the biosphere. It's easy to be soft-minded about this; perhaps it will harden our thinking to remember that these are the Malthusian questions taken to their extreme both in scope and consequence – and Malthus was an economist by trade.

Distinction 2: Examples from the history and epistemology of physics provide the standard models of scientific revolutions for historians, sociologists, and philosophers of science; but biology is different from and independent of physics in its patterns of historical development, and in epistemology as well.

The case for biology's independent standing was argued vigorously by Ernst Mayr, in the opening chapters of *The Growth of Biological Thought* (Mayr 1982). The distinction is relevant here because the underlying premise of this conference and this volume of papers is that a revolution is in fact going on – yet we speak of revolutions in biology as though we knew what they are like. In fact, most accepted notions about scientific revolutions derive from the history of physics; they derive, indeed, from a certain intellectual slant on certain selected episodes in that history. But revolutions in biology have not fit the models derived from physics. (Besides that, the patterns of change in physics are no longer what they were.)

Distinction 3: The relationship between science and technology, between so-called pure and so-called applied science, in the putatively most revolutionary branches of modern biology is not the same as it has traditionally been described in physics. One may say that the distinction was always a somewhat dubious intellectual construct; but my point here is that in precisely those branches of biology thought to be most revolutionary, discoveries and applications are so intimately related and the interchanges so rapid that they are indistinguishable – except for problems of scaling up for production.

REVOLUTIONS IN BIOLOGY TO THE PRESENT DAY

The argument runs that a revolution in biological science has been taking place this past forty years, that it is introducing remarkable new technologies, and that these present society with extraordinary opportunities and dangers. Many people, indeed, emphasize those dangers; they compare the revolution in biology to the coming of the atomic bomb and atomic power. Is any of this so? Have we not overstated the revolution absurdly? Does the history of science offer any comparable period?

A revolution in biology took place in the seventeenth century, which we can describe as the abandonment of Aristotelian and Galenic views of gross human physiology and their replacement by physiological reasoning and experiment that tended to ask questions more quantitatively and less teleologically. We can call this revolution after William Harvey; its emblem is his discovery of the circulation of the blood. The conceptual change was great, the practical consequences few. For example, blood transfusions were attempted, from dog to dog or from dog to man, but the recipients perked up only momentarily.

In the mid-nineteenth century and after, a great variety of foundation-laying developments took place in the biological sciences. The very term *biology* was coined. Cells, the basis of life in cells, and the beginnings of cellular physiology were discovered. Biochemistry began. The germ theory of fermentation and of disease was announced. The concepts of biological classification, set up by Linnaeus in the late eighteenth century, were elaborated, extended, refined. Mendel invented genetics, and after forty years his principles and his paper were rediscovered; in that interval Weissman and others put in place the chromosomal basis needed to make physiological sense of Mendel's principles. Darwin published the theory of evolution by natural selection. Immunology was launched, and by the end of the century its pioneers had achieved astonishing insights. By 1900, in fact, the roots of much of present-day biology – and of the supposed revolution in biology – were in place.

The conceptual transformation was tremendous, and in this case penetrated the popular consciousness deeply. Above the rest, the theory of evolution matters (and research in it should get government grants) for the same reason cosmology matters: it addresses questions of origins we find inherently fascinating however useless, which we want to understand for their own sake without thought of practical payout (Wilford 1991).

A little payout nineteenth-century biology did offer. Vintners, brewers, distillers, bakers, chicken farmers – as these became mass producers, they were the great beneficiaries of Pasteur's discoveries. Pasteur made himself a French national saint by showing how to intervene in rabies, but otherwise the germ theory of disease cured nobody. Surgeons discovered (but long resisted) asepsis; immunization was available against smallpox. Otherwise the Victorian strides in public health came through public works, starting with sewers and clean drinking water. Stockmen and plant breeders continued to rely upon the ad hoc rules that had fascinated and misled Darwin; genetics was of no agricultural use before the 1920s.

Medicine is a dramatic instance. Before the late 1930s, the greatest technological advance in medical history was the introduction, simultaneously in the mid-nineteenth century, of anesthesia and aseptic surgery. That was indeed a revolution. Except for that, though, physicians treated patients much as they had for two thousend years. Then beginning in the mid-1930s came antibiotics – the sulfa drugs, then penicillin. The doctor at the bedside could actually do something, and patients with dread, fatal diseases – syphillis, bubonic plague – could get well. This was the second medical revolution.

So far, this simplistic sketch of the history of biological science suggests that a couple of periods of great change did occur before the mid-point of this century – but that as revolutions they had three curious features. First, their practical effects were minimal. (This is in contrast to physics and chemistry, of which certain branches had an intimate and productive interaction with engineering and manufacturing.) Second, the undeniable importance, the excitement, of discoveries in biology lay, rather, in the conceptual transformations they produced. The scientists were rewarded by new understandings, new tools of inquiry, new or newly accessible problems – and by the esteem of their peers. The general public (of course including scientists) got apparent answers to impractical but deeply intriguing questions. Third, though more detail would be required to show this, the transformations in biology took place with little, and that local, scientific opposition. Not much was overthrown – no great scientific systems, no sets of ruling scientific preconceptions. The contrast to, say, the Newtonian revolution or the Einsteinian is great – and the opposition to Darwinism only reinforces the point, for that (though at its best systematic) was and is prescientific.

Now, the transformation in biological understandings that has been constantly accellerating for the past forty-odd years, the so-called biorevolution, though it is far greater in scope and detail than anything previous, seems to share those three curious features. Of course the practical rewards are mounting up. For example, organ and tissue transplants are a triumph of immunology and in the autumn of 1990 were once again the recipient of a Nobel prize in physiology or medicine. Again, the green revolution, high-yielding new crop strains for third-world agriculture, was a triumph of plant genetics (and of high-tech pesticides and fertilizers). Yet the developments that are called the present-day revolution in biology are centered on molecular biology and molecular cell biology – and while they have brought hundreds of new companies into the market, the companies have a dismal record in briding new products to market. Wheat or chicks are still hybridized, new medicines still found by methods long standard; they haven't "gone molecular". Even the birth control pill, as revolutionary in its social consequences as anything technological since the telephone, is premolecular pharmacology.

The shame is that biologists have trained legislators to put up money on the promise of payout – and this despite the overriding intrinsic interest of the subject. Fascination with the secret of life remains universal; the double helix is taught in elementary schools; the ground of beliefs is shifting again. The physicists have been more sophisticated in their fund raising. What practical payout justifies putting public money into the space telescope? The superconducting supercollider?

In the third respect, too, the present transformations in biological sciences continue the pattern of the past: although, of course, they threaten prescientific views of the world, they overthrow no vast scientific conceptual systems.

In sum, to speak of revolutions in biology, even recent biology, is awkward, glib, and in large part misleading. Yet I am sure that the situation of biology is now changing. We will soon be able to speak correctly of a revolution in biological science that has revolutionary practical consequences and revolutionary social effects. But this is a revolution to which history offers no guide.

THE REVOLUTION THAT IS IMMINENT

The second generation of molecular biology began about 1972, when molecular biologists turned their new methods and understanding, learned largely with bacteria and bacterial viruses, onto complex, multicelled creatures. The overarching problem was to understand how such an organism, starting with a single cell, the fertilized ovum, develops and differentiates into the adult with its many interacting types of cells. Molecular biologists thus returned to the classically intractable problem of embryology – but with a new approach, based on the idea that if you can add one or a few genes to a cell, and see how the cell changes what it does biochemically, you can figure out what the added genes are for. Science is perforce conservative: the new generation of molecular biologists began by adapting their methods of moving known snippets of genes into bacterial cells, but now trying to do the same with cells of higher organisms. They found this could be done surprisingly easily. To get the added genes to function in the host was more difficult but could also be done. Twenty years later, the problems of development and differentiation are far from solved, but molecular biologists have an armamentarium of tricks for moving foreign genes into cells of higher organisms and getting them expressed. These tricks collectively are called the technology of recombinant DNA, because pieces of the genetic material are taken from one organism and combined with that of another. The common term is genetic engineering.

Recombinant DNA and allied technologies are the basis of the revolution that is imminent but not yet here. This revolution will be followed by two more great transformations. The next is immunology, which has gone molecular with a vengeance, and which in the past four or five years has suddenly becom coherent and cohesive. Though many details remain to be filled in, leading immunologists believe that they are close to putting in place a complete framework of understanding. The third great transformation is just gathering momentum: a science that links neurobiology with behavior and with changes in behavior – again, at the molecular and cellular levels.

The question we desperately want to answer is what breakthroughs these three new sciences will produce. But by definition the actual content of a breakthrough to be made, say, a year from now cannot be predicted – or we would have it already. However, we can say something about the likely overall shape of discoveries to come, and something more about the practical consequences that may result as well as those much bruited that are not likely to result.

Though the biotech companies have been short of products so far, these will come. We will see new vaccines, new medicines, new pesticides designed to affect one species and nothing else. We will see new contraceptives, at last, including one that immunizes a female against her own ova. We will see new organisms – chiefly new kinds of food plants. We can be sure that for some time to come the greatest impact of genetic engineering will be in agriculture.

Gene therapy has attracted public as well as medical attention. This is the attempt to cure people who suffer a defect in metabolism because they were born without some particular gene – cure them by supplying that gene to their cells and making it function. Thousands of such genetic diseases are known, some dreadful, though most are extremely rare; a different gene defect is the cause of each. Gene therapy was a hope that occurred to many of the early workers in recombinant DNA. The difficulties turned out to be immense, and the first trial of it only began late in 1990. Gene therapy will not become widespread for many decades.

The ultimate gene therapy would add the gene not just to the body cells of the type that need it, but to the germline cells, sperm or ova, so that the cure would be hereditary. This prospect is remote, yet it fills some with horror.

Beyond that, in imagination at least, lies the idea of adding genes not to cure a defect but in some way to improve the organism, heritably. Yet imagination is a poor guide, because the general qualities of an organism are built and maintained by the interaction of many different genes – as we recognize at once when we consider, for instance, that a defect of any one of many hundreds of different genes will cause severe mental retardation.

Natural selection has already taken us far. We could not be much bigger, nor our muscles much stronger, without generating problems of engineering: bones would crack. We would like to live longer. A biologist friend recently suggested that if aging is caused in part by the accumulation of errors in the DNA of the body cells, then we need more copies of the gene that specifies the enzyme cells possess that repairs DNA. Then he recalled that many different enzymes repair DNA. We would like above all for our descendents to be more intelligent. But intelligence appears to be above all others the human quality that is most dependent upon the precise balance of the multiply interacting effects of scores of different genes – and of the interaction of all this, during development, with the social environment. Conceptually the problem is stupefying. Ethically the experiments, with the extreme risk of harm to the offspring, are horrifying. And anyway the extentions of intelligence provided by computers are becoming common – if not so common as spectacles for improving vision. In more than a decade of consideration and argument, I have thought of only one single-gene germ-line improvement that could be valuable to the human race – the gene for the enzyme that would have us make our own vitamin C, as our cousins the chimpanzees do.

In any case, molecular biologists experiment with bacteria, fruit flies, or mice because their generation time is short and the results of genetical experiments can be read in hours or days. Human generation times are very long. How would the experimenter known that he had found and successfully implanted the longevity gene into a germ line, even his own?

In immunology, the patterns of discovery and their consequences will be less contentious that those in molecular or cell biology. Those of molecular neurobehavioral science are too far off to justify speculation.

A MISLEADING ANALOGY; SOME UNFORESEEN CONSEQUENCES

Molecular biologists and their most bitter critics share an unthinking habit of comparing recombinant DNA and the issues raised by the science to the situation in the decades after the second world war when the science was nuclear physics. This analogy should be banished from the discussion. It is worn out, a forty-year-tired cliché, a stimulus that no longer elicits a jerk. It is out of date, for surely, in the past two years, fear of nuclear war has receded. Most fundamentally, the comparison of the dangers of DNA technology with the dangers of nuclear fission and fusion is not on all fours. Nuclear dangers were out there, separate from us, singular, massy, and isolable – peacetime accident or wartime desperation. The consequences, the supposed dangers, of DNA technology are inescapable, insidious, and omnipresent. They will enter the fabric of our daily lives – our standard of living and dying, our ideas of what it is to be human, our intuitions of our place in the universe. To take just one example, contrast the consequences for civil liberties of the two technologies. The nuclear complex indeed had nasty consequences for civil liberties, but these were not a direct product of the technology itself; rather, they were secondary effects of the secrecy and so on surrounding that technology. DNA technology is in and of itself a potential threat to civil liberties and rights, in ways that we can't yet fully foretell but that include such obvious possibilities as genetic screening of people applying for jobs or wanting to buy insurance.

Next, I come to one further prediction, of a novel and paradoxical kind. We must prepare ourselves to recognize and act early in cases that by definition cannot be specified in advance, where apparently benign discoveries in biology have unexpected consequences. Consider examples, not all biological.

Case 1: Atomic power was hailed in the late 1940s as the future of cheap, abundant energy. Nobody allowed himself to think ahead even a quarter century to the problem of radioactive wastes; nobody took seriously the statistical inevitability of accidents nor the possibility of popular protest.

Case 2: Electrical generating is now for the most part concentrated in large plants that also control immense distribution networks. Solar and other self-renewable energy sources for generating electricity will be widely dispersed and networked in a different way or not at all. Nobody has yet thought about the sociological and political consequences of dispersed power generation.

Case 3: In the summer of 1990, two scientists announced in a reputable journal a mid-size trial with elderly men that demonstrated convincingly that minute, precisely calibrated doses of human juvenile growth hormone over a six-month period reversed some effects of aging. In muscle tone, ratio of muscle to fat, and skin condition men of sixty became physiologically forty again (Rudman 1990; Vance 1990). Suppose this finding holds up. In the first place, human growth hormone is rare and expensive, although it has been produced successfully by recombinant DNA in bacteria. Who gets it? Deng Shiao-ping? Margaret Thatcher? The Ayatollah? A Stalin or a John D. Rockefeller? Beyond that, what if the stuff becomes cheap and abundant? The effect on population sizes would be catastrophic, on population structures still worse. What about inheritances? Retirement? What about such people as tenured professors in their seventies blocking promotion of younger academics?

Case 4: Suppose someone discovers a safe cheap pill that allows the choice of the sex of the child to be conceived. I would put this fairly high on the list of predictable breakthroughs. That pill, widely distributed in parts of the world where societies prefer boy children, would in one generation produce a great disproportion of males to females, and in the generation after that a population crash. (Is this one we should hope for?)

THE BIOLOGICAL REVOLUTION GOING ON NOW

A revolution is gathering speed in the human world and the living world generally that will transform the lives of the next generations more profoundly than anything since the invention of agriculture ten thousand years ago brought hunter-gatherers into settled villages. It can be described as the human domination of the world. As we all know, the human population has grown immense and will double again in the next twenty years if checks do not intervene. (They will intervene.)

The size of the population is only the start of the revolution in the way we live. The human species is in effect a domesticated species: we exhibit a range of variability that is not found in the wild but rather, say, in dogs. We are also taking on some of the characteristics of a monoculture, that is, a crop that is entirely of one variety planted densely without breaks. Monocultures are susceptible to pandemics, to extinctions. If the neolithic agricultural revolution put mankind in settled villages, the new revolution in concentrating us in cities. By the turn of the century, a good majority of the world's children will have been born in cities – itself a revolutionary shift – and will be growing up with the city as their natural environment. The other great change in the way we live is that intercommunication between human population groups is now unrestricted and fast. Geography no longer bars the spread of ideas, fads, and diseases.

The biological revolution of course has many other interrelated features; but I turn now to the next step in the argument. The consequences of this revolution seem to me entirely evil. The Malthusian checks on population are war, famine, pestilence, and vice. There's no cure for them, only postponement to a day that will be still worse. Here are illustrations.

About war: Entirely new types of weapon, many biological, are becoming possible and cheap.

About famine: The success of the green revolution is inherently a failure. In the first place, the technology relies massively on fertilizers and pesticides, high in cost and in destruction of soil and pollution of water. Secondly, it sows monocultures. Third, it allows the population to expand still more, up to a new but more dangerous stable plateau.

About pestilence: The outbreak of epidemic disease is the product not simply of a new or newly arrived disease organism, but of a change at the same time in the host population, usually a change in behavior. Changes in each of the two populations interact to touch off the epidemic. Many examples can be cited, beginning perhaps with the plague of Athens in the Pelopponesian War: In response to the threat from Sparta, Pericles ordered the building of the Long Walls between Athens and its port, Piraeus, and when war in fact came much of the rural population moved in from the countryside to concentrate between the walls. The disease appeared in Piraeus, necessarily with an arriving ship, and swept upwards, killing a third of the population. In sum, the preconditions of the plague of Athens were the new concentrations of population, and the mobility of trade and travel – and these are precisely the conditions that have built up today. AIDS is the prime modern instance, with the great changes in personal behavior that created the chief risk groups. More epidemics are sure to come.

About vice: The breakdown of families and of social control more generally in cities is notorious but true. Expressed most neutrally, vice represents behavior that has changed away from some norm: see pestilence, above.

Worse is to come. Malthus, writing before Pasteur, did not list the fifth and most dire check: wastes, pollution. Pasteur demostrated that organisms that cause fermentation of wine become inactive and die when the concentration rises of their own waste product, alcohol. Man at the end of this millennium is killing himself with the waste products of his standard of living. This check also cannot be cured, only postponed, for in a kind of second law of pollution dynamics, most proposed remedies have their own downwind effects.

RESPONSES

Our conference is aimed at discussion of a "new covenant" between biological science and society. I have no idea what that could mean: the language seems misty. What I do see is a need for hard-headed national and international science policy. We must learn from the cases where intervention of scientists worked – as it may have done, although it is too early to be sure, with chlorofluorocarbons and the ozone layer. We must heed the instances of failure – as in our apparent inhability in the United States to take the complex social measures, including the guarantee of universal medical care, that might have allowed the containment of AIDS. We must watch for benign discoveries that turn up societal side effects, as the sex-selection pill would do.

REFERENCES

Mayr, Ernst (1982) *The Growth of Biological Thought,* especially Chapters 1–3. Cambridge, MA: The Belknap Press of Harvard University Press.

Rudman, Daniel et al. (1990). Effects of human growth hormone in men over 60 years old. *New England Journal of Medicine* **233** (July 5, 1990: 1–6.

Vance, Mary Lee (1990). Growth hormone for the elderly? *New England Journal of Medicine* **233** (July 5, 1990): 52–54.

Wilford, John Noble (1991). Astronomers' new data jolt vital part of Big Bang theory. *The New York Times* from January 3, 1991, pp. 1–19 (reporting on Saunders, Will et al. (1991), The density field of the local universe. *Nature* **349** (Jan 3, 1991): 32–28; and comment by Lindley, David (1991). Cosmology: Cold dark matters makes an exit. *ibid.* 14–15).

Bioscience ⇌ Society
edited by D.J. Roy, B.E. Wynne, and R.W. Old, pp. 337–346
© 1991, John Wiley & Sons

Communication in Science / Communication of Science

Kenneth S. Warren

Maxwell Macmillan Group
866 Third Avenue
New York, NY 10022, U.S.A.

Abstract. Science is public knowledge, i.e., scientists must publish, thereby subjecting their work to scrutiny and repetition until it is "universally accepted", not only by their peers, but, eventually, by the general public as well. A fundamental problem of communication in science has been the exponential growth of the scientific literatures, which renders information retrieval both cumbersome and costly. Initially this was dealt with by the development of abstract publications. The next step should be concentration on the small proportion of scientifically sound information. Such a focus will also enhance communication of science to the public. The latter enterprise will be fostered by the development of effective means of communicating information of quality and by the growth of scientific literacy among the general public.

INTRODUCTION

Science is "public knowledge" in the sense that "its facts and theories must survive a period of critical study and testing by other competent and disinterested individuals, and must have been found so persuasive that they are almost universally accepted" (Ziman 1974). It goes without saying that the process by which science is made public is communication, and that acceptance involves not only peers but, in the final analysis, the general public. Thus, this paper discusses communication both *in* and *of* science. The former is essentially parochial, with professionals talking largely to small groups of their peers. Conversely, the latter is a catholic endeavor involving scientists, journalists, and even advertisers broadcasting to the general public. The audiences for these two modes of address vary strikingly. The critical difference is what Daniel Koshland (1985), editor of *Science,* has called "scientific literacy". He observes that the communication gap is "wide and serious" because "the fundamental concepts and methodologies of science are outside the understanding of the vast majority of the population, including its opinion-makers." Finally, the basic problem of communication *in* science is being overwhelmed by Lucullan feasts of information; that of communication *of* science is the inability to present information in an edible fashion, or to digest it upon ingestion.

My conceptions of communication in science stem from experience in biomedical, clinical, and information research. My concern for, and involvement in, the communication of science has been from the various perspectives of the academic, philanthropic, and business worlds.

COMMUNICATION IN SCIENCE

"The invention of a mechanism for the systematic publication of fragments of scientific work may well have been the key event in the history of modern science" (Ziman 1969). It began in 1665 with the publication of the *Journal de Scavans* by the French Academy and of the *Philosophical Transactions of the Royal Society* in England. There was a lag period in the growth of the number of journals, but since 1750 the 10 journals then extant has continued to double every 15 years, or by a factor of 100 every century. A consequence was that by 1830 "the process had reached a point of absurdity: no scientist could read all the journals or keep sufficiently conversant with all published work that might be relevant to his interest" (de Solla Price 1961).

The first attempt to solve this problem was the appearance of abstract journals at the beginning of the 19th century; they have also multiplied exponentially. The next stage in the attempt to deal with the plethora of information began with Charles Babbage's invention of an analytical machine in the late 19th century. One hundred years later, John von Neumann developed the idea of a computer system. Its subsequent development, particularly with the appearance of the personal computer in the mid-1970s, led to the almost universal belief that a third era in the handling of scientific information was imminent. Prescient organizations like the National Library of Medicine and the Institute for Scientific Information began to develop remarkable computerized databases called, respectively, MEDLINE and the Science Citation Index.

Unfortunately, this new technology has only added to the growth of scientific and medical information by providing computerized typesetters, laser printers, and desk-top publishing. It has not improved the ability of scientists to handle the information overload phenomenon. Satiation is not the only problem, although the ancient Romans learned how to solve it; the difficulty is the time and money involved in ingesting and digesting all that material. A recent report of the National Academy of Science entitled *Information Technology and the Conduct of Research – the User's View* (1989), stated that most searches of reference databases "are incomplete, cumbersome, inefficient, expensive, and executable only by specialists". The difficulties with factual databases are that "the researcher cannot get access to data; if he can, he cannot read them; if he can read them, he does not know how good they are; and if he finds them good, he cannot merge them with other data." A major obstacle in the path to more efficient information systems is that their development has been almost wholly technology-driven. Another is the lack of understanding of information systems that is rife among its scientist-users.

More than twenty years ago, Margit Kraft (1967), a renowned research librarian, succinctly described the technology problem. "The difference between a philosophy and a technique is that a philosophy sets goals, while a technique devises means to achieve these goals. If goals are nebulous, technique takes over and runs wild." The development of a "philosophy" in this context is dependent on knowledge of the environment of information, its ecology, epidemiology, demography, and sociology.

Theoretically, the concepts of the ecology and epidemiology of information are interlinked The epidemiological nature of the system was first described by André Siegfried (1960) in his *Itinéraires de Contagions: Épidémies et Idéologies*. In describing the routes of both epidemics and ideas he begins with caravans, but notes that while "the nature of communications does not alter, their intensity increases." This concept was later developed by William Goffman, mathematician and professor of library science, into a mathematical model based on a two factorial epidemiological process (tuberculosis, where there is direct transmission from person to person). He refined it to a three factorial process (malaria, where there is a mosquito vector), and, finally, to a four factorial process (schistosomiasis, where there is transmission between vertebrate and invertebrate hosts) (Warren and Goffman 1972). It was then observed that the third, and best, information model also described an ecological process that evolved over a long period of time and was well adapted to fulfilling its manifold functions. An awareness of the complex, interactive structure of our scientific communication system is necessary for its efficient utilization and to ensure that any significant changes are made safely.

The demography of the information system is of great importance in that it tells us that the "explosion" was not in information but in population. Studies of several major scientific literatures over a period of 80 years (1862–1962) have shown a constant paper-author ratio (Warren and Goffman 1972). Not only has the world's population been growing exponentially, but that of its scientists has been increasing at an even more rapid rate, owing to the expansion of higher education. It is even more important to take note of the structure of that population. First, it is essential to realize that the well-known hypothesis of Jose Ortega y Gasset that "experimental science has progressed thanks in large part to the work of men astoundingly mediocre, and even less than mediocre" (Cole and Cole 1972) is grossly wrong. In their paper, *The Ortega Hypothesis,* the Coles, sociologists of science, have reported that it is "a relatively small number of physicists [that] produce work that becomes the basis for future discoveries in physics." Sydney Brenner (1990) called attention to "the public myth that scientists are satisfied to be builders and that nobody's brick is better or bigger than anybody else's." What this means in demographic terms is that scientists are no different than those engaged in other endeavors, such as piano playing, basketball, carpentry, or ballet, i.e., that a good scientist is not common, and an outstanding scientist of the order of Magic Johnson or Margot Fonteyn is very rare indeed. It is obvious that the number of scientific papers of quality is proportional.

While the proportion of good scientists is small, de Solla Price (1963) has shown that they tend to be prolific. "It takes persistence and perseverance to be a good scientist, and these are frequently reflected in a sustained production of scholarly writing." Furthermore, "scientists tend to congregate in fields, in institutions, in countries, and in the use of certain journals. They do not spread out uniformly, however desirable that may or may not be." In other words, the best scientific papers tend to be produced by the best scientists, most of whom work in the better institutions and publish in the better journals. One way of objectively determining the better papers and journals is with the Science Citation Index, which records the number of times they are cited by their publishing peers. More than half of the published papers are never cited, and only 20% are cited more than once (Hamilton 1990); only a small proportion of journals have high impact factors, i.e., the number of citations divided by the number of papers they produce.

Another particularly revealing technique is to determine the use of journals in scientific libraries. While this approach is well known to librarians, it is not, unfortunately, known to scientists. Scientists do not know that more than

50% of the titles in the British Lending Library for Science and Technology are never consulted. Half of the demand is satisfied by 4% of the journals, and 80% is satisfied by 10% of them (Urquhart D. 1958). A recent study of the library of the Marine Biological Laboratory in Woods Hole yielded virtually identical results (Wolff 1986). Thus, it seems obvious that only a very small proportion of published scientific material is sighted or cited with any frequency. A large proportion is essentially ignored.

The environmental factors described above enable us to envision a third era in the retrieval of scientific information based on the small proportion, well less than 5%, of information of quality, i.e., scientifically sound. This is based on the identification of established scientists and their younger colleagues of quality who publish in journals of quality. In his article *Misinformation explosion: Is the literature worth reviewing?* the physicist Lewis Branscomb (1968) emphasized the crucial step of quality filtering of papers for review articles. He described an analysis of 30 independent reports on helium ionization cross-section in which only 10% of them had even rudimentary evidence concerning essential questions such as the prevention of secondary emission from the electron beam collector or the definition of the path length of the electrons in the ionizing region. Cynthia Mulrow (1987) recently pointed out that most medical reviews are "subjective, scientifically unsound, and inefficient." She then presented efficient strategies for identifying relevant material of substantive quality and excluding irrelevant or poor quality material. A company called *Current Science* has developed a unique series of "current opinion" journals which consist of concise, authoritative reviews with selected, annotated bibliographies, which are annually updated. In the process of providing comprehensive bibliographies for their reviewers they found that the high quality articles cited in their reviews usually came from fewer than 100 journals, enormously increasing the efficiency of their search mechanisms (V. Tracz, personal communication).

If scientists accept the fact that they are no different than craftsmen, athletes, and artists with respect to the presence of a pyramid of excellence, they will be able to develop much more efficient and cost-effective information systems. Furthermore, they will build into the ecological system of science that crucial element essential to evolution: natural selection. In the words of Charles Darwin, "under these circumstances, favourable variations would tend to be preserved and unfavourable ones to be destroyed", resulting in a natural decline in the quantity of information and an improvement in its quality. This in turn might feed back into a change in the perceptions of the public and a greater receptivity to what is clearly the best of science.

COMMUNICATION OF SCIENCE

There are certain values, crucial to the scientific enterprise, which must be understood in order to achieve scientific literacy. Most important was Francis Bacon's admonition to his colleagues to stop debating how many teeth a horse has, and actually go out and count them. Thus, while science is based on careful observations using the best available techniques, it begins by posing a question of substance and value. Then it performs a carefully controlled experiment which does the counting. The objective and truthful interpretation of the results then follows. Communication is the final and essential step, so that the results may undergo the annealing process of "public knowledge".

In recent years, the communication process, so necessary to science, has been institutionalized in a variety of forms, such as courses, departments, schools, and institutes, particularly in the U.S.A. My interest in the communication of science stems in particular from a review of a proposal to develop a major academic institute for communication studies in Europe. Its focus was on mass media, although it paid lip service to the sciences, arts, and humanities. While the proposal mentioned ethics it included a course on propaganda. This assignment, which came from an organization with a specific interest in science, led me to consider the idea of creating an Institute for Scientific Communication which would deal with the matters considered in this paper: communication among scientists and communication of science to the public. One clear-cut issue is that propaganda is the antithesis of science.

Some time later I was asked to help in the organization of an institute devoted to the communication of a major area of science, which impinges on public welfare. The plan was to develop a satellite-based global surveillance system which would input data into an advanced supercomputer center for analysis and synthesis. This information would be output to scientists within and outside of the institute, where it would undergo a variety of analyses, especially risk assessment. For significant risks, measures for prevention and control would be considered. If such measures were available, they would be examined for effectiveness and cost. If no measures existed, or those available were inadequate, research would be commissioned. A major function of this institute would be to communicate this value-added information to the public, using a variety of media for addressing a variety of different audiences, including the general public, schools, industry, and government.

The public is constantly bombarded with information concerning health and environmental hazards; most obtrusive and incessant are the daily newspapers and television. People's fundamental bodily functions are being assailed: from the air they breath to the food and water they drink, to the amount of exercise they perform, to the invisible hazards of radiation and the visible hazards of accidents. A major problem is that the information on all of these risks is constantly changing, so that the public is never sure where it stands. The difficulty is that the media are continuously reporting scientific work that has not passed Ziman's test of "public knowledge", i.e., repeated study and testing until it is almost universally accepted. Cigarette smoking has passed the test, and consensus is being achieved for the danger of high cholesterol levels, but there is still considerable controversy over the relative risk of a variety of different types of radiation, of different types of asbestos, of pesticides, of preservatives, etc. Things get far more difficult when dealing with micro-issues, such as whether olive oil, oat bran, margarine, and many other foodstuffs lower cholesterol levels, do not affect them, or raise them. With respect to the environment, we are constantly caught in controversies, such as the causes and effects of global warming, ozone holes, and acid rain.

In spite of all this confusion, the public remains fascinated by the development of new technologies. But even there it is bemused by matters such as hot fusion and cold fusion, since science accepts but cannot achieve the former, and does not accept that the latter has been achieved. The great achievement of the "double helix" and molecular biology is widely appreciated, but its practical outcome in genetic engineering has caused widespread fear. A case in point is concern about the use of bovine somatotropin, which increases milk yields up to 25% (Lesser 1990). Not only do small dairy farmers fear it for economic reasons, but neo-Luddite groups are using it as a major case in their ongoing battle against genetic engineering. The roles of the various parties in these situations was remarkably well described 300 years ago by Sir Francis Bacon (1597): "Crafty men contemne Studies; simple men admire them; and wise men use them: for they teach not their own use; but that there is a wisdome without them, and above them, won by observation."

Almost twenty years ago, the Scientists' Institute for Public Information (SIPI) was started in the U.S.A. Lewis Thomas, its chairman for most of the last decade, described SIPI in his usual elegant manner:

> There was a time, not long ago, when working scientists and professional journalists had as little as possible to do with each other. The scientists were apprehensive that whatever remarks they made about their research would be distorted or misunderstood, or worst of all, sensationalized. The journalists felt that scientists, as a group , tended to obscure the meaning of their findings deliberately, using impenetrable jargon to keep the press away.
>
> Things are beginning to change, and it is time. Science has become of critical importance to the very progress of society. The general public, knowing that virtually all basic science is dependent on public support, want to know what

> is happening in the nation's laboratories. The scientists themselves are worrying more than they used to about the need for public understanding of their work ... [SIPI] plays an important role in bringing these two communities together. It is developing novel mechanisms for doing this, and the quality of science reporting in the media is being enhanced by better sources of information.

These include the Media Resource Service (MRS), a free telephone referral service available to journalists working on science-related stories. Its computers contain information on more than 25,000 experts in science, technology, and medicine. On average they handle 90 calls a week from the media. In addition, they hold meetings, produce documents and a quarterly newsletter, and have a videotape referral service. The present chairman of SIPI is David Baltimore, the Nobel laureate president of Rockefeller University.

Five years ago Dr. David Evered of the Ciba Foundation founded an MRS for the United Kingdom. Its brochure stated:

> We live in a rapidly changing world. Scientific and technological changes have had an impact on almost every field of human endeavour, influencing the evolution of society and the conduct of our lives, but the benefits of some are controversial and others have had clear disadvantages for the community. The widespread impact and the growth of science emphasises its importance to us all, and provides the basis for the belief that an improved understanding of science and, no less important, an avoidance of misunderstanding, will be to the benefit of all, and will contribute to the effective operation of our democratic society. Many choices in society depend upon understanding the probable impact of scientific or technological change on the community and the individual.

The MRS has recently expanded to other Western European countries.

CONCLUSION

It has been clear for more than 150 years that the exponential growth of scientific information has made communication in science exceedingly cumbersome and costly. The technological revolution, which was perceived as a panacea, has as yet merely served to exacerbate the problem. As Robert Maxwell (1990) said in his Dainton Lecture at the British Library,

> Unless we do indeed learn in this country or in our civilized world how to harness modern information technology in order to package information in the form that individual research scientists and doctors can use to solve real time problems, then not only will our society continue to have to pay the penalty or re-invent

the wheel, but many, many problems at both micro and macro level will remain unsolved or will become even more unsolvable than they are already.

The editorial in *Science,* mentioned above, posed the crucial question for communication of science. Dan Koshland (1985) concluded,

> Scientists will be denounced for trying to introduce cold-blooded reason into an area in which warm-blooded humanity is supposed to reign supreme. But warm emotion frequently gives way to hot-headed anger and even bigotry. The scientific method has been the most effective means of overcoming poverty, starvation, and disease. Even those who are not professional scientists can understand its fundamental concepts, which will aid their decision-making in an increasingly difficult and technological world. It is time to bridge the "concept gap" by improving scientific literacy.

Science is "public knowledge", but the strange paradox is that the public has little understanding of the time-consuming process by which its ideas, concepts, and products are annealed into "almost universal acceptance". Improvement of communication in science and communication of science, which are interrelated phenomena, will do much to foster the well-being of mankind throughout the world.

REFERENCES

Bacon, Sir F. (1597). *Essayes. Religious Meditations. Places of perswasion and disswasion. Scene and allowed* London: Humfrey Hooper.

Branscomb, L. (1968). Misinformation explosion: Is the literature worth reviewing? *Scientific Research* **3:** 49–56.

Brenner, S. (1990). The greatest satisfaction (book review). *Nature* **345:** 675–676.

Cole, J.R. and S. Cole (1972). The Ortega hypothesis. *Science* **178:** 368–375.

de Solla Price, D. (1961). *Science Since Babylon.* New Haven: Yale University Press.

de Solla Price, D. (1963). *Little Science, Big Science.* New York: Columbia University Press.

Hamilton, D.P. (1990), Publishing by – and for? – the numbers. *Science* **250:** 1331–1332.

Koshland, D.E. Jr (1985). Scientific literacy. *Science* **230:** 391.

Kraft, M. (1967). An argument for selectivity in the acquisition of materials for research libraries. *Library Quarterly* **37:** 284–295.

Lesser, W. (1990). Technology and the family farm. *Nature* **347:** 1112.

Maxwell, R. (1990). *Information Technology as a Way of Reducing the Costs and Time in the Dissemination of Scientific and Technological Information.*Boston Spa: The British Library.

Mulrow, C. (1987). The medical review article: State of the science. *Annals of Internal Medicine* **106:** 485–488.
National Academy of Science. (1989). *Information Technology and the Conduct of Research: The User's View*. Washington, D.C.: National Academy Press.
Siegfried, A. (1960). *Itinéraires de Contagions: Épidémies et Idéologies*. Paris: Librairie Armand Colin.
Urquhart, D. (1958). Use of scientific periodicals. In: *International Conference on Scientific Information*. Washington, D.C.: National Academy Press. 277–290.
Warren, K.S. and W. Goffman (1972). The ecology of the medical literatures. *American Journal of the Medical Sciences* **263:** 267–273.
Ziman, J.M. (1969). Information, communication, knowledge. *Nature* **224:** 318–324.
Ziman, J.M. (1974). *Public Knowledge: The Social Dimension of Science*. London: Cambridge University Press.

Bioscience ⇌ Society
edited by D.J. Roy, B.E. Wynne, and R.W. Old, pp. 347–360
© 1991, John Wiley & Sons

Scientific Literacy: Can It Decrease Public Anxiety About Science and Technology?

Morris H. Shamos*

M.H. Shamos & Associates
6 East 43rd Street
New York, NY 10017, U.S.A.

Abstract. Any hope that increased scientific literacy in the general public will help decrease its anxiety about science and technology is ill-founded. We show a) that the literacy level required to make the public feel reasonably comfortable with these enterprises is so difficult to achieve as to be impractical, and b) the average educated individual sees no real incentive to spend the considerable effort needed to attain this level of literacy in science. The solution lies only partly in improved education, such as ensuring that the public understands the difference between science and technology, and knows the meaning of risk-benefit analysis, but mainly in finding ways to establish independent, knowledgeable (and credible) advisory groups that can gain the confidence of the public, assist it in the decision-making process, and form a new and productive alliance between the public and the science/technology community.

INTRODUCTION

World War II turned a spotlight on science, one that caught in its glare not only the scientific establishment, but also the science education enterprise as well as the changing interface between science and society. The rate of growth of scientific knowledge since World War II has been staggering, so much so that the average individual's knowledge of science today is proportionately far less than at any time since science became part of the traditional school

---

* Professor Emeritus (Physics), New York University

curriculum in the world's most highly developed nations. The same is true of professional scientists, who quickly found themselves hard-pressed because of sheer volume to keep pace with their own narrow specialities, let alone the overall field of science. Not only did the end of the war usher in an era of "big science", but the postwar drive for reindustrialization, coupled with a seemingly unlimited nuclear arms race, led many concerned individuals, scientists among them, to suggest early on that the utilization of science might be "getting out of hand". Thirty years ago Bertrand Russell warned scientists of their social responsibilities: "It is impossible in the modern world for a man of science to say with any honesty, 'My business is to provide knowledge, and what use is made of the knowledge is not my responsibility'." (Russell 1960). It was a theme to be repeated many times since, by scientists and nonscientists alike, but it failed to remedy the most obvious ills that much of the public saw as being visited by science upon a "helpless" society: industrial pollution of the air and water supplies, toxic wastes, devastation of the environment, and of course the implications of nuclear testing.

SCIENCE ILLITERACY AND ANTISCIENCE

In the nearly half century since the end of the war, despite intermittent appeals for more scientists and engineers, improved science education for all students, and most recently a renewed drive for widespread public literacy in science, the antiscience tide throughout the world appears to have grown stronger rather than weaker. It must be apparent to all concerned with public attitudes on science that large numbers of individuals are less enchanted with science and technology today than they once were. There appears to be more to it than a populace merely overwhelmed by the complexity of modern technology; one senses a deep concern, if not genuine fear, that unrestrained science and technology (the general public fails to distinguish these) will cause serious harm to the health and safety of present and future generations, or make the environment less hospitable. It may be noted that virtually all so-called "science-based societal issues" can be reduced in the final analysis to one or another of three fundamental human concerns: health, safety, or the preservation of the environment, whether for aesthetic reasons or for the effect that the environment may have on the health or safety of society. They may take on a variety of seemingly disparate forms, e.g., nuclear testing, toxic waste disposal, global warming, natural resources, food additives, even energy conservation, yet in the end all can be seen as affecting society through one or another of these basic concerns.

The latest target of the antiscience movement is the field of bioscience, where the fear of uncontrolled technology takes on new meaning, associated as it inevitably is in part of the public mind with the production of new organisms that may unleash untold disease and destruction, and in another part with the moral and ethical question of creating new life forms in the laboratory. Bioengineering, genetic engineering, gene therapy, genetic testing: all have evil connotations in the eyes of the self-styled neo-Luddites and antiscience proponents. Granted that there are legitimate moral and ethical issues involved with bioscience that call for reasoned debate, yet to many, including the "Greens", the answer is both simplistic and unrealistic: if they pose any element of risk at all to plants or animals (humans included) such activities should be prohibited. The fact that most of those who deplore the inroads of science and technology on modern life are basically illiterate in these fields is of little consequence, for so too is 90–95% of society. Hence the arguments used by such groups, unreasoned though these may seem to those trained in science, nevertheless carry more than a grain of truth behind their appeals to fear and emotion and strike a responsive chord in much of contemporary society. There is no question that many industries (and governments), knowingly or in ignorance, have acted recklessly with the environment; there is no question that the products of science have often been used to the detriment of society rather than to its benefit; and there is no question that scientists are often forced to dance to the tune of government funding. But to insist, as do many in the (science) counterculture movement, that "societal impact" studies be done before undertaking any frontier scientific research, or worse, to espouse the Aristotelian view that science and technology should in all respects be controlled by the state, would stifle enquiry altogether. When in 1905 Einstein elucidated his famous mass-energy equivalence ($E = mc^2$), he might have been appalled to think that some fourty years later it would play such a central role in nuclear explosives; but he had no way of knowing this at the time, and nor did anyone else. This inability, to forecast the uses to which scientific discoveries may be put, strikes at the heart of the problem, pointing up as it does the folly of attempting to control the growth of knowledge and directing us to seek other ways of allaying public concerns about science and technology.

We have already pointed out that most individuals are illiterate in science, which, as we shall see later, is perhaps understandable. Some social critics of science, like the political theorist Langdon Winner (Winner 1977), consider this a mark of objectivity, implying that becoming literate in science "brainwashes" one into a mold of conformity that lessens the individual's ability to criticize the establishment. This rationalization of scientific illiteracy on the part of some scholars in the newly labelled discipline STS (Science/Technology/Society) is specious. Yes, it is always easier to be critical from the outside looking in, but it does not follow that one can also be objective; the danger is that uninformed criticism may also be technically wrong and is rarely constructive. What society needs are fewer science illiterates who fashion their careers out of criticizing modern technology before receptive audiences among the equally illiterate public, and more critics who are also literate in science and technology and can thus target their criticism in a constructive manner. Preying on fears of the unknown and ignoring risk-benefit assessments are the stock-in-trade of most modern-day Luddites, and, unfortunately, the STS movement seems to attract more than its fair share of these individuals, who view the discipline as a convenient vehicle to attack government, big business, and science itself.

Of one thing we can be certain: the future of science and technology will be unlike the past. Russell's apprehension is perhaps more fitting today than it was three decades ago. All of us, scientists, engineers, and social scientists, have become sensitized to the concerns of society, whether we agree with them or not, and this is bound to affect what we do and how we view the fragile relationship between science and society. If indeed an appreciable fraction of society is losing faith in science and technology, where does the fault lie, and how might it be corrected? Is it a matter of trying to change certain mores within the scientific community? Or of doing battle against the neo-Luddite, antiscience, and counterculture movements in their attempts to bias the public? Or of developing some kind of covenant between science and society that might afford the latter a better understanding of what science is about and its role in contemporary civilization? The purpose of this paper is to explore one suggested approach; that is, whether increased scientific literacy in the general public, assuming this could be accomplished, might have the effect of decreasing its concerns about science and technology generally, and more specifically its apprehensions about bioscience.

THE DIMENSIONS OF SCIENTIFIC LITERACY

Attempts to fashion a simple definition of scientific literacy in absolute terms generally fail because a meaningful definition can only be expressed in terms of measurable outcomes, i.e., what should be expected of a science literate. At a symposium on measurements of public understanding of science and technology held during the 1989 Annual Meeting of the AAAS, Jon Miller, who has long sought to establish useful criteria for testing scientific literacy, suggested that "functional scientific literacy should be viewed as the level of understanding of science and technology needed to function minimally as citizens and consumers in our society." Requiring that the individual be able to "function minimally" begs the question, for as we shall see, many individuals actually believe that they function quite well despite their poor level of literacy in science. Miller fails to specify what he means by one functioning minimally, except to suggest that scientific literacy requires a basic vocabulary, an understanding of science process, and an understanding of the impact of science and technology on society. These are clearly among the earmarks of one who is truly literate in science, but it might be helpful to have a more comprehensive set of guidelines. Few educated individuals are totally illiterate in science; everyone knows some facts of nature and has some idea of what science is about, however naive or misconceived these notions may be. Thus, it is an oversimplification to assume that an individual is either literate or illiterate in science. Instead, one can distinguish forms or levels of literacy, levels that are normally attained sequentially by students in their formal exposure to science. The following are descriptions (definitions) of three such levels of literacy which build upon one another in degree of sophistication as well as in the chronological development of the science-oriented mind, and which, because of their vertical structure, may be more useful as criteria for judging scientific literacy.

Cultural scientific literacy

Clearly the simplest form of literacy is that proposed several years ago by Edward Hirsch (Hirsch 1987) in his highly popular book. Hirsch argues that "cultural literacy", by which he means a grasp of certain background information that communicators must assume their audiences already have, is the hidden key to effective education. He goes on to provide a list of several thousand names, dates, places, events, etc., illustrating the type and scope of knowledge that is shared by literate Americans. Included are several hundred science-related terms which, with appropriate definitions, are claimed to constitute a lexicon for the scientific literate – obviously a necessary, but hardly a sufficient, condition for scientific literacy beyond its most primitive meaning. Note the similarity between this and the basic vocabulary cited by Miller. Note, too, that if all one needed were such a lexicon, scientific literacy would be easy to acquire by rote; a companion dictionary, incidentally, has since been developed by Hirsch, Kett, and Trefil (Hirsch et al. 1988). Yet this is the only level of literacy held by most of the educated adults who believe they are reasonably literate in science. They recognize many of the science-based terms (the jargon) used by the media, which is generally their only exposure to science, and such recognition probably provides some measure of comfort that they are not totally illiterate in science. But for the most part this is where their knowledge of science ends.

Functional scientific literacy

Here we add some substance to the bare skeleton of cultural literacy by requiring that the individual not only have command of a science lexicon, but be able to converse, read, and write coherently, using such science terms in a nontechnical but meaningful context. This means using the terms correctly; e.g., knowing what might be called "some of the simple, everyday facts of nature", such as having some knowledge of our solar system, that the earth revolves about the sun once each year while rotating on its own axis once each day, that the moon revolves about the earth once each (lunar) month, and how eclipses occur. Or, to get a bit more sophisticated, identifying the ultimate source of our energy; or the "greenhouse effect"; or how we get the oxygen we breathe. To get still more sophisticated, knowing the difference between electrons and atoms; or knowing what DNA is and the role it plays in living things.

One could go on and on listing such simple facts about nature, and most objective tests of literacy are based upon just this kind of knowledge or recall. This is not a very demanding requirement, yet estimates of the number of adults in the U.S., for example, who might qualify at this functional level are distressingly low. Miller, in the symposium cited above, places the fraction of adult Americans who possess a "minimal" understanding of scientific terms and concepts at about 30%. This number seems rather high, but it is clearly a soft number, depending as it does on the sophistication of the test items designed to assess minimal understanding; obviously one could easily halve this figure by a more rigorous selection. Nonetheless, whatever the number, it does indicate that the science recall of most adults is very poor, since most of the terms and concepts tested would have been at their fingertips during their school years. It also shows that whatever subsequent exposure they may have to science through the mass media fails to reinforce such factual information about the natural world.

A useful distinction between these two levels of literacy is that the functional literate can engage in meaningful discourse on most science articles that appear in the popular press, at least posing intelligent questions, whereas the cultural literate is in the position of one who knows a smattering of a foreign language but can only stare blankly when one fluent in the language seeks to engage him in flowing conversation. Determining whether an individual qualifies for one or the other of these forms of literacy (functional literacy presupposes cultural literacy, of course) through objective tests is relatively easy, since both depend largely upon factual recall. But this is not what science is really about. Lacking in both these levels is literacy in: a) the process of science, a prescription for seeking out and organizing the factual information in the unique manner that is characteristic of science; and b) the fundamental role played by theory in the practice of science. Incorporating these brings us to the ultimate or true level of literacy, the kind most difficult to attain by students and to evaluate objectively by teachers.

True scientific literacy

At this level the individual actually knows something about the scientific enterprise. He understands (or at least is aware of) some of the major conceptual schemes (the theories) that form the foundations of science, how they were arrived at, and why they are widely accepted; how we make order out of a random universe, and the role of experiment in science. This individual also understands the elements of the so-called "scientific method", of proper questioning, of analytical and deductive reasoning, of logical thought processes, and of reliance upon objective evidence – the mental qualities that the famous educational reformer John Dewey called "scientific habits of the mind" nearly a century ago (Dewey 1909). He proposed these as the main rationale for compulsory science education, a rationale that today is usually called "critical thinking". Whatever term one chooses, it remains a mental state that has never come to pass in the general population. This is obviously a demanding definition of scientific literacy, and some may argue that it is designed to make such literacy unattainable by the public at large; but what it actually means is that the term has been used much too loosely in the past, and that this level of scientific literacy will likely be impossible to achieve in the foreseeable future. Notice that we do not require our scientific literate to have at his fingertips a wealth of facts, laws, or theories (the major conceptual schemes in science can virtually be counted on the fingers of both hands), or to be able to solve quantitative problems in science. Nor are advanced mathematical skills essential. All we should expect is that our literate person understand and appreciate the central role played by mathematics in science. Nevertheless, for reasons that will be discussed later, even this modest criterion puts scientific literacy beyond the reach of most educated individuals.

My estimate of the fraction of Americans who might qualify at this level is 4–5% of the adult population, nearly all being professional scientists or engineers. Miller places this number at about 6% by his benchmarks. It is worth noting that these low estimates of scientific literacy in the adult population are not confined to the United States; a companion study to Miller's conducted in the U.K. resulted in a 7% estimate.

FAILURE OF THE SCIENTIFIC LITERACY MOVEMENT

If we could achieve "true scientific literacy" in a reasonable fraction of the adult populations of our highly industrialized nations, say 20% rather than the present 5%, we should have much less concern about the public image of science and technology, or of society exercising sensible controls over these enterprises. With this core of knowledgeable adults participating in the democratic process we could have some confidence in the outcome. This is not to say that the problem would then be fully resolved, for so many of the so-called "science-based societal issues" are also emotionally burdened, which often moves them out of the arena of reasoned debate, but at least the science or technology involved should be better informed. The other two levels of literacy described above would not achieve the same purpose, for, in order to diminish public concern about science and technology, at least the educated public must learn to appreciate what these enterprises can do and how they work.

The reason for stressing adult literacy in science rather than student literacy is that good school performance, even a reasonably high level of scientific literacy while one is a student, is no guarantee that the individual will retain much science when he or she becomes a responsible adult in a position to contribute to the good of society. This is an important point, for whatever we do to create student interest in science, to make students aware of the world around them, and to get them to understand at least what the scientific enterprise is about, if not science itself, we are guided by immediate feedback rather than long-term retention. To be realistic about it, having literate students who turn out to be scientific illiterates as adults does not do society much good in a social sense. We know that the "staying power" of school science is very poor, but what is particularly sad about our failure to achieve any lasting form of scientific literacy is that all the surveys done on adult literacy in science not only prove that most adults are illiterate in science, but that they perceive themselves as being reasonably literate. Fully 70–80%

of American adults surveyed in a poll several years ago considered themselves interested in scientific and technological matters as well as concerned about government policy in science and technology, and rated their basic understanding of science and technology either as "very good" or "adequate" (Table 1). This great disparity between what the average adult really knows about science and what he/she thinks that he/she knows lies at the heart of the problem, for changing that perception is prerequisite to even thinking about ways to encourage adults that attaining scientific literacy may be worth the effort.

Table 1. Harris Poll (Poll conducted for U.S. Office of Technology Assessment in late 1986)

	very interested	somewhat interested	rather uninterested	not at all interested	not sure
How much interest do you have in scientific and technological matters?	23%	48%	11%	18%	1%

	very concerned	somewhat concerend	not very concerned	not at all concerned	not sure
How concerned are you about government policy concenring science and technology?	32%	50%	11%	7%	1%

	very good	adequate	poor	not sure
How would you rate your basic understanding of science and technology	16%	54%	28%	1%

Unfortunately, for reasons that have been explored previously, I think there is not the slightest possibility of achieving this level of scientific literacy in an appreciable fraction of the adult population (Shamos 1988a, 1988b). The basic reason is that becoming and remaining reasonably literate in science requires a special effort on the part of students, a commitment that very few nonscience students are prepared to make. Science is not easy; there is no getting around this. And the more we try to simplify it, the more we find ourselves moving away from those areas that comprise the essence of true scientific literacy. As far as the noncommitted student is concerned, science is difficult because of a) its cumulative nature, which makes it necessary to build one's understanding of it like a tall edifice, layer by layer; and b) its repeated failure to accord with "common sense", giving it a somewhat mystical quality. Indeed, in the view of many scientists and educators, it is its cumulative property that most distinguishes science from other forms of intellectual activity. As for common sense, while scientific inquiry generally begins with observations in the real, or everyday world and in the end returns to that world in the form of technology, i.e., in the form of practical things, the steps in between, where the real science is done, are largely unfathomable to all but specialists. We have no common-sense counterparts for photons, genes, cells, novas, or black holes. When scientists talk about these phenomena, they reason by means of models or abstractions: inventions that may seem reasonable to the scientist but which are more like "black magic" to the layperson.

Thus, taken together, the cumulative nature of science and its reliance on descriptions that often defy common sense make it necessary to devote extraordinary effort to the task of becoming scientifically literate. And we have said nothing yet about mathematics, the quintessential language of science, which so many students feel must have been invented to further complicate their studies rather than to make science more tractable. Nor have we said anything about the special mode of thought that is supposed to be characteristic of science. As we pointed out earlier, the habit of rational thought, or logical reasoning, or critical thinking, or "scientific habits of the mind", as John Dewey called it, is not inherent to our advanced societies – and never was. Hence, students discover still another source of discomfort in science: their fantasies, speculations, and ingrained misconceptions about the workings of the universe must now be replaced by factual observations, logical thinking, and reliance on sound evidence – a transition that most students find difficult, or are unwilling to make.

Coupled with the difficulty of learning science is the fact that there is no real incentive for students (or adults) to make the commitment necessary to become literate in science. Most individuals, certainly the vast majority of students who do not plan to become scientists, appear to be unwilling to make this effort. This should not be surprising, for, after all, what incentive do they have? As the sociologist David Harman points out, adult illiterates learn to read and write when they realize that being literate is in their own self-interest (Harman 1986). The same is certainly true of science; students and adults will evidence a genuine interest in scientific literacy only when they are convinced that becoming literate in science is for their personal good, not simply because science educators or commission reports tell them so. While enjoying the everyday comforts and benefits derived from science and technology, society has managed to insulate itself from any actual, or even perceived, need to understand its origins. The sad but simple fact is that one does not need to be literate in science (or mathematics) to be successful in most enterprises, or to lead the "good life" generally.

DIMINISHING PUBLIC CONCERN ABOUT SCIENCE AND TECHNOLOGY

We have shown that while there may be a level of scientific literacy that would cause the nonscientist to feel more comfortable about science and technology (true scientific literacy), the prospect of achieving this level in a reasonable fraction of the population is virtually nil. Education is not the only answer. Hence, we should look at the possibility of solving a subset of the overall problem, i.e., at specific failings in public understanding which appear to play a major role in fashioning public attitudes toward science, such as the following.

1) The public generally does not distinguish between science and technology. To identify its concerns properly, it must have a better understanding of the role of each of these enterprises in modern society.
2) The public has a very poor understanding of the meaning and use of risk/benefit analysis. Such understanding is essential to the decision-making process in a highly technical environment.
3) The public needs access to better advice and guidance on science/technology issues than that provided by the media or by self-serving individuals and organizations.

If these three relatively modest objectives could be achieved, substantial progress will have been made toward diminishing public concerns about science and technology, including bioscience. The first two are simple; they should be thoroughly treated in every science curriculum and constantly reinforced in the media through industry-sponsored public information campaigns. The last is far more difficult; for, while the public ultimately must turn to experts for advice on highly technical scientific matters, as it does in all other specialized fields such as law or medicine, establishing truly independent, unbiased panels of experts that could win the public trust, a sort of International "Science/Technology Watch", has thus far proved ineffective. There appears to be a mistrust of most scientists as advisors to the public. Table 2 shows some extracts from a recent study on public knowledge of science conducted in Australia by Schibeci (Schibeci 1990). Note that Aus-

Table 2. Public perceptions of its knowledge of science and technology (reproduced from *Bulletin of Science, Technology & Society*).

How well informed would you say most West Australian adults are in the general area of science and technology? Would you say it was . . .

Very well	0.8%
Fairly well	17.6
Neither well nor poorly	19.1
Not very well	48.4
Not at all well	12.5
Don't know	1.6

And how well informed would you say you yourself are?

Very well	9.0%
Fairly well	30.1
Neither well nor poorly	22.7
Not very well	48.4
Not at all well	4.7
Don't know	0.8

Generally do you think that science and technology do . . .

More good than harm	58.2%
Equal good and harm	34.0
More harm than good	5.1
Don't know	2.7

Table 3. Public perceptions of scientists (reproduced from *Bulletin of Science, Technology & Society*)

	agree strongly	tend to agree	neither agree nor disagree	tend to disagree	strongly disagree	don't know
Scientists often make statements that other scientists then oppose, so you don't know who to believe.	35.5%	44.1	5.5	9.4	3.9	1.6
Scientists are mostly concerned members of society who genuinely care about the future of the world.	9.4%	13.7	7.8	35.2	33.2	0.8
Scientists only make statements in newspapers to show people they are earning their salary.	3.5%	10.9	6.3	34.8	40.6	3.9
If a scientist is well known, or works for a well known organization, I would believe what he or she said.	14.1%	40.6	7.8	26.2	9.4	2.0

tralian adults hold a somewhat more realistic view than Americans of their shortcomings in science, but in common with most personal surveys they still believe that their own knowledge is better than that of their neighbors. The test sample, incidentally, did not have a university education, but most (62.5%) had gone beyond the minimum compulsory level of year 10. According to Schibeci, roughly the same results have been found in the U.K. But a more telling feature of this study is found in the last test item of Table 2 and in Table 3, dealing with public perception of science and scientists. What can be concluded from Table 3 is that most of the respondents were bewildered by what they perceive as "infighting" within the scientific community, and they tend to view scientists as uncaring about the future of the world. Also evident is the power of the press and public relations in shaping the image of science and scientists (last test item of Table 3).

Some way must be found to provide the public with sound advice on science-based questions, a reputable (and credible) forum that can help guide the public to informed choices rather than leave it at the peril of well-organized vested interests and scientific frauds. Developing such a vital interchange would serve not only the best interest of the public at large, but that of the science/technology community as well. This is by no means a simple problem, as all should appreciate, but it is one on which the scientific community (and this workshop) must focus its attention if we wish the public to feel more comfortable with what we do.

REFERENCES

Dewey, J. (1909). Symposium on the purpose and organization of physics teaching in secondary schools; Part XIII. *School Science and Mathematics* **9:** 291–92.

Harman, D. (1986). *Illiteracy: A National Dilemma.* New York: Cambridge Book Company.

Hirsch, E.D. 1987. *Cultural Literacy: What Every American Needs to Know.* Boston: Houghton Mifflin Company.

Hirsch, E.D., J.F. Kett, and J. Trefil, (1988). *The Dictionary of Cultural Literacy.* Boston: Houghton Mifflin Company.

Russell, B. (1960). The social responsibilities of scientists. *Science.* **131:** 391–92.

Schibeci, R.A. (1990). Public knowledge and perceptions of science and technology. *Bulletin of Science Technology and Society* **10:** 86–92.

Shamos, M. (1988a). A false alarm in science education. *Issues in Science and Technology,* **IV(3):** 65–69.

Shamos, M. (1988b). The lesson every child need not learn. *The Sciences (July/August): 14–20.*

Winner, L. (1977). Autonomous Technology. Cambridge, MA: MIT Press.

Bioscience ⇌ Society
edited by D.J. Roy, B.E. Wynne, and R.W. Old, pp. 361–376
© 1991, John Wiley & Sons

Private Enterprise Involvement in Bioscience and Technology: Liability or Asset?

Margaret N. Maxey

Biomedical Engineering Program
Murchison Chair of Free Enterprise
Petroleum/CPE 3.168
The University of Texas at Austin
Austin, TX 78712, U.S.A.

Abstract. As its underlying theme, this essay maintans that problems besetting bioscience and biotechnologies should be viewed as the offspring of conflicting risk perceptions derived from competing world views and preferred forms of social organization, rather than from measurable risks "out there" in a physical state-of-being. Despite dramatic bioscientific advances that have lengthened life expectancy and conquered major diseases, affluent citizens of Western industrialized societies have selected the risks induced by science and technology for paramount concern. Two distinct yet compentary explanations of this phenomenon have been proposed. One approach uses historical analyses of similar occurrences in the past; the other approach analyzes the cultural biases present among diverse constituencies, each with its preferred form of social organization. The involvement of private enterprise in bioscience and technology development will be an asset rather than a liability to the extent that its practitioners not only disengage its role from defenders and detractors alike, but also recapture its moral foundations expressed in a commitment to serve "a new social covenant".

SIGNS OF OUR TIMES

As we enter the final decade of the twentieth century and approach the end of the second millennium, two recent publications dramatize the pervasive influence exerted by conflicting worldviews now competing for our allegiance. The authors of *Megatrends 2000* invite us to conclude from the evidence that we are standing on the threshold of a global renaissance and that good economic times lie ahead (Naisbitt and Aburdene 1990). Megatrenders Naisbitt and Aburdene remind us that it is the optimists who "get things done", while cynics sneer and skeptics fantasize about "what-ifs". A second publication, by Hillel Schwartz, *Century's End: A Cultural History of the Fin de Siecle from the 990s through the 1990s,* examines evidence that the historical record is hostile and downright discordant with irrepressible optimism of this sort (Schwartz 1990). Nostradamus and other dead prophets of doom have eloquent counterparts in the 1990s.

Discussions of the present uneasy relationship between bioscience and society will be foreshortened if they do not take account of the origins of a deepening conflict between competing worldviews. The reassurance that humans in the past have always "muddled through" does little to diminish a premonition that the 1990s present stark choices etched far more antithetically "as good will or holocaust, ecology or extinction, higher consciousness or the end of (Western) civilization" (Schwartz 1990). Believe as we may that the end of the millennium does not mean the end of the world, it cannot be denied that a sense of impending doom seems stronger now than ever before in human memory. Neither can we deny evidence that this decade may culminate in a decisive passage from a natural world to a human world (Singer 1987). In a natural world mankind remains vulnerable to such ravaging afflictions as AIDS and Alzheimer's disease; in a human world bioscientific research and biotechnologies are expected to decisively pierce the veils of ignorance surrounding the genetic makeup of living cells and tissues, and reveal the root causes of human disease and death.

Whatever their enduring merit, these publications interject a salutary reminder that the quest for scientific truth as the reward of unbiased research cannot claim exemption from public trial-by-ordeal. Scientific truth must inevitably be filtered through a dominant cultural climate of opinion, variously exploited for political ends. We may prefer to believe that the objective quality of scientific fact is self-justifying, but we ignore at our peril the role now played by societal risk perceptions.

AMBIVALENT SOCIETAL RISK PERCEPTIONS

It would be a serious error to suppose that ordinary people exhibit risk perceptions dominated exclusively by expectations of impending doom. On the contrary, ambivalence reigns.

As a result of bioscientific research having achieved unprecedented breakthroughs in biotechnology, the way has been paved for a new quantum leap: nanotechnology. Segments of the public are aware that bioscientific research has resulted in striking innovations in five categories: human diagnostics; human therapeutics; agricultural biotech products; contaminant-monitoring devices; and specialty products for pollution control, bioremediation, and enzymes used in vitamins and detergents. Whereas biotechnology creates these so-called new life forms designed for specific purposes, the focus of molecular technology (nanotechnology) is upon the direct manipulation of individual atoms and molecules of matter. The Human Genome Project's goal is not only to create a precise "map" of the 40–100 thousand genes in human chromosomes, but also to "sequence" the three billion DNA molecules that comprise these genes. This project expects to yield our first detailed understanding of some four thousand genetic diseases, making a quantum leap in mastering their detection and treatment. Bioscientists envision microscopic computers comprised of proteins injected as tiny "robots" into the bloodstream, cleansing cholesterol deposits from arteries. Ultimately, molecule-sized mechanisms of cell repair could eradicate the root causes of all human disease, and even the aging process itself.

It might seem inevitable that this impressive array of achievements in bioscience and technology would assure public acceptance of innovations yet to come. Such is not the case. Social analysts express fascination with a puzzling anomaly: affluent citizens, whose quality of life and standard of living rank among the highest in the world today and who enjoy political freedoms and social amenities unparalleled in history, have nonetheless developed strongly negative attitudes toward bioscience and genetic engineering in particular, and modern high technology in general. All manner of social complaints are surfacing, leveled against those experts who are presumed to be in charge of, and responsible for, a modern "technological leviathan" regarded as out of control. They are referred to as a new technocratic elite who have a vested interest in pushing society toward ever more complex, sophisticated, and esoteric systems in which ordinary human values have become eclipsed (Winner 1977).

Time was when those skilled in the principles of science and engineering were esteemed for exerting control over a hostile world of nature (Florman

1976). For centuries these forms of knowledge were regarded as time-honored sources of safety, but in a remarkably brief period of thirty years, popular confidence in science has been transformed into skepticism and doubt. This puzzling cultural phenomenon has attracted a number of explanations, each falling within two distinct, yet complementary, approaches. One examines the historical record and extracts lessons with important implications for current and future strategies to cope more effectively with societal risk perceptions. The other approach analyzes differing types of social organization that are preferred by diverse constituencies, whose members embrace divergent interpretations of social justice and economic interests.

After examining each approach in turn, this essay will offer proposals for optimizing the role private enterprise should play in securing its essential contribution to bioscientific research and biotechnology.

RESCOPING THE PROBLEM OF SOCIETAL RISK PERCEPTION

In Western industrialized societies, life expectancy since the early 1900s has almost doubled; infant mortality has dropped dramatically; major causes of death are no longer communicable, parasitic diseases. These achievements raise troubling questions about risk perceptions among affluent citizens in technological societies. What underlying social values can account for a widespread political campaign to eliminate synthetic pesticides and petrochemical fertilizers which, in a brief 50 years, have tripled crop yields, reduced food costs more than 30%, and cleansed the environment of disease-bearing organisms? What social philosophy attempts to justify, with claims about protecting public health from toxic wastes, a virtual elimination of essential chemicals and proven radiotechnologies (CAT, SPECT, NMRI, PET) which enable medical professionals to detect, prevent, and cure major causes of human disease (Ray 1990)?

A Historical Analysis Approach

The most influential response to these philosophical questions appeared over a quarter-century ago. In the final chapters of Rachel Carson's *Silent Spring* we find five assumptions which have decisively shaped public risk perception (Carson 1962). She assumes that the magnitude and character of scientific innovations since World War II have given modern man unprecedented power to cause planetary impacts with adverse effects on generations yet to come. By their universal contamination of a natural environment, sinister chemicals in partnership with radiation are allegedly changing the very nature of the world and life as we know it. Second, nature has slowly, over eons of unhurried time, reached a fragile, precarious "balance" which, she claims, is increasingly vulnerable to the rapidity of man's technological encroachments. Consistent with her second assumption, Carson also asserts that existing carcinogens are not natural but man-made, "... for man, alone of all forms of life, can create [*sic*] cancer producing substances". Her fourth assumption maintains that exposures to repeated small doses of carcinogenic agents are far more dangerous than a single large dose, since a large dose may kill cells outright, whereas small doses allow cells to survive in a damaged condition, causing them to mutate and proliferate as alien cells. Since there is allegedly "no safe dose" of any exposure, morality dictates that the only safe dose must be "zero dose". Consequently, Carson concludes, substances which man has introduced into the environment not only can, but must, be removed as a moral imperative; the moral goal must be "zero pollution" so as to achieve zero exposure to man-made agents of disease and death.

These assumptions underlie an entire ethical framework governing societal perceptions and moral imperatives for social policy. Bioscientific research reveals, however, that this framework has been erected upon a scaffolding of beliefs which are seriously at odds with the actual status of scientific evidence. To be sure, there is an important distinction between "evidence" which persuades and "proof" which convinces. Nevertheless the evidence cannot be denied.

Whereas Carson told her readers that any natural carcinogens remaining in nature were "few in number" and, anyway, humans had long ago adjusted to their adverse effects, bioscientist Bruce Ames has produced overwhelming evidence of the existence, magnitude, and pervasive ubiquity of natural carcinogens (Ames 1983). The human diet contains a plethora of natural mutagens and carcinogens. Current preoccupation with relatively small amounts of man-made effluents from various industrial processes has diverted scientific research away from the most health-effective and cost-effective ways of reducing the perceived burden of cancer, namely, by making dietary and lifestyle changes. Renowned scientists have produced evidence that it is not among man-made agents that one should look for primary carcinogens (Handler 1979; Higginson 1979). Instead, the all-pervasive agent appears to be oxygen: it is a recognized mutagen; experiments have shown that it causes tumors in fruit flies; in the Ames assay test for screening carcinogens, it shows up positive (Totter 1980).

What a different world we would inhabit if regulatory scientists since the 1960s had rejected Carson's assumptions, and instead, had applied to natural carcinogens the same standards of research that were applied to industrial by-products. And what a radically different world we would now inhabit if the law of diffusion had been enshrined in the pantheon of regulatory science as an antidote to the quest for zero pollution. This law of physics – which governs the movement and intermingling in nature of all molecules of liquids, gases, and solids – requires the conclusion that nothing is completely uncontaminated by anything else. Hence the zero-threshold/one-molecule/one-hit theory becomes intellectually bankrupt when confronted with reality governed by the law of diffusion (Efron 1984).

Bioscientific research exposing the fallacy of popular "received wisdom" about public health has subsequently led William Clark to examine the psychology of public risk selection through the lens of witch hunting during two centuries spanning the Renaissance and Reformation (Clark 1980). His premise is that people and their fears have always been at the heart of the risk problem – fears of loss, fears of injury, above all, fears of the unknown. Ironically, in our much-heralded "Information Age", unknowns and uncertainties reign supreme.

Not unlike today's litany of sorrows, says Clark, the 14th and 15th centuries have recorded urgent societal pressures to explain wheat rotting in fields, sheep dying, vineyards afflicted with unseasonable frost, and a rising incidence of human disease and infertility. These phenomena precipitated the medieval church, as the locus of institutionalized expertise, to initiate the Inquisition, which was mandated to create a witch-free world. Through its creative and energetic efforts, Clark observes, "the rates of witch identification, assessment and evaluation soared. By the dawn of the Enlightenment, witches had been virtually eliminated from Europe and North America. Crop failures, disease, and general misfortune had not." In the interim, half a million people had been burned at the stake.

What Kefauver's Hearings achieved for drug regulators, and what Rachel Carson's *Silent Spring* has provided for environmentalists, witchcraft inquisitor Jacob Sprenger's *Malleus Maleficarum,* published in 1486, provided for the witch hunting business (Sprenger 1486). It mobilized individualized complaints into a collective consciousness that witches did in fact exist, wielding real power for evil. "As agents of Satan," observes Clark, "they were a heresy – a dangerous sin in need of eradication. Not individuals, but society as a whole was in peril as long as witches remained at large." Once the witch risk was "socialized" as a common threat, collective action taken against an individual was justified in the name of the common good. Ironically, the carbonization of half a million people was inflicted by the most elite and educated institution of its day.

As for the causal relation between assessment and risk, asks Clark, "Which is driving which? A strong case can be made for the notion that search effort creates the thing being sought." As a result, the higher discovery rate of witch risks justifies more search effort, in turn creating a self-amplifying process, with no "natural limitation" imposed by rules of evidence to verify an objective frequency of witch risks in the environment. As its principal tool for identifying witches, the Inquisition used torture. If an accused witch said she was not one, she was tortured until she confessed the truth. Not until a perceptive Inquisitor in 1610 ordered that the Spanish Inquisition could no longer use torture under any circumstances, and that accusations could not be considered unless supported by independent evidence, did the number of witch trials drop precipitously.

Rejecting historical determinism, Clark nonetheless draws instructive parallels between past witch hunting and today's risk assessment movement, urging us to pay more attention to "its earthly results rather than its heavenly intentions". First, just as the Inquisition was the growth industry of its day, offering exciting work, rapid advancement, and career opportunities, so today, the risk assessment movement exhibits elements of opportunistic careerism. Second, it is much more significant, though less firmly established, that the witch mania happened to coincide with a social crisis in late medieval society. It enabled authorities in elite institutions to shift responsibility for the crisis from both church and state to imaginary demons in human form. Not only were these institutions exonerated of blame for human suffering; they managed to transform themselves into indispensable protectors of mankind against a sinister enemy alleged to be omnipresent, yet difficult to detect.

The most significant insight gained from Clark's analysis of parallels between witch hunting and our current risk assessment movement can be gleaned from M.K. Matossian's recent publication, *Poisons of the Past: Molds, Epidemics, and History* (Matossian 1989). Her research in medical history and historical demographics reveals persuasive evidence of a strong correlation between food poisoning in cereal grains from fungal poisons or mycotoxins formed when climatic conditions are cold and wet, and such localized phenomena as mass hallucinations, witch persecutions, and panics. The most common form of fungal poisoning (ergotism) has two variants. Convulsive or "dystonic ergotism" is characterized by nervous disfunctions such as tremors, writhing, wry neck, muscle spasms, delusions, and hallucinations. "Gangrenous ergotism" may result in loss of fingers, toes, and limbs caused by a vasoconstrictive chemical. Using arguments from incidence, from regional prevalence, from exposure resulting from class differences in diet, and from elimination due to dietary changes, Matossian uncovers persuasive evidence for her position that "ergot poisoning was entirely responsible for the appearance of abnormal central nervous system symptoms" in communities of early modern Europe where bizarre behavior was interpreted as bewitchment. "To blame witch accusers and courts for witchcraft persecution," she insists, "is to mistake an effect for a cause. Witches were persecuted because harm had befallen a community, not just because there were people vulnerable to indictment and other people prone to indict them."

The significance of Matossian's timely research lies in a fundamental lesson. In a world devoid of bioscientific research bounded by rules of evidence, in a society searching for causal explanations among nonnatural agents over which some forms of social control could be exerted, witch hunting and persecution became a means of escape from an admission of powerlessness. Health effects which gave rise to witch hunting were not without foundation in reality, but because no rigorous scientific explanation of natural causes was available or socially acceptable, demonic causes were substituted. Nature's agents and processes remained concealed. Scapegoats prevailed. As Clark explains, medieval witch hunters were public demagogues, not scientific experts, and witch hunting had no rules of evidence permitting closure to public debate over the causes of bizarre behavior. "An accused witch was a convicted witch because of the methods adopted (torture) to find out whether she was." Similarly today, " a possible risk is an actionable risk" as soon as someone raises the possibility that it might be. This state of affairs continues because legislative and regulatory institutions have failed to promulgate laws, rules, or tests for identifying nonriskiness or "safety".

A Cultural Analysis Approach

When attempting to explain why educated people today differ in their perceptions and in the selection of risks they are disposed to accept or avoid, we might easily assume without much reflection that personalities account for different perceptions. However Douglas and Wildavsky in *Risk and Culture* (Douglas and Wildavsky 1982) suggest that the reason people pay attention to certain risks, and ignore others, is for the purpose of reenforcing and conforming to a preferred way of life and environmental quality already selected on other grounds. Cultural bias dictates risk selection.

In the current dispute about hazards from biotechnologies and perceived threats from bioscientific research, partisans accuse each other of serving the vested interests of the social institution they serve. One side accuses the other of being a lackey of the "industrial establishment", which is presumed to be pursuing profits at the expense of public safety now threatened by death-dealing contaminants. The other side responds by accusing its critics of joining the ranks of the "danger establishment" dedicated to destroying the economic foundations of material well-being that afford them the leisure time to be prophets of doom. Each side accuses the other of irrational bias, of misperceptions of "real risks", of subversion of the public interest.

Current disputes over the discrepancy between actual vs. perceived risks cannot be resolved if we cling to the view that real knowledge of an external world should be allocated to experts in the physical and biosciences, while mistaken perceptions pertain to the realm of personal psychology. According to this faulty division, dangers are assumed to be inherent in a physical state of affairs; hence the risks posed appear objectively ascertainable only by experts. This division attributes a bias toward selecting certain types of risk and not others to flaws in subjective personality traits, i.e., an individual is either a bold risk-taker or a timid risk-avoider. But a subjectivist view is quickly nullified by two questions: Why is it that experts disagree? Why does one and the same individual fear minute traces of alar in apple juice measured in parts-per-billion, to the exclusion of those dangers more immediately life-threatening, such as addiction to cocaine among children, or auto accidents, fires, and falls?

Only a cultural theory of risk selection as a product of cultural bias and social criticism can account for the anomalies we now encounter in affluent technological societies. The risks we select to control or mitigate, individually or collectively, are integral to the choices we make with respect to the best way to organize social relations, to protect shared values, and to devise institutional processes for achieving consent in the formulation of social policy.

Michael Thompson offers an enlightening illustration of how pervasively cultural bias leads to diverse ways of managing technological risk (Thompson 1986). In Britain a cultural bias in favor of compromise and negotiation identifies it as a "consensus culture", whereas in the United States a cultural bias in favor of intransigence and steadfastness entails the approach of an "adversary culture". This basic polarity is reflected in stylistic differences for managing risks of high technology. The British favor a "consultative" process conducted among scientific/engineering experts who alone can anticipate and prevent whatever might go wrong. Americans favor a legalistic process of writing statutory regulations which are expected to prevent something from going wrong – a process accessible to any nonexpert. Whereas a consensus culture will tolerate closure on decision making, an adversary culture demands openness and virtually endless challenges in the court system. As for differing relationships between leaders and those led, the political regime prevailing in Britain and many other Western European countries is one in which a deferential populace trusts its experts and permits its government to keep individuals and groups from getting out of line; but Americans can only seem to tolerate a "bottom up" leadership in which a truculent populace is relentlessly engaged in whistleblowing directed at government officials and other authority figures.

To understand how cultural biases influence consent in risk debates, such as those emerging from bioscientific research, Thompson describes distinct kinds of social individuals. The "sectist type", such as members in voluntary associations, are distinguished from the "casteist type", typified by those who function in bureaucratic settings. Both of these are distinct from the entrepreneur, representing a bias in favor of "wealth creation" and "employment opportunities" pursued by risk taking. More or less peripheral to any debate are "ineffectual" individuals and "autonomists", such as New York taxidrivers and British owner-driver haulage contractors. Two kinds of rationality are adequate for the British political regime of a "top down" consultative process: bureaucratic rationality (casteist bias) and market rationality (entrepreneurial bias), but in the American political regime, bureaucratic and market rationalities must strike a balance with the rationality of truculence (sectist bias) to garner consent. Thompson concludes that the iron law of consent ("you can't fool all of the people all of the time") holds true in risk debates for political regimes on both sides of the Atlantic. However, "in Britain you can fool the sects, the ineffectuals, and the autonomists all the time and get away with it, while in the United States you can fool only the ineffectuals and the autonomists".

Questions raised by cultural analysis of this kind are fundamental. Are public officials today regulating what is actually harming people, or only what appears to frighten people? Should hypothetical risks receive the same priority attention as empirically verifiable risks of harm? Should perception be regarded as equivalent to reality? Or does bioethics require a more rigorous method of determining appropriate risk management strategies?

In a cultural analysis similar to that of Thompson, Rayner and Cantor focus upon a different facet of these problems (Rayner and Cantor 1987). They call in question three dominant assumptions about societal risk management. First, they challenge an engineering concept which reifies "risk" as a measurable entity "out there" in a physical state-of-affairs. Second, while recognizing the necessity of assessing probabilities and magnitudes of undesired outcomes essential to a sound engineering decision, they insist that these considerations are irrelevant to societal choices about scientific and technological innovations. Third, they maintain that the critical question facing risk managers is not "How safe is safe enough?" but rather, "How fair is safe enough?"

In contrast to engineering concerns, Rayner and Cantor have rephrased the question to suggest that what ordinary people care about are trustworthy institutions, procedures to assure informed consent, and adequate compensation for harmful effects, should a project fail. Different constituencies have a cultural bias toward preferred forms of social organization with differing goals; hence, members of each type choose a different liability distribution. Each type entails vastly different expectations about the proper procedures for securing consent, for judging adequate compensation and liability, and for eliciting trust in institutions responsible for managing potential threats to societal health and safety.

Rayner and Cantor outline a typology of four ideal types of social organization. First, entrepreneurial individualism functions in a competitive market system which keeps to a minimum the rules or prior claims of others restricting social behavior. Hierarchical or bureaucratic forms of organization function with restrictive controls exerted by formal systems of accountability. A third type, often in opposition to both competitive individualism and bureaucratic hierarchy, is the collectivist egalitarian constituency, typified by religious sects or revolutionary political groups. A fourth constituency is comprised of atomized and often alienated individuals.

In the debate over bioscience and biotechnology options, a deeper understanding of risk conflicts among these constituencies requires us to recognize that no one risk management solution can satisfy those already divided because of cultural biases toward social justification, diverse liability demands, and the divergent goals they pursue and people they trust. Each constituency exhibits differing risk management strategies.

Competitive-individualist risk takers depend upon social consequences-based justifications for achieving their goal of market success in responding to revealed preferences of clients and consumers. They invest their confidence in troubleshooting individuals (e.g., Red Adair and Lee Iaccoca). They favor a loss-spreading approach to liability.

Members of bureaucratic-hierarchical organizations rely on a social contract principle of justification to achieve their goal of system maintenance. Their trust is placed in long-established formal organizations (AMA, FDA). Bureaucracies use redistributive mechanisms, e.g., taxation, to apportion costs and spread liabilities among those constituencies whose stability is essential to bureaucratic survival.

Since egalitarians distrust both bureaucracies and entrepreneurial individuals, they favor rights-based justifications for decisions they hope will achieve their goal of a "new social order". Their trust is in town meetings and affinity groups. They favor the moral calculus of a strict liability fault system, rather than market or distributive methods of loss-spreading favored by entrepreneurs or bureaucrats.

A cultural analysis approach reinforces the view that risk conflicts among diverse constituencies are not rooted in physical entities. They emerge from competing worldviews which entail basic disagreements about ethical principles that should govern consent, trust, and compensatory liability. This insight has practical implications for our examination of the appropriate involvement of private enterprise in bioscientific research and its technological applications.

PRIVATE ENTERPRISE INVOLVEMENT IN BIOSCIENCE: LIABILITY OR ASSET?

It is fair to say that the political agenda of Western industrialized societies in the past quarter century has been driven by opponents of economic growth who advocate a "wholly new ethic". Its advocates assert that a violent, plundering mankind must abandon its destructive ways, propelled by an economic system of private enterprise, and begin forging a sustainable society. The ideological vision of the environmental or green movement, now entering its third stage with deep ecology goals, has pronounced nature sacred and untouchable. To make the desert bloom is the greatest sin. Its credo is succinctly captured by V. Postrel: "Technology is too complicated, work too demanding, communication too instantaneous, information too abundant, the pace of life too fast. Stasis looks attractive, not only for nature but also for human beings" (Postrel 1990). Its sectist prescriptions are straightforward: restrict trade to local areas. Eliminate markets and end specialization. Relocate individuals in "bioregions" defined by environmental features. Shrink population. Restore life to simple, small, self-contained enclaves. Abandon the spirit of inquiry that drives the scientific process. The world must be kept simple, simply understood, or "so subjective," says Postrel, "that it cannot be understood".

The underlying theme of this essay maintains that problems besetting bioscience and biotechnology should be viewed as the result of conflicting risk perceptions dictated by competing worldviews, rather than by measurable risks "out there" in a physical state-of-being. Bioscientific research, by its very nature, must engage in penetrating so-called sacred regions of life and death, evoking reminders of the symbolic fates of Prometheus and Pandora, Faust and Frankenstein. It is to be expected that those committed to an apocalyptic worldview, focused on survival in a sustainable society, vigorously oppose bioscience and applied technologies. They exhort us to adopt a new environmental ethic which derives its ethical norms for social policy from nature's ecosystemic harmonies. An old ethic of utility with its anachronistic pursuit of "the greatest good for the greatest number", appears to sustain a political agenda in disrepute.

The cultural climate of opinion confronting us in the 1990s stands in sharp contrast to the momentum of achievements and future promise of bioscientific research. However, the lessons of history and cultural analysis offer useful conceptual tools. For two centuries practitioners of private enterprise have been the catalyst for social change induced by technological innovations, as well as the driving force "riding point" in the struggle to overcome the facticities of nature. Private enterprise has, with uneven success, pursued its social role in the face of counterforces protecting the status quo – chiefly, bureaucratic-hierarchical organizations sustained by a casteist rationality, favoring inertia and complacency as antidotes to the uncertainties of risk-taking.The social role of private enterprise can, and must, continue to be an asset, despite opposing forces who regard it as a liability.

The current ascendancy of the green movement represents still another opposing force, typified by a variety of egalitarian-collectivist groups imbued with a sectist rationality and reformist social goals of an ill-defined "sustainability".

The involvement of private enterprise in bioscientific research is not a question of "whether", but rather of "how" and "why". Because of its market rationality, its goals of serving individual needs and preferences, and its social justification through tangible consequences, private enterprise functions at the effective center of society, where it alone embodies the cultural bias capable of fulfilling the promises of bioscientific research. Its social role will be a liability if it fails to understand the true nature of opposing forces, and trivializes the symbolic power of discourse based on philosophical assumptions held as tenaciously as articles of faith.

Private enterprise must disengage its identity from defenders and detractors alike, who have reduced it to an economic system capsulized in academic economic theory and teachable as a behavioral science. Private enterprise must recover its moral foundations, directly traceable to the linkage between the dawn of conscience, individual freehold, and the spirit of individual enterprise. Paul Johnson has eloquently articulated these foundations (Johnson 1980). Rooted in Greek and Judeo-Christian concepts implying that every individual is responsible for his or her actions, hence, "free" in a moral sense, the ideas of freehold property coupled with inalienable rights of individuals have slowly matured over time, induced by the notion of equality before the judgment of God and, by derivation, before the rule of law. After Erasmus had linked the individual conscience to the spirit of individual enterprise for the first time, the grip exerted by a frozen status society has slowly given way to economic emancipation for ordinary working people, who began to be rewarded for their fundamental freehold property – energy and skill. In short, private enterprise has continued to create a multiplicity of rival power centers, keeping in check not only statist bureaucratic powers, but also oppressive economic monopolies. By virtue of the self-correcting mechanism of individual freehold, the cultural bias of private enterprise empowers a democratic society with its greatest asset.

Apocalyptic articles of faith, attempting to re-locate "the sacred" in an inviolable nature, prohibiting scientific intrusions, must be repudiated as a return to primitive animism. Theologian Richard Neuhaus is unequivocal: "... the location of the sacred in man is the greatest achievement of human history. Upon this foundation rests the whole construct of humanism, including Christian humanism" (Neuhaus 1971). A refusal to recognize nature's causality – genetic, physiological, cosmological – in inflicting widespread disease and suffering on entire populations is not only fallacious but ethically untenable because of its consequences for human rights.

To remain an enduring asset, practitioners of private enterprise must not only articulate, but also demonstrate, a moral commitment to serve "a new social convenant" (Sturm 1986). This commitment transcends the goals of sectist egalitarian-collectivities, of casteist bureaucratic hierarchies, and of atomized-alienated individuals. Market-oriented enterprise alone can serve the needs and preferences – even the vices and virtues – of all ordinary citizens. Private enterprise should not be expected to produce a perfect society, much less a perfectly just one. It can only provide an abundance of material goods, a conquest of human disease, and a range of freedom for coping with human problems according to each individual's capacity (Kristol 1978).

REFERENCES

Ames, B. (1983). Dietary Carcinogens and Anticarcinogens. *Science* **221:** 1256–1264.

Carson, R. (1962). *Silent Spring*. Greenwich, CN: Fawcett.

Clark, W.C. (1980). Witches, floods, and wonder drugs: Historical perspectives on risk management. In: *Societal Risk Assessment: How Safe Is Safe Enough?* eds. R. Schwing and W. Albers, pp. 287–318. New York: Plenum Press.

Douglas, M. and A. Wildavsky (1982). *Risk and Culture*. Berkeley, CA: University of California Press.

Efron, E. (1984). *The Apocalyptics: Cancer and the Big Lie*. New York: Simon and Schuster.

Florman, S. (1976). *The Existential Pleasures of Engineering*. New York: St. Martin's Press.

Handler, P. (1979). *Dedication Address*. Evanston, IL: Northwestern University Cancer Center.

Higginson, J. (1979). Cancer and environment: Higginson speaks out. Reported by T. Maugh. Research news. *Science* **205:** 1363–1366.

Johnson, P. (1980). Is there a moral basis for capitalism? In: *Democracy and Mediating Structures,* ed. M. Novak, pp. 49–58. Washington, D.C.: American Enterprise Institute for Policy Research.

Kristol, I. (1978). A capitalist conception of justice. In: *Ethics, Free Enterprise, and Public Policy,* eds. R.T. DeGeorge and J.A. Pichler, pp. 57–69. New York: Oxford University Press.

Matossian, M.K. (1989). *Poisons of the Past: Molds, Epidemics, and History*. New Haven, CN: Yale University Press.

Naisbitt, J. and P. Aburdene (1990). *Megatrends 2000*. New York: Morrow.

Neuhaus, J.R. (1971). *In Defense of People*. New York: MacMillan Company.

Postrel, V. (1990). The green road to serfdom. *Reason* **22/2:** 22–28.

Ray, D.L. (1990). *Trashing the Planet*. Washington: Regnery Gateway.

Rayner, S. and R. Cantor (1987). How fair is safe enough? The cultural approach to societal technology choice. *Risk Analysis* **7/1:** 3–9.

Schwartz, H. (1990). *Century's End: A Cultural History of the Fin de Siecle from the 990s through the 1990s*. New York: Doubleday.

Simon, J. (1981). *The Ultimate Resource*. Princeton, NJ: Princeton University Press.

Singer, M. (1987). *Passage to a Human World*. Indianapolis, ID: Hudson Institute.

Sprenger, J. (1486). *Malleus Maleficarum. Hammer of Witches,* trans. by M. Summers, London, 1928.

Sturm, D. (1986). Toward a new social covenant: From commodity to commonwealth. In: *Christianity and Capitalism: Perspectives on Religion, Liberalism and the Economy,* eds. B. Grelle and D.A. Krueger, pp. 91–108. Chicago, IL: Center for the Scientific Study of Religion.

Thompson, M. (1986). To hell with the turkeys! A diatribe directed at the pernicious trepidity of the current intellectual debate on risk. In: *Values At Risk,* ed. D. McLean, pp. 113–135. New Jersey: Rowman & Allanheld.
Totter, J. (1980). Spontaneous cancer and its possible relationship to oxygen metabolism. In: *Proceedings of the National Academy of Sciences.* Washington, D.C.: National Academy of Sciences.
Winner, L. (1977). *Autonomous Technology. Technics-out-of-Control as a Theme in Political Thought.* Cambridge, MA: MIT Press.

Standing, left to right: Kenneth Warren, Wilhelm Girstenbrey, Patrick Vinay
Seated, left to right: Alexander Keynan, Margaret Maxey, Morris Shamos, Sheila Jasanoff
Not shown: James Wyngaarden, Horace Judson, Robert Anderson

Bioscience ⇄ Society
edited by D.J. Roy, B.E. Wynne, and R.W. Old, pp. 379–390
© 1991, John Wiley & Sons

Group Report
What Actions Are Required to Improve the Present Uneasy Relationship Between Bioscience and Society?

Rapporteur: S.S. Jasanoff
R. McC. Anderson
H.F. Judson
W. Girstenbrey
A. Keynan
M.N. Maxey
M.H. Shamos
P. Vinay
K.S. Warren
J. Wyngaarden

INTRODUCTION

The task of our group was to explore how, and whether, we can give operational meaning to the concept of a new covenant (or, more neutrally, relationship) between bioscience and society. Specifically, taking into account recent advances in the life sciences and their technological applications, what recommendations could we make to improve the "present uneasy relationship" of bioscience with society? Of necessity, our discussion of these questions had to encompass the descriptive as well as the prescriptive. We ranged from observations about the nature of the science-society relationship to possible strategies for alleviating present – and, we hope, future – tensions. Our conclusions with regard to the problems and opportunities confronting bioscience and society are subsumed under three headings: 1) public understanding of science; 2) public role in the direction of science; and 3) social responsibility of scientists. The dominant themes that emerged from our discussion under all three headings were *pluralism* (that is, the existence of multiple voices and viewpoints) and *dialogue*.

PUBLIC UNDERSTANDING OF SCIENCE

Negative public reactions to biotechnology, and more broadly, to the biosciences, are frequently blamed on deficiencies in the public understanding of science. Accordingly, recommendations are often made to educate the public, so as to increase its knowledge, awareness, or appreciation of science. We attempted to refine these assumptions based on a consideration of three aspects of public understanding: scientific literacy; communication of science; and public perceptions of risk.

Scientific literacy

The notion that a greater level of "scientific literacy" (as conventionally defined) among the general public would diminish public concern about bioscience was questioned by members of our group, though not necessarily on the same grounds. Some group members saw the goal of scientific literacy as important in principle but unattainable in practice. Others wondered whether the goal itself was relevant to the problems confronting this conference. The following positions were articulated.

a) According to one view, chiefly represented by M. Shamos, developing scientific literacy is impractical because it would be difficult, if not impossible, to persuade members of the public that it is in their best interest to acquire and maintain a high level of literacy throughout their adult lives. Even if it were possible to achieve widespread literacy, such a massive undertaking would not have noticeable impact for about two generations, i.e., in less than 40–50 years. Improvements in science education are, of course, essential to the welfare of any technologically advanced nation, most notably for the sake of meeting society's need for scientific and technical manpower. These goals, however, should be independently addressed; they are not directly related to the present tensions involving the biosciences and society.
b) Another view, expressed by R. Anderson among others, was that the emphasis on scientific literacy, if it is too narrowly conceived, may be misplaced, because it encourages us to focus on factual knowledge and knowledge claims, whereas public concerns are almost invariably directed either at the applications of knowledge, or else at the impact of the scientific worldview on personal values and religious beliefs. The questions that the public asks about bioscience would not be asked by children in

the schoolroom; they are adult questions about the economic, social, and ethical implications of the use of knowledge.

c) In any event, evaluating what the public knows or understands calls for caution. The "89 percent story" from the European community is instructive. A recent Danish survey sought to replicate an earlier survey of public attitudes to biotechnology by the U.S. Office of Technology Assessment. Asked to rate how well informed they were about biotechnology, 89% of the respondents said they were well or very well informed. In a roughly contemporaneous survey done in Ireland, the question of understanding was put differently. Respondents were asked to state in their own words what they understood by "biotechnology". This time 89% confessed ignorance, saying they knew nothing about biotechnology; only 2% said it had something to do with genes.

Recommendations. Although our group differed in its assessment of the problems, we nevertheless agreed on certain broad principles that should guide future efforts to increase the understanding of science in society. We also agreed that such efforts should be directed not only toward the general public but also toward scientists themselves.

a) Efforts to educate the public should focus on the "awareness" of science. Science education should seek to impart not merely facts, but an appreciation of the methods and limitations of science, the differences between science and technology, and the risks and benefits of technology. Such awareness has to begin in the classroom with basic science courses. But awareness in adults can and should be fostered through channels in addition to formal education – such as newspapers, science journals, science museums, and television programs. These avenues should be actively pursued, even though their reach may be limited to attentive and interested members of the public.

b) The adult audience for science includes several different publics, and educational efforts should be targeted accordingly. For example, awareness of science is important to public interest groups, legislators and regulators, the mass media, and the scientific community itself. Each of these relevant publics has its distinctive knowledge needs and capacity to use knowledge.

Communication of science

The public understanding of science has sometimes been construed too narrowly as "scientific literacy". Similarly, the need to communicate science to the public has been equated all too often with experts making scientific facts available to the public. This view of communication is based on an oversimplified and positivistic views of "facts" (Douglas and Windavsky 1982; Johnson and Covello 1987; Jasanoff 1990) as well as an exaggerated appraisal of the authority of "experts" (Nelkin 1984; Collingridge and Reeve 1986). It also underestimates the "social illiteracy" of scientists and the need for scientists to learn from the nonscientific public (Krimsky and Plough 1988).

Recommendations. We agreed that the notion and practice of communicating science should be broadened in the following ways.

a) Communication about science and its uses should be a two-way street – a continuous dialogue rather than an authoritative monologue, which J. Wyngaarden described as "dropping leaflets" from a propaganda airplance. Scientists should learn to listen to public fears, wishes, and concerns. Like medicine, science should strive to be more "caring". Through this dialogue, the public in turn should learn more about the aspirations, ideals, potential, and limitations of science.
b) Communicating science requires facts to be understood in context. Science has to be seen in the round, as a nested construct consisting not only of claims and counterclaims, but also of the methods and techniques that led to their production, the organizations of scientists and scientific institutions that carried out the underlying research, and the social understandings and imperatives that created the demand for new knowledge (Barnes and Edge 1982). Effective communication of science requires public discussion of all of these dimensions, both between science and society and within the scientific community.
c) Information should be backed up by responsible governmental action. As discussed further below, public confidence depends not only on the availability of facts, but perhaps even more on the existence of trustworthy institutions capable of processing and responding to facts.

Public perceptions of risk

Our discussion of public perceptions of risk represented in microcosmic form a larger and still unresolved societal debate. Some in the group were frustrated by the public's apparent disregard for expert assessments of risk and its seemingly irrational tendency to focus on eliminating smaller risks before larger ones (Zeckhauser and Viscusi 1990). Others saw the problem of risk management more in political than in psychological terms, attributing past failures to such factors as the public's lack of trust in experts and government agencies.

According to an account offered by M.N. Maxey, risk management efforts in the U.S. have repeatedly run afoul of the public's general commitment to a zero risk policy, which impedes rational comparison of risks and cost-effective policy-making. Seeking explanations for this alleged state of affairs, some group members suggested that the public, and even segments of the scientific community, are inadequately trained in probabilistic thinking and are unable to draw meaningful comparisons or otherwise distinguish among extremely low-probability events. However, B. Wynne and others challenged the underlying assumption that the public routinely demands zero risk. Research on risk perception suggests, on the contrary, that the public is prepared to accept technological hazards, provided that risk management practices satisfy requirements of openness, candor, and accountability (Wynne 1989).

We agreed, in any event, that if the public feels unable to evaluate a risk, it will resort instead to evaluating the regulator. To maintain public confidence, regulatory institutions must be capable of making credible decisions in the face of uncertainty. The controversy over the proposed Hoechst insulin plant in Germany and the early controversies over the Ice Minus and Frostban bacteria in the U.S. (Krimsky and Plough 1988) illustrated the importance of adopting timely regulatory mechanisms. In each case, the absence of such a structure left scientists vulnerable to public distrust and rejection.

In contrast, the establishment of the NIH Recombinant DNA Advisory Committee (RAC) and the subsequent adoption of containment guidelines for DNA research were in many respects a regulatory success story (Swazel et al. 1978). The NIH's action effectively defused pressures for federal legislation which the scientific community might well have found too restrictive. As J. Wyngaarden noted, even a decade later the NIH director reaped good will from Congress because of this highly visible act of scientific self-regulation. Indeed, the RAC mechanism may have been perceived as almost too successful, for it lulled many scientists and policy-makers into thinking that the regulatory problems of bioscience had been solved, when in fact only a narrow band of issues (laboratory research involving genetic engineering) had been adequately addressed.

These considerations led our group to conclude that, as regards the risks of bioscience, regulatory agencies are the key mediators between science and the public. In most of the industrialized world, these agencies are the entities that bear primary responsibility for developing methods of risk assessment, weighing risks against benefits, and eliminating unreasonable risks.

Recommendations. We therefore agreed that scientists' efforts to improve the quality of risk decisions should focus on the mediating role of regulatory agencies and on the legislators to whom the agencies in turn are accountable. This communication should be an essential part of the dialogue between bioscience and society. The scientific community can make a positive difference by providing policy-makers with good technical analyses, including comparative assessments of risks and improved methods of risk assessment. However, in a democratic society it is ultimately the policy-maker's duty to devise informed risk reduction strategies and to persuade the public, where necessary, to change its risk preferences in the light of new scientific knowledge.

PUBLIC ROLE IN DIRECTING SCIENCE

Unease about the biosciences is conditioned in large part by misgivings about the uses to which scientists will decide to put their knowledge. We therefore thought it important to look at mechanisms for increasing the public's control over – or even ownership of – new knowledge. We concluded that the issues of greater public empowerment were substantially different, depending on whether research is publicly or industrially funded and depending on the realtionship between the producers and the consumers of research.

Publicly funded research

In an era of heavy public investment in scientific research, particularly in the life sciences, it is scarcely controversial to suggest that the public should have a large role in determining research priorities (Chubin and Hackett 1990). Thus, it is appropriate for the public to indicate which problems it wants science to address – cancer, AIDS, overpopulation, environmental pollution – and with what level of monetary support. Scientists, too, have an indispensable part to play in setting priorities, principally by selecting feasible and interesting research projects within broadly defined subject areas. It is generally accepted that scientists should be free to determine their own research practices and protocols except where ethical considerations intervene, as in studies involving human or animal subjects.

There are substantial cross-national differences in the research funding system's responsiveness to the public will. At one end of the scale is the U.S. process for allocating funds to the National Institutes of Health (NIH) for biomedical research. In this highly porous system, groups of virtually any composition, from powerful lobbies like the American Cancer Society to single, concerned individuals, can gain the ear of Congress and have an impact on appropriations. The case of Irene Plescher is illustrative, though undoubtedly extreme. Driven by concern for her two children, who were suffering from a rare and debilitating skin disease, this determined mother persuaded the U.S. Congress to approve $1 mllion to study the condition. Rather exceptionally, as noted by J. Wyngaarden, this investment led to a scientific discovery and some improvement in treatment.

Funding for research in Germany is significantly less responsive to individuals and groups in society. Scientists determine spending priorities with respect to the DM 1 billion budget administered by the Deutsche Forschungsgemeinschaft. The DM 1.1 billion budget for the Max Planck Institutes is also insulated from public control. (These funds are used, in any case, to support individual researchers rathers than projects.) Finally, the DM 6 billion budget of the Federal Ministry for Research and Technology (BMFT) is subject to parliamentary control, but is far more shielded from direct public input than the extraordinarily open, 8–12 week-long NIH hearings.

British biomedical funding, channeled in part through the Medical Research Council and in part through private foundations, likewise provides only minimal opportunities for public influence.

Participants in our discussion, regardless of their national origin, generally favored the NIH funding model in spite of its flaws. The relatively high level of political support for biomedical research in the U.S. was thought to be a direct consequence of the openness and responsiveness of the process. Nonetheless, the NIH model has drawbacks that were explicitly noted. The principal "warts" in the system are as follows:

Waste: New programs require 4–5 years to produce work that is up to the NIH level; money is wasted when the public sets problems that are far beyond science's current ability to answer.

Rigidity: Administrative costs and burdens increase along with levels of public funding; programs once instituted are virtually impossible to eliminate.

Fickleness: Result-oriented funding can be withdrawn if the public's demands are not met in a timely fashion; political rather than scientific considerations determine what counts as "timely".
Institutional hypocrisy: Since the public supports results, not basic research, NIH is in the business of funding science while selling technology (e.g., in testimony to its congressional sponsors). This creates a potentially troubling distance between what bioscientists really do and what they are perceived to be doing, engendering unrealistically high expectations concerning basic research.

Industrially funded research

A substantial proportion of bioscientific research and development around the world is funded by industry. In the U.S., in particular, about 40% of the $20 billion spent annually on health-related research and development comes from industry and is directed primarily toward product development. Accordingly, the fruits of bioscientific research that the public sees are often products of investment by industry. Yet there are few formal channels by which the public can participate in directing where and how industry makes expenditures for research and development.

To date, market research has been the primary means by which industry seeks to incorporate the public's views and values into developing biotechnological products. This approach has resulted in several notable commercial and public relations failures. Examples include the early U.S. attempts to field-test genetically engineered microbial pesticides and the more recent effort to market bovine somatotropin (BST) in Europe and the U.S. Such failures should lead scientists working in industry to consider seriously whether there are more effective ways of increasing public participation at the development stage of biotechnological research and development.

Participation by developing countries

It is widely believed that foreseeable advances in the biosciences will have their most dramatic impacts in developing countries. Members of our group strongly urged that the power of bioscience should be used to alleviate the problems of the developing world. Any new understanding between science and society should include ways of giving the currently disempowered end-users of bioscience a real voice in directing the course of research and development.

A major obstacle to such participation is the dearth of institutional relationships between bioscientists working at the frontiers of knowledge in industrial societies and the potential beneficiaries of their enterprise in developing countries. UNIDO's coupled biotechnology centers in New Delhi and Trieste provide a possible model for future efforts. Another positive example is the recent collaboration between the World Health Organization's Tropical Diseases Research program and the U.S.-based Rockefeller Foundation's Great Neglected Diseases program; this initiative paired 16 of the best laboratories of nations in the South with outstanding counterparts in the North.

SOCIAL RESPONSIBILITY OF SCIENTISTS

In light of the preceding analysis, our group concluded that the overriding new responsibility for bioscientists in the coming decades will be to foster, and participate in, constructive dialogue with the public about scientific research and its social applications. We identified several areas where there is need for scientists to reassess their social reponsibilities.

Interaction with the media

Whether because of national traditions or professional acculturation, many bioscientists are reluctant to leave the quiet of the laboratory to communicate with journalists about what the scientists regard as unfounded fears or fictional scenarios. As a result, marginal scientists or false prophets may dominate communications with society – leading to the distortions that many U.S. bioscientists have attributed to biotech adversary Jeremy Rifkin or, in Germany, to Jost Herbig, author of *Die Gen-Ingenieure.*

Our group felt that scientists should accept as a matter of principle that they have an obligation to communicate with journalists and the general public about the implications of their research. To this end, the scientific community should participate more actively in existing networks of communication, such as the Scientists' Institute for Public Information (SIPI) and its Media Resource Service (now operating in Britain as well as the U.S.). Other countries should explore the possibility of establishing similar networks. Scientists should also train themselves to address public concerns in generally understandable language.

A minor caveat was expressed with respect to younger scientists, whose already difficult career paths could be intolerably burdened by more extensive obligations to interact with the public. Efforts to raise the social and ethical awareness of young scientists should instead be incorporated into graduate and postgraduate training programs – for example, through required or strongly recommended courses in science writing, research ethics, or social studies of science and technology.

Uncertainty and provisionality

Policy-makers and the public frequently desire certainty, but science in most cases can deliver only provisional knowledge. The gap between expectation and reality creates tensions for all scientists, including those in the biosciences. Beset by impossible social demands and seemingly wilful misinterpretations of data, many scientists have expressed a longing for authoritative institutions, such as "science courts" (Kantrowitz 1967), which could provide the public with dispassionate and fair readings of the state of scientific knowledge. An idea proposed during our discussions by M. Shamos was for one or more national "science watches", perhaps constituted under an international umbrella, whose function would be to monitor activities in the biosciences and to keep the public informed, in as impartial a manner as possible, of the limitations and significance of new technologies. At a minimum, such bodies should be able to project sufficient expertise and independence to win the confidence of the media and of policy-making bodies.

The feeling of the group was that all efforts by scientists to address questions of trust and credibility should in principle be applauded. At the same time, the vision of detached , impartial bodies "speaking truth to power" was viewed as unrealistic (Jasanoff 1990). The experience of consensus development conferences at the NIH shows how difficult it is to achieve unanimity, even when technical issues are carefully delimited and participants share a common level of expertise. It would be foolhardy to expect any single institution to prevail as the sole voice of reason in the complex, morally charged debates between bioscience and society. In a pluralistic world, bioscientists should be prepared to accept conflict and dissent as the most likely pathways to a robust agreement.

Technology transfer

Science sometimes claims to be pure knowledge, shining above the realm of politics, apparently free of social, moral, or ideological commitments. However, science impinges upon society primarily through the development of technologies that may modify the environment, alter our ways of life, challenge our personal and collective ethics, and perhaps create as many new problems as they solve.

Technologies, moreover, are products of social interaction and are shaped by contextual factors – economics, law, political organization – that tend to make them culturally specific. A technological solution for a problem in one society (e.g., birth control) may thus be wholly unacceptable in a different cultural context. Bioscientists accordingly should take care not to use technology as an indirect means of imposing their own social and cultural views on other societies. Technology transfer should be grounded in a respect for cultural differences.

Furthermore, technologies chosen (or not chosen) by one society may have a profound impact on other societies and even on the entire planet. Such effects are especially likely in connection with the ecological applications of bioscience. Accordingly, there is a need for establishing institutional networks that would allow countries with diverse cultural traditions to establish and maintain continuing dialogue with respect to each other's scientific and technological choices.

Setting science's house in order

It is important, finally, to recognize that public anxiety about modern bioscience is closely linked to several structural and philosophical changes that are fundamentally reshaping this area of scientific and technological activity. We considered the following to be worthy of special attention: the growing interpenetration of academic and industrial research, with potentially negative consequences for openness in scientific communication; the commodification of knowledge as intellectual property; increased opportunities for conflicts of interest, particularly for researchers using public funds; visible and recurrent episodes of fraud and misconduct in the biomedical sciences (Broad and Wade 1982; Kohn 1986); and the use of bioscience for military purposes. All of these developments present scientists in a very different guise from that of disinterested seekers after truth or bringers of Promethean fire.

We were unable to explore in detail how bioscientists should confront and deal with these negative attributes, images, and associations. It seems clear, however, that if these transformations persist, then bioscientists will increasingly be challenged in public forums to account for real or perceived deviations from the norms of science as commonly understood by the public.

In such confrontations, humility and an acknowledgment that modern science must be answerable to society will serve bioscientists better than arrogance.

REFERENCES

Barnes, B. and D. Edge (1982). *Science in Context.* Milton Keynes, U.K.: Open University Press.

Broad, W. and N. Wade (1982). *Betrayers of the Truth.* Oxford: Oxford University Press.

Chubin, D.E. and E.J. Hackett (1990). *Peerless Science: Peer Review and U.S. Science Policy.* Albany, NY: State University of New York Press.

Collingridge, D. and C. Reeve (1986). *Science Speaks to Power.* London: Pinter.

Douglas, M. and A. Wildavsky (1982). *Risk and Culture.* Berkeley, CA: University of California Press.

Jasanoff, S. (1990). *The Fifth Branch: Science Advisers as Policymakers.* Cambridge, MA: Harvard University Press.

Johnson, B. and V. Covello, eds. (1987). *The Social and Cultural Construction of Risk.* New York: Reidel.

Kantrowitz, A. (1967). Proposal for an institution for scientific judgment. *Science* **156:** 763–764.

Kohn, A. (1986). *False Prophets: Fraud and Error in Science and Medicine.* Oxford: Basil Blackwell.

Krimsky, S. and A. Plough (1988). *Environmental Hazards: Communicating Risks as a Social Process.* Dover, MA: Auburn House.

Nelkin, D., ed. (1984). *Controversy,* 2nd edition. Beverly Hills, CA: Sage.

Swazey, J.P., J.R. Sorenson, and C.B. Wong (1978). Risks and benefits, rights and responsibilities: A history of the recombinant DNA research controversy. *Southern California Law Review* **51:** 1019.

Wynne, B. (1989). Frameworks of rationality in risk management: Towards the testing of naive sociology. In: *Environmental Threats: Social Sciences Approaches to Public Risk Perceptions,* ed. J. Brown, pp. 33–45. London: Belhaven.

Zeckhauser, R.J. and W.K. Viscusi (1990). Risk within reason. *Science* **248:** 559–564.

THE DAHLEM WORKSHOP MODEL

MONDAY	TUESDAY	WEDNESDAY	THURSDAY	FRIDAY
A. OPENING (P) B. INTRODUCTION (P) C. SELECTION OF PROBLEMS FOR THE GROUP AGENDA (S) 1 2 3 4	1 2 3 4	1 3	F. REPORT SESSION (S) 1 2 3 4	G. DISTRIBUTION OF THE GROUP REPORTS H. READING TIME I. DISCUSSION OF THE GROUP REPORTS (P)
D. PRESENTATION OF GROUP AGENDA (P) E. GROUP DISCUSSIONS (S) 1 2		2 4 Publication Session		J. GROUPS MEET TO REVISE THEIR REPORTS (S) 1 2 3 4

Key: (P) = Plenary Session

(S) = Simultaneous Session

◯ = Each circle represents one discussion group

Explanation of the Dahlem Workshop Model

A. Opening
Background information is given about the Dahlem Workshop Model.

B. Introduction
The goal and the scientific aspects of the workshop are explained.

C. Selection of Problems for the Group Agenda
Each participant is requested to define priority problems of his choice to be discussed within the framework of the workshop goal and his or her discussion group topic. Each group discusses these suggestions and compiles an agenda of these problems for their discussions.

D. Presentation of the Group Agenda
The agenda for each group is presented by the moderator. A plenary discussion follows to finalize these agendas.

E. Group Discussions
Two groups start their discussions simultaneously. Participants not assigned to either of these two groups attend discussions on topics of their choice.
The groups then change roles as indicated on the chart.

F. Report Session
The rapporteurs discuss the contents of their reports with their group members and write their reports, which are then typed and duplicated.

G. Distribution of Group Reports
The four group reports are distributed to all participants.

H. Reading Time
Participants read these group reports and formulate written questions/comments.

I. Discussion of Group Reports
Each rapporteur summarizes the highlights, controversies, and open problems of his group. A plenary discussion follows.

J. Groups Meet to Revise their Reports
The groups meed to decide which of the comments and issues raised during the plenary discussion should be included in the final report

S. Bernhard

List of Participants / Fields of Interest

Robert McC. Anderson
European Foundation for the Improvement of Living and Working Conditions, Loughlinstown House, Shankill, Co. Dublin, Ireland

Attitudes to Biotechnology of the Public and Consumer/Environmental Interest Groups

Erik-Olof Backlund
Dept. of Neurosurgery, Haukeland Hospital, N–5021 Bergen, Norway

Neurosurgery

Deborah M. Barnes
The Journal of NIH Research, 2101 L. Street, NW, Suite 207, Washington, D.C. 20037, U.S.A.

AIDS, Neuroscience, Signal Transduction, Cell Biology, Molecular Biology

Walter Bechinger
Evangelische Akademie Arnoldshain, D–6384 Schmitten 1, Germany

Social Science, Social Ethics, Biotechnology, Gene Technology

Michael Bernhard
ENEA, P.O. Box 316, I–19100 La Spezia, Italy

Ecology and Environmental Pollution

Silke Bernhard
Koenigsalle 35, D–1000 Berlin 33, Germany

Communication in Science, Models for Scientific Meetings

Diana Brahams
15, Old Square, Lincoln's Inn, London WC2A 3UH, U.K.

Medicine and Law; Medical Science, Law and Society

Baruch A. Brody*
Center for Ethics, Baylor College of Medicine, One Baylor Plaza,
Houston, TX 77030, U.S.A.

Ethical Issues Raised in the Development and Testing of New Drugs

Jeremy Cherfas
Top Flat, Woodlands, Bridge Road, Bristol B58 3PB, U.K.

Biologist and Journalist Interested in Genetic Resources

Patricia Smith Churchland
Dept. of Philosophy, B-002, University of California,
San Diego, CA 92093, U.S.A.

Philosophy of Neuroscience

---

* unable to attend

Michael J. Crawley
Centre for Population Biology, Dept. of Biology, Imperial College,
Silwood Park, Ascot, Berkshire SL5 7PY, U.K.

Population Ecology

H. Tristram Engelhardt, Jr.
Center for Ethics, Medicine and Public Issues, Baylor College of Medicine,
One Baylor Plaza Houston, TX 77030, U.S.A.

Bioethics, Health Care, Policy and the Philosophy of Medicine

Hans Günter Gassen
Institut für Biochemie der Technischen Hochschule Darmstadt,
Petersenstr. 22, D–6100 Darmstadt, Germany

Neurochemistry, Molecular Genetics

Wilhelm Girstenbrey
Birkenallee 9, D–8131 Pentenried, Germany

Journalism

Robin A. Gordon
Department of Applied Mathematics, Schering AG, P.O. Box 650311,
D–1000 Berlin 65, Germany

Mathematical Modelling, Theory of Risk, Operations Research

Youssef Halim
Faculty of Science, University of Alexandria, Moharram Bay, Alexandria, Egypt

Marine Ecology, Pollution

Barbara Hobom
Arndtstr. 14, D–6300 Gießen, Germany

Science Writer; Molecular Biology, Genetic Engineering

Kenneth J. Hsü
Geologisches Institut, ETH-Zentrum, CH–9092 Zürich, Switzerland

Social Studies of Science and Technology, Science Policy, Comparitive Regulatory Policy

Sheila S. Jasanoff
Cornell University, Department of Science and Technology Studies, 632 Clark Hall, Ithaca, NY 14853, U.S.A.

Social Studies of Science and Technology, Science Policy, Comparitive Regulatory Policy

William R. Jordan III
University of Wisconsin – Madison Arboretum, 1207 Seminole Highway, Madison, WI 53711, U.S.A.

Ecological Restoration as a Model for Relationship Between Human Beings and Nature

Horace Freeland Judson
807 West University Parkway, Baltimore, MD 21210, U.S.A.

History of Recent Science: Molecular Biology, Evolutionary Theory, Immunology

Peter Kafka
Max-Planck-Institut für Astrophysik, Karl-Schwarzschild-Str. 1, D–8046 Garching, Germany

Astrophysics, Political Ecology in Connection with a General Theory of Evolution

Alexander Keynan
The Hebrew University, Jerusalem, Israel

Science Policy, Government-Science Relations, Application of Science for the Development of Less-developed Countries, Extra-mural Research Funding for Universities

Roger Lewin
3802 Ingomar Street NW, Washington, D.C. 20015, U.S.A.

The History of Molecular Evolution

Margaret N. Maxey
Biomedical Engineering Program, Murchison Chair of Free Enterprise, Petroleum/CPE 3.168, The University of Texas at Austin, Austin, TX 78712 U.S.A.

Cultural, Ethical, and Religious Origins of Policy Disputes and Conflicts in Risk-Selection Occurring in Affluent Industrialized Societies

Maurice McGregor
Conseil d'Evaluation de Technologies de la Santé du Québec,
800, Place Victoria, Tour de la Bourse, Bureau 42.05, C.P. 215,
Montréal, Québec H4Z 1E3, Canada

Evaluation of Technology in Health Care

Dietmar Mieth
Katholisches Theologisches Seminar, Abt. Ethik II, Liebermeisterstr. 12,
D–7400 Tübingen, Germany

Theological Ethics, Ethics in the Natural Sciences, Genome Analysis, Genetic Engineering in Plants

Jonathan D. Moreno
Division of Humanities in Medicine, SUNY Health Science Center at Brooklyn, 450 Clarkson Ave., Box 116, Brooklyn, NY 11203-2098, U.S.A.

Biomedical Ethics

Benno Müller-Hill
Institut für Genetik der Universität zu Köln, Weyertal 121,
D–5000 Köln 41, Germany

Genetics

Horst Nöthel
Institut für Genetik, Freie Universität Berlin, Arnimallee 7,
D–1000 Berlin 33, Germany

Genetics and Evolution, Mutagen-Mutation Equilibria

Robert W. Old
Department of Biological Sciences, University of Warwick,
Coventry CV4 7AL, U.K.

Embryology / Early Development of the Amphibial Xenopus, Molecular Biology

Daniel M. Pauly
International Center for Living, Aquatic Resources Management,
MC P.O. Box 15 01, Makati / Metro Manila 1299, Phillipines

Tropical Fisheries and Aquatic Ecology, Fisheries and Ecosystem Modelling

Marcus E. Pembrey
Paediatric Genetics, Institute of Child Health, 30 Guilford Street,
London WC1N 1EH, U.K.

Clinical Genetics

Elliot E. Philipp
46, Harley Street, London W1N 1AD, U.K.

Gynecology, Surgery, Infertility, Medical Ethics, the History of Medicine

Stuart L. Pimm
Department of Zoology and Graduate Program of Ecology, The University
of Tennessee, Knoxville, TN 37996, U.S.A.

Conservation Biology

Andrew N. Rowan
Tufts Center for Animals and Public Policy, Tufts School of Veterinary Medicine, 200 Westborough Road, N. Grafton, MA 01536, U.S.A.

Animals and Public Policy, Particular Pmphasis on Animal Research, Companion Animals, and Ethics

David J. Roy
Center for Bioethics, Clinical Research Institute of Montreal, 110 Ave. des Pines Ouest, Montreal, Quebec H2W 1R7, Canada

Bioethics, Clinical Ethics, Ethics of Research, Ethics and Public Policy Relevant to Bioscience

Gary S. Sayler
Center for Environmental Biotechnology, 10515 Research Drive, Suite 200, Knoxville, TN 37932, U.S.A.

Environmental Biotechnology for Degradation of Hazardous Wastes and Environmental Pollutants, Gene Transfer in the Environment

Morris H. Shamos
M.H. Shamos & Associates, 6 East 43rd Street, New York, NY 10017, U.S.A.

Biomedical Diagnostics, Scientific Literacy and the Interaction of Society with Science and Technology

David W. Sharp
Deputy Editor, The Lancet, 42, Bedford Square, London WC1B 3SL, U.K.

Medical Journalism

Robert C. Solomon
Philosophy Department, University of Texas, Waggener Hall 316,
Austin TX 78712, U.S.A.

Ethics, Existentialism, Theories of Human Nature, Philosophical Psychology

Wolfgang Van den Daele
Wissenschaftszentrum Berlin, Reichpietschufer 50, D–1000 Berlin 30,
Germany

Sociology of the Environment

Patrick Vinay
Fond de la Recherche en Santé du Québec, 550, Rue Sherbrooke Ouest,
Bureau 1950, Montréal, Québec H3A 1B9, Canada

Renal Metabolism and Physiology

Kenneth S. Warren
Maxwell Macmillan Group, 866 Third Ave, New York, NY 10022, U.S.A.

Information Sciences

James Wyngaarden
National Academy of Sciences, 2101 Constitution Ave.,
Washington, D.C. 20418, U.S.A.

Medicine, Genetics, Science Policy, Interaction between Science and Society

Brian E. Wynne
Centre for Science Studies and Science Policy, Lancaster University, Lancaster LA1 4YN, U.K.

Sociology of Science, Technology and Risk, Public Understanding of Science

Rolf Andreas Zell
Journalistenbüro KLARTEXT, Osianderstr. 13, D–7000 Stuttgart 1, Germany

Journalism, Biotechnology

Subject Index

Author Index